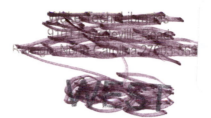

GLOBAL ISSUES

BIOTECHNOLOGY AND GENETIC ENGINEERING

GLOBAL ISSUES

BIOTECHNOLOGY AND GENETIC ENGINEERING

Kathy Wilson Peacock

Foreword by Charles Hagedorn, Ph.D.
Professor, Environmental Microbiology, Virginia Tech

Facts On File
An imprint of Infobase Publishing

GLOBAL ISSUES: BIOTECHNOLOGY AND GENETIC ENGINEERING

Copyright © 2010 by Infobase Publishing

Facts On File, Inc.
An imprint of Infobase Publishing
132 West 31st Street
New York NY 10001

Library of Congress Cataloging-in-Publication Data

Peacock, Kathy Wilson.
    Biotechnology and genetic engineering / Kathy Wilson Peacock; foreword by Charles Hagedorn.
        p.; cm. — (Global issues)
    Includes bibliographical references and index.
    ISBN 978-0-8160-7784-7 (alk. paper)
    1. Biotechnology—Popular works. 2. Genetic engineering—Popular works. I. Title. II. Series: Global issues (Facts on File, Inc.)
    [DNLM: 1. Biotechnology. 2. Genetic Engineering. 3. Organisms, Genetically Modified—genetics. QU 450 P352b 2010]
    TP248.215.P43 2010
    660.6—dc22   2009025794

Facts On File books are available at special discounts when purchased in bulk quantities for businesses, associations, institutions, or sales promotions. Please call our Special Sales Department in New York at (212) 967-8800 or (800) 322-8755.

You can find Facts On File on the World Wide Web at
http://www.factsonfile.com

Text design by Lina Farinella
Illustrations by Dale Williams
Composition by Mary Susan Ryan-Flynn
Cover printed by Art Print, Taylor, Pa.
Book printed and bound by Maple Press, York, Pa.
Date printed: May 2010
Printed in the United States of America

This book is printed on acid-free paper.

CONTENTS

PART II: Primary Sources

PART III: Research Tools

Foreword

In September 1992, the University of California, San Francisco and the Exploratorium, housed in San Francisco's Palace of Fine Arts, presented a public symposium entitled "Winding Your Way through DNA." This program was designed to educate the public and to encourage a dialogue about the scientific possibilities and associated problems with recombinant DNA technology. Following the symposium, a team of high school and college teachers, ethicists, historians, and scientists created a series of documents and videos that they hoped would establish a long-term discussion about this technology and the ethical, legal, and societal issues that have emerged since its inception. This symposium fully recognized that the first century of the new millennium would belong to biotechnology, among all the biological sciences, and that biotechnology could bring unprecedented advances in human and animal health, agriculture and food production, manufacturing, and sustainable environmental management. Also recognized was the need to exercise caution and judgment in its application to ensure that the potential risks to human health and the environment arising from the commercial use of genetically modified organisms in food production were properly managed. Such management included continuous assessment of biotechnology programs, establishment of suitable regulatory systems to oversee biotechnology products, and efforts to increase public awareness and acceptance of these products.

What the symposium participants could not have foreseen in 1992 was the incredible speed by which biotechnology would develop. This speed resulted in more than 115 million hectares of biotechnology crops in 2007, compared to none in 1992; completion of the Human Genome Project in 2003, two years ahead of schedule and under budget; and progress in technologies such as cloning, stem cell research, processes to increase longevity, intelligence, and physical abilities—the potential for human enhancement—termed by many as the most fundamental social and political issue facing the world today. The 1992 symposium participants also could not have foreseen how controversial every development in biotechnology would become: that efforts to increase public understanding would result in whole societies rejecting certain technologies (for example, the European Union ban on importation of genetically engineered food products); that regulatory systems would vary so widely from stringent in the United States and Europe to almost nonexistent in many developing countries; that growth in food production would remain largely within developed and wealthy countries; and that developments in the potential for human enhancement would provide hope to many but dread and horror to others.

Biotechnology dates from the dawn of civilization, where the earliest farmers selected edible plants to grow as crops and saved some of the seeds for the next season, and domesticated cattle, pigs, sheep, and goats. Over the years, farmers bred both the plants and animals they liked and learned how to best produce them with irrigation and weed control for plants and growing grain and forages for the animals. Early civilizations around the globe also used yeast to make alcohol and bread—yeast being a living microorganism (a fungus)—long before its role in fermentation was understood. One of the primary goals of biotechnology is to feed the world's 6 billion people, but there is substantial disagreement over the best way to accomplish this. Some people think genetically altered crops are the best answer because they allow farmers to grow more food less expensively than ever before. Others see these crops as being available only in wealthy countries and not accessible to much of the developing world, places where such crops would do the most good (for example, 73 percent of all biotechnology crops are grown in the United States, Canada, and Argentina).

Most of the controversy surrounding biotechnology involves drawing distinctions between what is acceptable and what is not. Is it wrong to modify germ-line cells (those that can be passed on to future generations) to ensure that a child has blue eyes? Is that more wrong than aborting a fetus with grave medical defects? Clearly, people have differing ideas about where such lines should be drawn. In practice, a double standard is sometimes evident. For example, growth hormones are illegal in the United States for bodybuilders and athletes, but the U.S. meat supply is heavily dependent on these same hormones to promote rapid growth in food animals.

The development of biotechnology in the United States has been similar to that of other countries that have embraced it. The earliest policies were designed to help farmers because the United States was primarily a rural, agrarian nation well into the 20th century. Much research focused on crop science and development of new varieties. As immigration and urbanization began to change American society, eugenics arose as a "scientific" approach to cure the social problems such as poverty, overcrowding, and crime. Eugenics serves as a good example as it was embraced by many of the country's intellectuals at the time, and their decisions created a legacy that casts a dark shadow over bioethics in the 21st century. As the promises of the Human Genome Project become closer to reality and scientists identify the genes responsible for specific and treatable conditions, those highly concerned with bioethics will continue to question the morality of rearranging the building code of life.

The biotechnology debate has essentially settled into two opposing sides, with rather extreme positions prevalent in both. On the advocacy side are the proponents, perhaps best exemplified by Ronald Bailey in his book *Liberation Biology: The Scientific and Moral Case for the Biotech Revolution.* This book eloquently describes the benefits of biotechnology: curing diseases and disabilities for millions of sufferers, producing more nutritious food with less damage to the natural environment, enhancing human physical and intellectual capacities, and retarding the onset of the ravages of old age—all of which are at least possible in the not-too-distant future. Bailey aims to use biotechnology to dramatically boost people's physical and intellectual capacities, eradicate diseases and cancers, restore the natural environment, and make death optional. Opposing the advocates is best expressed by Fritz Allhoff et al. (editors) in *Nanoethics: The Ethical and Social Implications of Nanotechnology.* This book describes the issue of relinquishment—the wholesale abandonment of certain fields of research as ethically unacceptable. This is the most controversial recommendation in the book, and the editors advocate that relinquishment at the right level is part of a responsible and constructive response to genuine perils. The issue, however, is this: At what level are we to relinquish new and developing technologies?

Biotechnology and Genetic Engineering, part of Facts On File Global Issues series, is designed to place itself outside of the wide-ranging biotechnology debate and to provide factual information on the current status of the science and its potential. Written for students and general readers, it is one of the most comprehensive and accessible introductions to biotechnology that I have read. Part I begins with an introduction that defines what is at issue with biotechnology, outlines the global challenges involved, and provides a brief history of the subject. Following the introduction are detailed case studies of the United States and four other countries (Japan, India, Germany, and South Africa) that explore how biotechnology has affected each one and what strategies and perspectives each has pursued in response. Part II draws together significant U.S. and international primary source documents on biotechnology, such as excerpts from essays, speeches, newspaper articles, relevant treaties and other legal documents, and scientific reports. Part III gathers useful research tools for biotechnology, such as brief biographies of key international players, facts and figures, an annotated bibliography, and a list of relevant international organizations and agencies. A chronology, glossary, and index provide additional help.

In many respects we are fortunate: New advances in biotechnology promise to make the path of progress a great deal easier and shorter. However, much is at stake—from stem cells to cure many diseases to the open release of engineered microbes designed to clean up the environment to feeding the world's population with engineered foods. It is imperative that all students and citizens alike learn about biotechnology so they can make informed choices. We stand at a crossroads and our responses to these opportunities will shape our future.

—Charles Hagedorn
Professor of Environmental Microbiology and
Outreach Specialist in Biotechnology, Virginia Tech

List of Acronyms

A*STAR	Agency for Science, Technology and Research (Singapore)
AAT	alpha-1 antitrypsin (glycoprotein)
ACLU	American Civil Liberties Union
ADD	attention deficit disorder
AES	American Eugenics Society
AIDS	acquired immune deficiency syndrome
AMA	American Medical Association
ART	assisted reproductive technologies
ASPM	abnormal spindle-like microcephaly-associated (gene)
AT	adenine and thymine
BCCST	Bioethics Committee Council for Science and Technology (Japan)
BGH	bovine growth hormone
BSE	Bovine spongiform encephalopathy
BST	bovine somatotropin
Bt	*Bacillus thuringiensis*
CAFO	concentrated animal feeding operation
CAPRISA	Centre for the AIDS Programme of Research in South Africa
CBEL	Center for Biomedical Ethics and Law (Japan)
CDC	Centers for Disease Control and Prevention
CG	cytosine and guanine
CGIAR	Consultative Group on International Agricultural Research
CIIFAD	Cornell International Institute for Food, Agriculture and Development
CODIS	Combined DNA Index System
CSIR	Council for Scientific and Industrial Research (South Africa)
DDT	dichlorodiphenyltrichloroethane

DEFRA	Department of Environment, Food and Rural Affairs (U.K.)
DNA	deoxyribonucleic acid
DOE	Department of Energy
ELCA	Evangelical Lutheran Church in America
ELSI	ethical, legal, and social issues
EPA	Environmental Protection Agency
EU	European Union
FAO	UN Food and Agriculture Organization
FDA	Food and Drug Administration
FIFRA	Federal Insecticide, Fungicide and Rodenticide Act
GDP	gross domestic product
GE	genetically engineered
GLC1A	glaucoma 1A (gene)
GM	genetically modified
GMO	genetically modified organism
GURT	genetic use restriction technology
HBB	hemoglobin beta (gene)
HGP	Human Genome Project
HHS	Department of Health and Human Services
HIPAA	Health Insurance Portability and Accountability Act
HIV	human immunodeficiency virus
HUGO	Human Genome Organization
IFN	interferon
IFPRI	International Food Policy Research Institute
IGSP	Influenza Genome Sequencing Project
IJC	Islamic Jurisprudence Council
iPSC	induced pluripotent stem cells
IPM	integrated pest management
IVF	in-vitro fertilization
MCD	mad cow disease
MDMA	3,4-Methylenedioxymethamphetamine (Ecstasy)
mRNA	messenger ribonucleic acid
MSG	monosodium glutamate
NGO	non-government organization
NIH	National Institutes for Health
PCBE	President's Council on Bioethics
PCR	polymerase chain reaction
PGD	preimplantation genetic diagnosis
PGS	preimplantation genetic selection
PMP	plant-made pharmaceuticals

List of Acronyms

rBGH	recombinant bovine growth hormone
rBST	recombinant bovine somatotropin
rDNA	recombinant deoxyribonucleic acid
RNA	ribonucleic acid
SAA1	serum amyloid Al (gene)
SAAAS	South African Association for the Advancement of Science
SAAVI	South African AIDS Vaccine Initiative
SANBI	South African National Bioinformatics Institute
SCID	severe combined immunodeficiency
SCNT	somatic cell nuclear transfer
SRI	system of rice intensification
SV40	simian virus 40
TB	tuberculosis
TOCP	tri-o-tolyl phosphate
tRNA	transfer ribonucleic acid
UN	United Nations
UNESCO	United Nations Educational, Scientific and Cultural Organization
USAID	United States Agency for International Development
USDA	United States Department of Agriculture
vCJD	variant Creutzfeldt-Jakob disease
WHO	World Health Organization
WMA	World Medical Association

List of Maps and Graphs

PART I

At Issue

1

Introduction

In Michael Crichton's novel *Jurassic Park*, bioengineers use DNA trapped in amber to clone 15 species of long-extinct dinosaurs. The creatures become the star attraction at an isolated island theme park, where they are safely segregated from other animals to prevent their DNA from infiltrating the modern gene pool. As added insurance, the scientists have genetically engineered only females, thereby eliminating the possibility of the dinosaurs reproducing on their own. Nevertheless, trouble ensues: Not only do rival geneticists get their hands on the DNA technology, but escaped dinosaurs embark on a rampage, killing people and setting off a chain of events that the park's scientists are incapable of stopping. Technology is no longer contained within a controlled environment, which has devastating consequences for humanity.

Jurassic Park is a compelling fictional thriller because it explores the moral consequences of scientific achievement. Should humans tinker with the building blocks of life? What could happen if we do? While most everyone wants to benefit from scientists' medical breakthroughs and to end world hunger and disease, most people agree that scientists must operate within ethical boundaries. Take genetically modified food: If genes from a fish are inserted into a tomato plant to make it resistant to colder weather, have biologists solved a problem by allowing nutritious food to be produced under harsher conditions or have they tampered with an ecosystem that has achieved a delicate balance through eons of evolution? And while welcoming the advent of gene therapy for a debilitating disease like Parkinson's, we may remain uneasy about the process, which involves attaching the corrected gene sequence to a virus and injecting that virus into the patient's body.

Biotechnology is the manipulation of living organisms for purposes other than their original intent. While the term may sound futuristic, biotechnology is nearly as old as civilization itself. It began with food; agriculture in its most basic sense is biotechnology. The first farmers selected particular plants to grow as crops and saved their seeds for the following season. Over the years, they bred the varieties of seeds they liked best and learned how to grow them more efficiently through irrigation and weed control. Emerging cultures around the world also used yeast to make alcohol and bread—yeast being a fungi, a living organism—long before they understood its role in the process of fermentation.

One of the main goals of biotechnology is to feed the world's 6 billion people, but there is disagreement over the best way to accomplish this. Some say genetically modified organisms (GMOs) are the answer because they allow farmers to grow more food less expensively than ever before. Genetically modified (GM) or genetically engineered (GE) corn can be grown without with the use of harmful pesticides; "golden" rice contains beta carotene, a nutrient that could save the lives of some of the world's most malnourished citizens.

When it comes to livestock, the same principles apply. Ranchers struggling to raise more cattle on less land inject antibiotics and hormones into their livestock to reduce disease and increase body mass. Poultry farmers build warehouses and mechanize the life cycles of chickens to increase their productivity and lower overhead costs.

THE ISSUE

Biotechnology is a huge topic; it is hard to define its exact boundaries. Some scientists (particularly in Europe) divide the field into red biotechnology and green biotechnology. Red biotechnology relates to medicine, and green biotechnology relates to food.[1] Some subdivide biotechnology into white and blue. White biotechnology, also called industrial biotechnology, uses natural processes such as fermentation and enzymes to create products formerly made with chemicals. Bioplastics made with vegetable oil and starches instead of petroleum are examples of white biotechnology. Blue biotechnology encompasses all aspects of marine biology and genomics (the application of biotechnology through gene mapping, DNA sequencing, and other techniques). Color coding this vast subject is a handy way to break it into more manageable chunks, but these categorizations are mainly cosmetic. All the categories share two elements: Research is driven by the desire to find solutions to modern-day problems, and opposition is driven by concern about the unknown effects of altering the natural world and the ethics of doing so, be it the genome of corn or people.

Academically, biotechnology falls under many umbrellas. It is generally considered a natural science, and more specifically, a life science. The life sciences include biology, which is the study of living organisms and their environments. Biology encompasses botany, the study of plants, and zoology, the study of animals. Beyond these classifications are numerous overlapping categories including cell biology, microbiology, molecular biology, physiology, ecology, embryology, genetics, population genetics, epigenetics, proteomics, and bioinformatics. The list will continue to grow as new fields emerge. Whatever the discipline, if it has to do with altering living organisms for any purpose, it can be considered biotechnology.

This book breaks biotechnology into two imprecise categories—agriculture and medicine—because much of the genetic engineering taking place today is aimed at solving issues related to food and health.

Biotechnology and Agribusiness

Since the end of World War II, nationwide patchworks of multigenerational family farms have gradually disappeared due to urbanization and the rise of large corporate farming operations. Agribusiness encompasses both crops and livestock, and biotechnology is central to both as corporations develop ways to increase output and maximize profit. Transgenic crops (those altered by the insertion of DNA from another organism) yield greater harvests and require fewer pesticides than those grown from seeds that have not been genetically altered. Livestock are injected with growth hormones to produce more meat at a faster rate, and antibiotics keep animals free of disease in crowded confines. The result is that more people are now fed with food raised on less land than at any time in history.

BIOTECHNOLOGY AND CROPS

In 2007, some 250 million acres of genetically modified crops—mainly corn, soybeans, cotton, canola, and alfalfa—were planted worldwide, and more than half of these were in the United States.[2] The remainder were primarily in Canada, Argentina, Brazil, China, and South Africa.[3] GM crops have become popular in recent years for three main reasons. First and foremost, they have higher yields than non–GM crops. For example, prior to 1950, farmers typically harvested 40 bushels of corn per acre, whereas GM corn produces up to 180 bushels per acre.[4] This has important ramifications in a world that is increasingly urban, where fewer farmers supply food for a growing population. Second, GM seed is designed to resist pests and herbicides. GM corn crops, such as Monsanto's Roundup Ready Bt corn, can be sprayed with Monsanto's Roundup herbicide and continue to grow unabated while weeds shrivel and die. Bugs and diseases that have been the bane of farmers since time immemorial leave the crops unscathed. Third, GM foods can be manipulated to withstand the rigors of long-distance shipping and to ripen more slowly. This allows food to reach more people in more locations, ensuring variety in people's diets.

The labeling of GM foods is a controversial issue in the United States and Europe. Consumers generally want GM foods to be labeled as such, while companies have been reluctant to do this. In the United States, the Food and Drug Administration (FDA) does not require labeling unless the GM food differs significantly from its unmodified counterpart. This is the concept of "substantial equivalence," which is the belief that GM foods are equal in nutrition and safety to those that are not modified. Under this definition, few GM foods are labeled in the United States. However, in Europe the same definition does not apply, and public sentiment has resulted in widespread labeling of GM foods.

Another contentious issue relates to genetic use restriction technologies (GURTs), or terminator technology. Embedded in GM seeds, GURTs cause crops to kill off their own seed before germination. Farmers are therefore prevented from harvesting and replanting patented seeds the following season, requiring them to purchase new seed from their suppliers each year. As of 2009, no GURT products were available in the United States or elsewhere, and the opposition to GURTs seems likely to keep it that way for the foreseeable future. Genetic contamination is a key concern of GURT opponents, who fear that the engineered genes could enter the genomes of other plants through cross-pollination or other fertilization practices and could have adverse consequences on the ecosystem.

Many believe that GM crops could cause a loss in biodiversity. Biodiversity is the variety of plant and animal life within a given ecosystem, with a high level of biodiversity corresponding to a healthy ecosystem. Monoculture, or the practice of growing one crop over a large area, characterizes much commercial farming in the United States and around the world. It tends to damage the land and leave crops vulnerable to pests, which leads to heavy applications of pesticides. While commercial farmers generally attempt to mitigate some effects of monoculture, some people fear they do not do enough to ensure food security or a community's ability to obtain the nourishment necessary for peaceful survival. The widespread failure of a monoculture GM crop on which a community has relied could be both economically and politically destabilizing (as happened during Ireland's Great Famine in the 1840s when a blight destroyed the potato crop).

In 1994, Calgene, Inc.'s Flavr Savr tomato became the first genetically engineered food marketed for human consumption. Launched as a hearty, durable, flavorful fruit that could withstand the rigors of cross-country transport because it ripened slowly, the Flavr Savr tomato underwhelmed consumers and quickly fell victim to bad management decisions. It was taken off the market after only a few months.

Nevertheless, GM food soon gained a solid foothold as the agribusiness giant Monsanto pioneered herbicide-resistant Roundup Ready soybeans and obtained contracts with significant numbers of farmers to plant the seed exclusively. Rather than marketing GM food to consumers, Monsanto worked with farmers whose crops were used mainly as animal feed or as raw ingredients for processed food. By the early 2000s, Monsanto's Bt seeds for corn, canola, cotton, and soybeans—genetically engineered to produce the *Bacillus thuringiensis* (Bt) toxin, which is lethal to some insect pests—represented a substantial portion of all seeds planted by U.S. farmers. Other companies, including Dow AgroSciences, Syngenta Seeds, Aventis CropScience (which manufactured StarLink Bt corn in 2000), and Bayer CropScience (maker of Liberty Link rice), pursued GM seed development.

BIOTECHNOLOGY AND LIVESTOCK

The demand for meat and dairy products has skyrocketed in recent years due to increasing prosperity in developing countries such as China and India, whose populations previously subsisted mainly on grains, vegetables, and fruits. This has prompted a livestock revolution in the meat industry, with ranchers, farmers, and abattoirs all changing the way meat is brought to market. Historically, meat has been expensive. It takes more land, time, and resources to raise a cow than it takes to grow a bushel of corn, for example. Thus, many people around the world could afford to eat meat only rarely, if at all. Now that more people can afford meat, cattle ranchers need to raise more animals on less land. The result is an intensification of the livestock-breeding process in which, for example, free-range poultry farms have been replaced by factory farms known as concentrated animal feeding operations, or CAFOs. Pigs, veal calves, turkeys, and other livestock can be raised in CAFOs. The crowded conditions provide fertile ground for disease, which is often staved off by injecting animals with antibiotics. The waste from CAFOs can pollute nearby land and waterways.

The intensification process is similar for cattle. At a young age, cattle are shipped to feedlots, where they are confined in a small area and fed large amounts of corn. The corn fattens them up faster than if they grazed in an open field. The massive amounts of corn tend to cause the animals to suffer acidosis, or an excess of stomach acid. To combat this, they are injected with antibiotics before they are sent to the slaughterhouse. Some believe that these processes will reduce the genetic diversity of cattle and lead to problems in the food supply.

Bovine Growth Hormone

Bovine growth hormone (BGH; sometimes called bovine somatotropin, or BST) is a protein generated in the pituitary glands of cows. It can also be produced artificially using DNA containing genetically engineered *E. coli* bacteria, in which case it is known as rBGH or rBST. The agriculture giant Monsanto has been producing and marketing rBGH under the trade name Posilac since 1994 as a product that increases milk production in cows by preventing the death of mammary cells. Cows injected with Posilac lactate longer, producing 10 percent more milk in the second half of their cycles. The injections, however, tend to cause mastitis (a painful infection of the udder), reduce fertility, and increase lameness.[5] While no study has found detrimental effects in humans who consume milk from cows injected with rBGH, both the European Union (EU) and Canada no longer allow its use because of its detrimental effects on dairy cattle. Australia and New Zealand have also banned rBGH. It has been claimed that rBGH is dangerous to human health because it increases levels of IGF-1, a hormone associated with breast, prostate, and colon cancer. Consumer concern over Posilac has resulted in a decline in its use in the United States: In 2002, 22.3 percent of dairy cows were treated with Posilac, and by 2007 that number had dropped to 17.2 percent.[6]

Some farmers do not believe that rBGH increases milk production significantly and that its harm outweighs its benefits. For instance, cows that develop mastitis from rBGH are given large doses of antibiotics as a remedy. While this may help the cow, from a public health standpoint the practice is questionable because the more widely antibiotics are used the less effective they become for both cows and humans.[7] There is also some evidence that antibiotics can be passed along to humans who consume the milk from treated cows.

The FDA does not require milk from hormone-treated cows to be labeled as such. This angers some consumers, who believe they have the right to know how their milk is produced. The FDA and Monsanto, however, maintain that milk derived from hormone-treated cows is not significantly different from that derived from untreated cows. Furthermore, when Oakhurst Dairy of Portland, Maine, labeled its milk as having been produced with "no artificial growth hormones"—employing reverse labeling (which lists what is *not* in a product rather than what is in a product)—it was sued by Monsanto. The settlement required Oakhurst's statement to be accompanied by a disclaimer stating that the FDA has found no difference between milk produced with rBGH and milk produced without it.[8] Nevertheless, many consumers are willing to pay more for organic and rBGH-free milk. In 2007, though only 2.7 percent of all milk sold was organic, it represented a $1.3 billion industry, with consumer demand rising 20 percent annually.[9] Major retailers, including Starbucks, Wal-Mart, and Kroger, began selling non-rBGH milk in 2008.[10]

8

Introduction

Antibiotics

Antibiotics are a modern miracle: They kill bacteria, cure sickness, and prevent millions from dying of diseases that used to be fatal. They work on people as well as animals, and of course no one wants to eat meat derived from a sick animal. Since the 1990s, however, ranchers have been administering antibiotics to prevent illness in their herds, with the unintentional result of lessening those antibiotics' overall effectiveness, even among humans. This happens because bacteria consistently exposed to a certain antibiotic can mutate and become resistant to it. Over the long term, this will render antibiotics useless. According to the Union of Concerned Scientists (UCS), "antibiotic-resistant bacteria are on the rise. Patients once effectively treated for pneumonia, tuberculosis, or ear infections may now have to try three or more antibiotics before they find one that works. And as more bacterial strains develop resistance, more people will die because effective antibiotics are not identified quickly enough or because the bacteria causing the disease are resistant to all available antibiotics."[11] This phenomenon is well established and has prompted both the U.S. Centers for Disease Control (CDC) and the World Health Organization (WHO) to condemn the use of antibiotics in the food supply.

Overall, 70 percent of the antibiotics used in the United States each year (amounting to 25 million pounds) are given to chickens, pigs, and cows to prevent disease caused by CAFO practices, such as mastitis and acidosis.[12] According to journalist Ken Midkiff, "It is the concentration of animals and consequent stress on them that prompts the use of antibiotics to enhance appetite and to promote more efficient conversion of feeds to weight gain."[13]

The Nonpolluting Pig

Pig meat in the form of pork, ham, and bacon is a source of protein, but pig manure contains phosphorus, which is highly polluting to the environment. Nevertheless, pig farming is on the rise in China and elsewhere as part of the livestock revolution. As a result, the lakes, rivers, and streams that receive runoff from pig farms are becoming polluted as tons of phosphorus-laden waste deplete oxygen, kill fish, and emit greenhouse gases. Geneticists at the University of Guelph in Canada developed the Enviropig as a partial solution to the problem. Enviropigs are genetically altered with mouse genes. The altered pigs' waste contains 75 percent less phosphorus than that of regular pigs because the mouse genes endow the pigs' saliva with enzymes that extract phosphorus from their feed and prevent it from passing through the digestive system and into the environment.[14] Critics maintain that even pig waste without phosphorus is harmful because its high nitrogen levels cause algal blooms in waterways, which suffocate marine life. The Enviropig has not been approved for the market as of 2010.

Biotechnology and Medicine

Medicine is the science of healing. Biotechnology has brought medicine out of the age of potions and into the era of genetic manipulation. Diseases that were once incurable might now be remedied by tinkering with a person's DNA. Many people would welcome this tinkering if it could cure debilitating diseases such as cancer, Alzheimer's, or multiple sclerosis (MS). Others question where to draw the line between healing and tampering. As scientists unlock the secrets of the human genome and learn how to manipulate sequences of DNA to modify everything from eye color to memory skills, the hope is that people will live longer and enjoy better health than ever before. Since the first bioengineered medicine—synthetic human insulin—arrived on the market in 1982, more than 100 such medicines have withstood rigorous testing and entered the marketplace. Beyond new treatments for existing disease, recombinant DNA technology holds great potential for developing vaccines that could prevent illnesses that still routinely devastate parts of the globe.

UNDERSTANDING DNA AND RNA

When Francis Crick and James D. Watson unveiled their double helix model of deoxyribonucleic acid, or DNA, the world was astounded. Their article "Molecular Structure of Nucleic Acids," which appeared in the academic journal *Nature* in 1953, was accompanied by a simple sketch of a spiraling, ladderlike design and claimed the DNA "structure has novel features which are of considerable biological interest."[15] A subject that previously left laypeople scratching their heads now had a visual cue that helped them understand the complexity and beauty hidden in each living cell. DNA is a long chain of genetic material organized into base pairs of either adenine and thymine (AT) or cytosine and guanine (CG) in such a way that the helix can "unzip" itself and make copies. Watson and Crick's discovery was a monumental insight into the nature of life itself; it was as important as Darwin's theory of natural selection or Einstein's theory of relativity.

Previously, no one had known how genetic material was inherited or even where the information was stored or coded in the body. Yet it took just over 50 years for scientists to catapult from Watson and Crick's initial discovery to completing the Human Genome Project, which created a public database of all the sequenced chemical base pairs (some 3 billion combinations of AT and CG) that comprise the 46 human chromosomes and roughly 25,000 genes in each human being.

Introduction

DNA is a chain of nitrogen-containing molecules called nucleotides that stores all of an organism's genetic information; each cell in an organism contains a copy of the organism's DNA. Think of DNA as a recipe and ribonucleic acid (RNA) as a chef. The RNA follows the DNA recipe to create the proteins that combine into enzymes, which in turn become the compounds that comprise an organism. Understanding the structure and components of DNA is the work of molecular biologists. Their hope is that by mapping the building blocks of genes, they will discover, and possibly be able to change, the genes that are responsible for certain diseases.

RNA is similar to DNA, but it has a slightly different chemical structure, and its molecules of adenine are paired with uracil instead of thymine. RNA is usually single-stranded and plays an important role in the synthesis of proteins. There are different types of RNA that fulfill different roles within an organism. For example, messenger RNA (mRNA) carries the blueprint for a protein to its construction site in the ribosome. Transfer RNA (tRNA) conveys an amino acid to the construction site as part of the translation process.

Each cell in the human body contains the individual's 23 pairs of chromosomes (46 altogether), which collectively house between 20,000 and 25,000 genes. A gene is "a union of genomic sequences encoding a coherent set of potentially overlapping function products."[16] Think of a gene as a discrete section of your genomic sequence that is inherited. A chromosome is a defined portion of DNA that contains many genes, nucleotide sequences, and protein packages. Each chromosome from number 1 through 22 is paired with a copy. The 23rd pair of chromosomes is the sex chromosomes, with males having an X and a Y, and females possessing two X chromosomes.

Chromosome 1 is the largest human chromosome. It is comprised of 247 million base pairs of nucleotides and contains 3,148 known genes, including the gene *ASPM*, which helps determine brain size, and *GLC1A*, a gene for glaucoma. Diseases associated with genes on chromosome 1 include Alzheimer's, breast cancer, congenital hypothyroidism, and Parkinson's. Chromosome 11 is smaller; it contains 134.5 million base pairs and approximately 1,500 genes, including *HBB* (hemoglobin, beta) and *SAA1* (serum amyloid A1), which relate to autism, breast cancer, sickle cell anemia, and many other conditions. Scientists have also identified the genes responsible for many other diseases on the remaining chromosomes, although much more work is required before the picture is complete.

Recombinant DNA
Recombinant DNA is produced when scientists add DNA to an organism's genome to code it for a new trait or to alter an existing trait. The technique

was pioneered in 1973 by the National Medal of Science recipients Stanley Norman Cohen, a Stanford genetics professor, and Herbert Boyer, a cofounder of the biotechnology company Genentech. The first recombinant DNA substance approved by the FDA was synthetic human insulin, which was created to treat diabetes. Since then, recombinant DNA has been used for a host of other inventions, including the GloFish, a zebrafish altered by the addition of proteins that cause it to become fluorescent. Initially designed to act as a warning signal for water pollution, the GloFish instead became a novelty item on the tropical fish market.

GENETIC ENGINEERING AND PROTEOMICS

Genetic engineering is any process in which an organism's genome is intentionally altered. Genetic engineering does not encompass traditional breeding techniques because it requires manipulation of an organism's genes through cloning or transformation via the addition of foreign DNA. This process has five steps:

1. Isolation of the genes
2. Insertion of those genes into a transfer vector (a virus or a plasmid used as a conduit)
3. Transfer of the vector to the organism to be modified
4. Transformation of that organism's cells
5. Separation of the genetically modified organism (GMO) from organisms that have not been successfully modified

Proteomics is the study of proteins, and it represents the other side of genomics. Proteins are the substance within cells that allow for the cells' various functions. They are organic compounds comprised of a chain of molecules called amino acids that are joined by a chemical bond. While an organism's genome remains constant across the body's many different types of cells, its proteome, the particular proteins expressed by the genome, shifts from cell to cell, making proteomics more complicated than genomics.

The field of proteomics is headed by an international consortium of scientists and organizations known as the Human Proteome Organization, which was formed in 2001 to capitalize on the discoveries of the Human Genome Project. The science of pharmocogenomics, which combines the fields of genomics and proteomics, may lead to custom-made treatments for individuals based on their genetic analysis. Scientists hope that pharmocogenomics will eliminate much of the trial-and-error process patients currently undergo to find optimal medication dosages.

THE STORY OF INSULIN

Diabetes mellitus is a disease in which the body cannot produce enough insulin to regulate blood sugar, or glucose, levels. Blood sugar levels that are either too high or too low can result in significant health problems, including loss of consciousness and brain, kidney, eye, and nerve damage. Synthetically produced insulin was one of the earliest success stories of molecular medicine. Type 2 diabetes, the most common variety, is usually caused by obesity, a high-fat diet, and a lack of physical activity. Symptoms can include an impaired cardiovascular system, blindness, and renal failure. The CDC has declared Type 2 diabetes to be an epidemic in the United States, and the disease is being diagnosed more frequently in young people than it used to be. Individuals may also inherit a genetic predisposition toward Type 2 diabetes.

The treatment for diabetes includes insulin shots. Insulin is a hormone normally produced by the pancreas that allows the body's cells to obtain glucose from blood cells and store it as glycogen in the liver and muscles. Without insulin, a person will die. Historically, diabetics received shots of insulin derived from cows, horses, pigs, or fish. But in 1982 the first synthetic insulin, Humulin, came on the market; it was genetically engineered by Genentech and marketed by Eli Lilly. Humulin was the first medication genetically engineered by inserting human DNA into a host cell, which then reproduced with the correct form of insulin to be injected into the diabetic. This synthetic insulin has proven invaluable as rates of diabetes skyrocket.

GENETIC SCREENING: ELIMINATING TAY-SACHS DISEASE

Genetic screening is the process of testing a person's DNA to see if he or she carries a gene that may lead to a certain disease. The first successful, wide-scale genetic screening took place in 1971, when Michael Kaback, a pediatric neurologist at Johns Hopkins University, gave 1,800 people of Ashkenazi Jewish ancestry an enzyme test to determine if they carried the gene for Tay-Sachs disease. Tay-Sachs disease is a debilitating, fatal genetic disorder that causes atrophy of the nervous system and usually results in death by age five. It afflicts Ashkenazi Jews (German Jews) more frequently than other ethnic groups, and the people tested volunteered to find out if they carried the recessive gene that could be passed on to a child. The test was both inexpensive and reliable and was soon widely adopted; 1.3 million people were tested, and nearly 50,000 carriers of the gene were identified within a few years. When both prospective parents were identified as carriers, pregnancies were tested with amniocentesis or chorionic villus sampling to determine whether the child had Tay-Sachs. Out of the 604 cases where the diagnosis was confirmed, 583 couples chose to abort the fetus. Almost all of the remaining children were born with Tay-Sachs and died in early childhood.[17] Essentially, Kaback's test meant that Tay-Sachs became exceedingly rare among Ashkenazi Jews.

GENE THERAPY

Gene therapy is the process of treating a genetic disease by replacing a person's defective gene with a new, normally functioning one. Gene therapy has a much shorter history than genetic engineering, dating back only to 1990, when doctors performed a groundbreaking procedure on four-year-old Ashanthi DeSilva. DeSilva suffered from severe combined immunodeficiency (SCID), which meant her immune system lacked the ability to fight off even mild infections. People born with SCID typically die in childhood. Researchers took some of DeSilva's white blood cells, grew them in a laboratory, inserted a missing gene into them, and then put the modified cells back into her bloodstream. While the treatment was not a cure, it did boost DeSilva's immune system enough for her to lead a normal life, although she must receive treatment every few months as the modified cells begin to die off and need to be replaced.[18]

Researchers have high hopes for developing gene therapies for single-gene diseases, such as cystic fibrosis, hemophilia, muscular dystrophy, and sickle cell anemia. Progress is slow because of the size of the human genome; finding where exactly to insert a modified gene is a daunting task, as is finding an appropriate vector, or carrier. Viruses are often used as vectors; modified genes are inserted into the virus, which is then introduced in the patient with the hope that it will infect the target cell with the altered gene.

Gene therapy comes in two forms: germ line gene therapy and somatic cell gene therapy. Germ line therapy focuses on reproductive cells—sperm or eggs—that are altered by inserting engineered genes. Changes made this way would alter a genome forever and be passed on to future generations. As of 2009, germ line therapy is theoretical. Somatic cell gene therapy involves altering the body's somatic cells, or organ and tissue cells that are not involved in reproduction. This type of gene therapy would cure or treat a person's condition, but the changes would not be passed on to future generations.

The success of DeSilva's gene therapy encouraged scientists, but their optimism proved short-lived. The next major clinical trial of gene therapy in 1999 ended in the death of 18-year-old Jesse Gelsinger, who was afflicted with ornithine transcarbamylase deficiency, a liver disease that usually results in death at birth. Gelsinger's version of the disease, in which his liver could not metabolize ammonia, resulted from a genetic mutation rather than hereditary factors, and he had survived on a strict diet and regimen of drugs until he underwent gene therapy. Doctors at the University of Pennsylvania created an adenovirus that contained a corrected gene and injected it into Gelsinger. His body initiated a massive immune response to the virus used to insert the new gene, which led to organ failure and then death five days later. The trial was a huge setback for gene therapy researchers.

In 2008, gene therapy was used to slow the progression of a rare, deadly form of Batten disease, a severe neurological condition that leaves children unable to see, speak, or breathe on their own. Ten children were treated with a corrective gene that was placed in their brains via a virus. The corrective gene was adopted by the mutant cells, which then began working normally. Eight of the children showed signs that the disease was slowing down, although they were not cured. Two children died, one from complications of the procedure.[19]

VIRUSES, RETROVIRUSES, AND ANTIRETROVIRAL DRUGS

A virus is an infectious agent that grows or reproduces by attaching itself to a host cell. Viruses have their own genetic material—either DNA or RNA—but they do not have cells. Thus, some scientists believe that viruses are living organisms, while others do not. Viruses can be described as existing on the cusp of life, with the ability to evolve as their environment warrants. Viruses are abundant in the world; scientists had discovered and classified more than 5,000 of them by 2010. The common cold, influenza, cold sores, Ebola, rabies, polio, and chicken pox are all viruses. Viruses cannot be killed with antibiotics, but they can be controlled with antiviral medication and vaccines.

A retrovirus is a type of virus that contains RNA instead of DNA. It replicates by using the enzyme reverse transcriptase to change its RNA into DNA. The DNA then integrates itself into the host genome. The virus then continues to replicate via the host's DNA. Human immunodeficiency virus, or HIV, is a retrovirus, and so are some types of leukemia. Antiretroviral drugs combat retroviruses, but because retroviruses mutate quickly and often, antiretroviral drug therapy usually involves taking several medications at once in high doses. The course of treatment may change frequently as the retrovirus mutates in the body.

FERTILITY

When Louise Brown was born via caesarean section in Manchester, England, on July 25, 1978, she became the first live-born baby to be conceived through in vitro fertilization (IVF), a process whereby an egg is removed from a woman's body and fertilized with a man's sperm in a laboratory, with the resulting zygote then implanted in the woman's womb. Many people in 1978 thought the process was highly unethical and should not be allowed.

They feared the rise of a new eugenics movement, or even human-animal hybrids. Others, such as Louise's parents, who had struggled with infertility for years and desperately wanted a child, considered the procedure a miracle. Louise Brown proved to be a normal, happy baby girl, and her doctors, Patrick Steptoe and Robert Edwards, became heroes to those who clamored for the procedure in the succeeding years. In fact, the Browns' second daughter, Natalie, was also conceived via IVF and became the first test-tube baby to become a mother herself, in 1999.[20] As the number of IVF births rose, the procedure gained acceptance. By 2006, more than 3 million IVF babies had been born worldwide.[21]

Several years after the first successful IVF, genetic engineering made preimplantation genetic selection (PGS) possible. This process tests embryos for genetic diseases such as Down syndrome, Tay-Sachs, and cystic fibrosis before they are implanted into a woman's womb. If an embryo is found to contain such a disease, it is not implanted.[22] While PGS is fairly uncontroversial when used to test for serious diseases, it can also be used to test for gender and tissue matching, which is of greater ethical concern. According to an article in the *Houston Chronicle,* 9 percent of all embryo screenings in U.S. fertility clinics in 2005 were for sex selection, without a specific medical reason.[23]

Tissue matching allows parents to choose to implant an embryo that contains the same tissue as a child with a genetic disease; some find this so-called spare parts argument morally reprehensible. Such was the case with Charlie Whitaker, who was born in 1999 with a form of anemia that left his body unable to produce red blood cells. He took drugs every day and received a blood transfusion every three weeks, and chances were good that the condition would eventually prove fatal. The British government denied Charlie's parents permission to genetically screen IVF embryos to find a match for him, so they found a doctor in the United States who would do so. An embryo found to be a tissue match with Charlie was implanted into Michelle Whitaker. She gave birth to a baby, named Jamie, whose umbilical cord cells were used to cure Charlie of his debilitating disease.[24]

Now that the artificially created human embryo is commonplace, the debate centers around how those embryos are used. IVF often results in the creation of more embryos than can be implanted in a woman's uterus. Usually the extras are destroyed, but some researchers would like to harvest stem cells from those embryos for use in research devoted to curing disease. An embryonic stem cell is valuable because it can be turned into any type of cell that researchers need, such as a brain cell, a skin cell, or a liver cell, and it can reproduce itself. To start a line of stem cells for research, an embryo must be destroyed or cloned. Many people believe both processes are unethical and could lead to an industry in which embryos and clones of embryos are created and destroyed in the name of science.

DESIGNER BABIES

In the 1997 movie *Gattaca,* society embraces eugenic practices to the point that middle- and upper-class parents use biometrics to create children of superior intellect and talent who are called Valids. For example, some children are genetically engineered to have six fingers, which makes them better pianists than their five-fingered cohorts. Such children are highly valued by society and given advantages not enjoyed by those who are in-Valid, or the product of traditional, nongenetically altered reproduction. The in-Valids believe they are more than the arrangement of their nucleotides and fight against being pigeonholed because of their ordinary DNA.

Some fear that life could imitate art with the creation of designer babies, or children whose traits have been chosen by their parents through genetic testing or alteration. The possibility brings eugenics—the practice of improving the human race by placing conditions on reproduction—into the 21st century. Supporters believe that coercing parents into negative eugenics (prohibiting them from reproducing, usually through forced sterilization) is wrong, whereas giving parents a choice about the characteristics of their offspring will be beneficial to society. Others believe that such privatized eugenics, despite good intentions, is immoral. "Some will curse these new technologies," writes Stephen L. Baird, "sounding the death knell for humanity, envisioning the social, cultural, and moral collapse of our society and perhaps our civilization. Others see the same technologies as the ability to take charge of our own evolution, to transcend human limitations, and to improve ourselves as a species."[25]

The term *designer baby* was invented by the media; doctors and scientists prefer advanced reproductive technologies (ART), which includes everything from IVF to human germ-line engineering (genetic alterations that can be passed on to future generations). While PGS simply allows parents to choose which of several embryos contains the tissue or genes they would like their offspring to possess, germ-line engineering involves changing the genetic makeup of an embryo's germ cells—egg and sperm cells—so the embryo will possess engineered traits that he or she will be able to pass along to his or her own offspring. These changes would take place at the blastocyst stage (around five days, or 150 cells) via a viral vector that would insert altered genes into the blastocyst's genome. As of 2010, such technology is in its infancy, but bioethicists acknowledge that guidelines need to be developed in preparation for the day it becomes reality.

PHARMING: GROWING MEDICINE

A growth area of biotechnology variously called pharming, biopharming, or molecular farming involves inserting genes with pharmaceutical benefits into crops that do not normally contain those genes. The crop is then harvested and either consumed for its pharmaceutical properties or refined into a marketable pharmaceutical product. Much of the new genetic material introduced into these organisms is protein. As with other forms of biotechnology used in agriculture, some worry that the genetically altered material may contaminate the general food supply. Thus, various agencies in the United States and abroad have strict regulations on pharming processes. Many scientists are optimistic about pharming because the genetic alterations target the causes of a disease, not just the symptoms. This could result in cheap, highly effective treatments.

Plant-made pharmaceuticals (PMP) could also help malnourished people in developing countries. Golden Rice 2 is a pharming product that contains the provitamin beta carotene, which can be converted by the body into vitamin A. In parts of the world where diet staples are low in beta carotene, this rice may alleviate this one deficiency. Researchers are working on a host of other pharming applications, including safflowers that contain insulin, alfalfas that contain an influenza vaccine, and potatoes that contain a vaccine for hepatitis C.[26] ToBio, a Virginia-based group of tobacco farmers, has teamed up with CropTech to produce tobacco that will contain altered proteins that can be used in pharmaceuticals and vaccines.[27] Instead of growing tobacco to sell to cigarette makers, the farmers are growing it for pharmaceutical companies.

Pharming processes that result in biopharmaceuticals—a field sometimes called biotherapeutics—in the milk of cows, sheep, or goats is another growth area. The process involves genetically altering an animal, then cloning that animal to take advantage of its modified genome. Recombinant human antithrombin (ATryn) is a protein with anticoagulant and anti-inflammatory properties that has been manufactured from the milk of transgenic goats. The drug can be used to treat people with a hereditary antithrombin deficiency or acquired antithrombin deficiency, which can result from sepsis, liver failure, and cardiopulmonary bypass surgery.[28]

Regulatory agencies have been cautious when it comes to approving biotherapeutics. The European Medicines Agency blocked the marketing of ATryn in 2006 but reversed its decision shortly thereafter. Each country has review processes to determine if a drug maker can market a substance in that country. As of 2010, ATryn is undergoing trials for FDA approval in the United States.[29] Biopharmaceutical companies are developing a host of other biotherapeutics. Dow AgroSciences hopes to market a vaccine for West Nile virus,[30] while Medicago, Inc., is developing a flu vaccine that is grown and harvested from plants.[31]

THE CHALLENGES
Food Testing: When and Who?

In the United States, the FDA is responsible for testing and approving food additives before they enter the marketplace. The FDA does not require GM seed to be tested however. It did not test seed modified with the toxic protein Bt, which kills the European corn borer, because the Bt toxin is considered a pesticide, not a food additive. The Environmental Protection Agency (EPA) is responsible for testing pesticides, but it uses a different yardstick to establish safety. By definition, a pesticide is toxic, so evaluating human consumption requires finding an acceptable level of the pesticide via a risk-benefit analysis.[32] When the science writer Michael Pollan investigated the development of Monsanto's New Leaf potato, modified with the Bt protein, he found that because the EPA had previously determined that Monsanto's potatoes were safe and that the Bt protein added to them was safe, no further studies were conducted to determine if the New Leaf potato modified with the Bt protein was also safe.[33]

Initially, GM food consisted of crops that had new genes inserted into their genomes, but increasingly scientists are finding ways of "turning off" seeds' existing genes. This could result in a "nothing added" product such as a hypoallergenic peanut, according to Alan Orloff, that stops the expression of the gene that causes life-threatening allergic reactions in some people.[34] These nothing-added products are indeed genetically modified, but by the definitions of the FDA and the EPA as they stand in 2010, the products would not need to be tested or labeled as such in the marketplace.

Bioethics

The main debate on GM food is between those who believe that the health and environmental effects of transgenic crops are not known and those who believe they have been adequately tested for safety. "Many scientists who've worked in this field take great offense at the accusation, often repeated, that genetically engineered crops harbor unknown and unresearched risks to ecosystems," wrote Daniel Charles in the book *Lords of the Harvest*. "They insist that these issues have, in fact, been researched intensively, but that their opponents studiously ignore the results."[35] Consumer advocates state that regardless of the research, GM foods should be labeled in the marketplace and that testing should continue in order to determine long-term effects of the technology.

BIOTECHNOLOGY AND GENETIC ENGINEERING

The main issues relating to biotechnology and livestock concern maintaining the health of animals and making sure the meat they provide does not adversely affect human health. People are also concerned about animal rights as agriculture becomes increasingly mechanized with factory-like conditions that disregard animals' health and well-being in favor of obtaining as much product out of them as quickly as possible. Cattle are force-fed food they would not normally eat (e.g., corn instead of grass) and injected with antibiotics that may lessen their effectiveness for human use and with hormones that may cause painful diseases. Their movement is often severely restricted, and the waste they generate pollutes the environment.

When it comes to genetic research and ethics, the issue often boils down to resources. Millions of people die each year in the developing world from influenza, malaria, and other diseases. Fewer, but still a substantial number, die from cancer in the developed world. Yet because the developed world controls the majority of research funding, more money is invested in research to cure cancer than to combat malaria and AIDS in the developing world. Columbia University professor Philip Kitcher spoke about the ethical implications of this situation:

> *It seems to me a very, very serious question whether we should be pouring as much money as we are into the molecular genetics of cancer when we could be sequencing the genomes of known pathogens and working very hard to develop vaccines that could possibly prevent diseases that kill millions and millions of children annually. Surely there are some serious moral issues here. We can certainly justify the expenditure of some funds on these molecular tools against cancer, but from the figures I've gotten so far the disproportion between what we do with respect to cancer and what we do with respect to diseases like malaria is extraordinary.*[36]

GENETIC DISCRIMINATION

Genetic testing can easily identify cystic fibrosis or muscular dystrophy. Newborn babies are routinely tested for phenylketonuria and prospective parents for Tay-Sachs disease. This causes no real problems—a positive diagnosis will help a person get the treatment he or she needs as quickly as possible. But what about genetic testing to find out if a person is at increased risk for heart disease, Alzheimer's disease, diabetes, or colon cancer? This might be welcomed by those who could use the information to head off a health crisis. Genetic testing, however, is controversial. For example, what if a person is denied health insurance because such tests show that he or she may develop a serious illness?

Introduction

So far, genetic screening is most acceptable and most accurate for those who have a family history of a disease with a known genetic component. But soon geneticists will be able to test for a myriad of predispositions, rather than actual diseases. Having a predisposition to a disease is information, but murky information, some bioethicists say. Why should health insurers pay for expensive tests that are not diagnostic in nature? Patients may demand costly procedures to ensure they do not get a disease, such as a woman with a family history of breast cancer who elects to undergo a mastectomy. Such a procedure would be costly and not entirely risk-free, and there is no guarantee she would have developed breast cancer in the first place. Finally, the expense could all be for naught if the patient simply develops a different kind of cancer.

SHOULD GENES AND DRUGS BE PATENTED?

Many drugs developed by private companies are patented, meaning the companies own the exclusive right to manufacture and sell them. This situation has given rise to some ethical questions. While some believe that the technology to make possibly lifesaving drugs should be made available to all, others say that the enormous amount of money it takes to research and develop new drugs requires companies to protect their investments with a patent. "No patents, no cures" goes the slogan. Yet when someone asked Jonas Salk in 1952 why he did not patent his vaccine for polio, he replied, "Would you patent the sun?"[37] Salk meant that something so crucial for so many people should not be selfishly protected by corporate interests. The ethicist Philip Kitcher takes a similar position: "It's inappropriate for medicine to enter this kind of free-market mentality."[38] Nevertheless, in the United States and most other countries, that is exactly what has happened. The lively public debate often extends to Capitol Hill, where industry lobbyists go head-to-head with citizen activists. For example, the Greater Access to Affordable Pharmaceuticals Act was introduced in the U.S. House of Representatives in 2002 but never made it to a vote. The bill would have prevented pharmaceutical companies from extending their patents after they expire, thereby allowing people greater access to less expensive, generic prescription drugs.

The ramifications of genetic testing on people's ability to obtain health insurance are huge. If people discover they have genes that predispose them to a serious disease, they may be denied insurance for themselves or their families, even if they are still healthy. While test results may prompt people to make lifestyle changes to lessen their chances of developing a disease, this would not necessarily change insurance companies' decisions because the business model relies on calculated risk. In a 2007 letter to the Senate Health, Education, Labor and Pensions Committee, representatives of the American Civil Liberties Union (ACLU) summarized the issue:

21

Genetic information may identify an individual's predisposition to de-velop certain diseases, allowing for early diagnosis and treatment and ensuring that people can make informed decisions and retain maximum control over their health. This information, however, can also be misused to deny individuals health care and employment. Some cases of discrimi-nation have already been documented. For example, one woman was denied health care for her children because they carry a gene for alpha-1 antitrypsin deficiency (AAT), even though her sons are merely carriers and will never develop the condition. The occurrence of such cases is certain to increase as genetic testing becomes more common in the fu-ture. Fear of such discrimination will also have a chilling effect, causing individuals to refuse potentially lifesaving testing due to fear of how the results will be used by employers, insurers, or the government.[39]

Biotechnology and Religion

A country's regulations regarding technology often partially depend on its predominant religious tradition. In western countries that are home to many Catholics and Protestants, religious beliefs often curtail support for embryonic stem cell research, cloning, and genetic engineering of animals and people. The Roman Catholic Church has published statements outlining its position against cloning and cautions against biotechnology. Archbishop Gianfranco Girotti, a Vatican spokesperson for matters of conscience, stated in 2008 that within bioethics, "there are areas where we absolutely must denounce some violations of the fundamental rights of human nature through experiments and genetic manipulation whose outcome is difficult to predict and control."[40] GM food, however, is condoned by the Catholic Church as a way to feed those most vulnerable to starvation and malnutrition in the world.

Protestant denominations have similar views. The Evangelical Lutheran Church in America (ELCA), for example, endorses biotechnology in general but opposes research that creates human embryos for the purpose of destroy-ing them. It acknowledges that scientific progress may reach the point where cloning is an acceptable practice and is in the process of developing a state-ment on genetics. The Southern Baptist Convention on numerous occasions has opposed embryonic stem cell research, citing the Nuremberg Code, the Declaration of Helsinki, and the UN Declaration of Human Rights, but it sup-ports adult stem cell research because it does not destroy a potential life. It also opposes fetal tissue experimentation and any form of human germ-line genetic modification.[41]

Jewish rabbis, both Reform and Orthodox, have determined that biotech foods are kosher[42] and support stem cell research and therapeutic cloning on the basis of the doctrine that states a child does not receive his or her soul until eight days after birth. Such biotech and gene research, because it aims to save lives, is welcomed among people of the Jewish faith.

Islam, the world's second-largest religion, has specific guidelines for dealing with issues of biotechnology and engineering, which are rooted in the teachings of the Quran. Islamic law differentiates between things that are halal, or lawful, and those that are *haram,* or forbidden. In the case of biotechnology, Muslim scholars use four guidelines to determine whether a practice is halal or *haram:*

1. Necessity overrules prohibition.
2. Choose between the lesser of two evils if a situation cannot be avoided.
3. Actions are accepted or allowed depending on a person's original intention.
4. Everything is halal unless it has specifically been deemed *haram.*

Using these guidelines, the U.S. Islamic Jurisprudence Council (IJC) has determined that all GM foods available as of 2004 are halal. This may not always be the case; Muslim leaders caution that genetically altered food should not contain genes from animals. Any genetic engineering involving pigs, which are forbidden in the Muslim diet, is hotly debated. Biotechnology in general, however, is encouraged by Muslims when it aims to improve human well-being and eliminate suffering. Under certain conditions, it is even obligatory. Under others, however, it is unequivocally forbidden. The use of reproductive technology is almost always forbidden. IVF is permissible only if a woman is able to receive her husband's sperm and the couple is still married. Donor sperm and surrogate mothers are not allowed under any circumstances. If the husband is sterile, IVF is not an option. However, because Muslims sometimes practice polygamy, IVF with a husband's sperm and a wife's egg implanted into a second wife is permitted. Spare embryos created during the IVF process are a difficult issue. They may be used by the couple, but if they are not, scholars debate whether they should be used for research purposes. Certainly they are not allowed, under Islamic law, to be donated to an infertile couple. Most believe that destruction of the embryo is acceptable because the Quran teaches that ensoulment takes place on the 120th day of gestation. Some believe that destruction of embryos for the sake of research that could lead to cures for human disease is obligatory because to destroy them would be wasteful.

Genetic screening is allowed under Islamic law, as is prenatal testing to save a mother's life. Abortion for genetic abnormalities is not allowed. Gene therapy, like IVF, is conditional. Somatic gene therapy, which would alleviate a person afflicted with a disease, is allowed and even encouraged because it relieves human suffering. However, germ-line therapy is not acceptable because it aims to make permanent changes in genes that will be passed to future generations and could be used for nonessential treatments such as improving intelligence or enhancing beauty. Such interference in Allah's creation, Muslims believe, will upset the balance of the universe.[43]

When it comes to cloning, Muslim law forbids creating a genetic copy of a person in part because of the way it could affect the relationships between man and woman and parent and child. On the other hand, animal cloning is allowed if it is done with an eye toward improving human lives, and cloning human genes is permissible if it alleviates human suffering.

Hinduism is the world's third-largest religion, with roughly 800 million adherents. It has no single written document of law but consists of a wide variety of beliefs and traditions formulated over the centuries. Its various sects practice in numerous ways around the globe, so making pronouncements as to what Hindus believe is difficult. However, most Hindus believe in the eternal soul and reincarnation, and this significantly affects their views on biotechnology. "Asian religions worry less than Western religions that biotechnology is about playing God," according to Cynthia Fox, a writer on stem cell research. "Therapeutic cloning in particular jibes well with the Buddhist and Hindu ideas of reincarnation."[44]

Ideas about reincarnation prompted the South Korean researcher Hwang Woo-Suk, who claimed to have cloned human embryonic stem cells but was later found to be a fraud, to state that his work honored Buddhist beliefs. The U.S. geneticist and author Lee Silver has looked at biotechnology's religious divide in his book *Challenging Nature*. As he explained to the journalist John Tierney, "Most people in Hindu and Buddhist countries have a root tradition in which there is no single creator God. Instead, there may be no gods or many gods, and there is no master plan for the universe. Instead, spirits are eternal and individual virtue—karma—determines what happens to your spirit in your next life. With some exceptions, this view generally allows the acceptance of both embryo research to support life and genetically modified crops."[45]

Green Revolution to Gene Revolution

After World War II, as the world's population spiraled out of control, many governments became concerned that they would not be able to feed everyone within their borders. The specter of mass starvation loomed. Mexico was one of the first countries to sound the alarm. In 1944, it imported half the wheat it needed but wanted to become self-sufficient. The government hired the agronomist Norman Borlaug, who had worked with the Civilian Conservation Corps during the Great Depression, the U.S. Forestry Service, and at the DuPont Chemical Company, to find a solution. Funded with grants from the Ford Foundation and the Rockefeller Foundation, Borlaug turned his attention to the problem, and by 1956 Mexico was producing enough wheat to feed its population. Furthermore, within a few short years it was exporting wheat to other countries. Mexico's quick turnaround combined with that seen in other populous countries was dubbed the Green Revolution.

Borlaug was a plant geneticist and microbiologist for DuPont when he accepted the position in Mexico. In the first few years, he experimented with 6,000 crossbreeds of wheat, producing varieties that were high yield and disease resistant. More important, he instituted two growing seasons per year, thereby doubling the amount of wheat the country produced. With Mexico as a template for success, Borlaug assisted other countries such as India and Pakistan in attaining food security. In 1970, Borlaug received the Nobel Peace Prize for his role in alleviating world hunger. He is often credited with saving more than 1 billion people from starvation through his work. Although Borlaug's initial success predated the introduction of GM foods, as a plant geneticist he has always been a supporter of biotechnology as a solution to world hunger.

Among the critics of GM foods are those who do not object to them on principle but on the circumstances of their development. For instance, most GM crops have been developed to benefit the bottom line of large agricultural corporations. While GM seed may yield a greater harvest and prevent pest damage, it is also more expensive than non–GM seed, promotes environmentally damaging monoculture, cannot legally be reused by farmers, and sometimes requires the application of a corresponding herbicide (such as Monsanto's Roundup). These conditions bode well for those who hold the patent on the seed but are not necessarily conducive to helping those who need help the most—subsistence farmers in the developing world who cannot afford the high-priced seed and herbicides. In fact, Vandana Shiva, a respected agriculture activist from India, believes that conditions surrounding the distribution of GM seed are destroying subsistence farmers' ability to feed themselves.[46] What smallholder farmers need, Shiva believes, is access to a variety of crops suitable for the land they cultivate and the promotion of time-honored traditions that protect the environment, such as crop rotation and natural pest control. What they do not need, she says, is GM seed that is grown specifically to become feed for livestock or as an ingredient in another product in which the farmer has no stake.

The Food and Agriculture Organization (FAO) of the United Nation's 2004 annual report, *The State of Food and Agriculture 2003–2004,* called for a "gene revolution" on par with the Green Revolution of the 1960s that would bestow GM seed on those who would most benefit. In the next 30 years, the population of the poorest countries is forecasted to swell by 2 billion. The gains of the Green Revolution have made continued population growth possible, and a new solution is needed. However, most GM seed planted worldwide is corn, along with lesser amounts of soybeans, canola, and cotton. The gene revolution needs to apply biotechnology to a wider range of crops. According to the FAO Director-General Dr. Jacques Diouf, "Neither the private nor the public sector has invested significantly in new genetic technologies for the so-called 'orphan crops' such as cowpea, millet, sorghum and tef that are critical for the food supply and livelihoods of the world's poorest people."[47]

Furthermore, according to the FAO:

> [GMOs] can provide farmers with disease-free planting materials and develop crops that resist pests and disease, reducing use of chemicals that harm the environment and human health. It can provide diagnostic tools and vaccines that help control devastating animal diseases. It can improve the nutritional quality of staple foods such as rice and cassava and create new products for health and industrial uses.[48]

India could also greatly benefit from a gene revolution. Predicted to overtake China as the world's most populous country by 2050, India must attain food security by addressing the problems of soil erosion, water shortages, and rural poverty. According to C. S. Prakash, the director of the Center for Plant Biotechnology Research at Alabama's Tuskegee University, "India also has serious problems of blast in rice, rust in wheat, leaf rust in coffee, viruses in tomato and chilies and leaf spot in groundnut across the country. These problems can be significantly minimised in an ecologically-friendly manner with the development of genetically reprogrammed seeds designed to resist these disease attacks, while minimising or even eliminating costly and hazardous pesticide sprays."[49]

Even in the United States, some farmers are transitioning from traditional crops, such as wheat and corn (whose markets have fluctuated unfavorably) to transgenic crops that can benefit third world countries. Rice genetically modified with proteins from human milk, saliva, and tears is being test-grown in Missouri in the hopes that it may be consumed by at-risk populations in countries that suffer from high death rates due to diarrhea.[50] These GM crops could produce food that is medically beneficial in areas that have inadequate health care or sanitation systems. It could also help domestic farmers gain a better foothold in an industry that suffered an almost total collapse in the 1980s.

But would a gene revolution be overkill? Some believe the food problem requires a more simple solution. A low-tech agricultural practice called the system of rice intensification (SRI) could produce greater yields and require little in the way of scientific intervention. More than half of the world's population depends on rice, and between 2007 and 2008 its price tripled, laying the groundwork for a possible humanitarian crisis in some of the world's most fragile economies. Norman T. Uphoff, the former director of the Cornell International Institute for Food, Agriculture and Development (CIIFAD), developed the SRI as a way to help solve the global food crisis. No genetic engineering is necessary, according to Uphoff. Farmers simply plant rice early, give seedlings more room to grow, water them less, and rotate crops annually. Fewer seeds and deeper roots make for harvests roughly two to three times larger than traditional cultivation practices allow.[51] If such processes are so easy, then why have farmers not adopted them sooner, ask critics of SRI. They believe the claims made for SRI are exaggerated and that the system cannot be replicated on a wide scale. While basic, it also requires much old-fashioned weeding by farmers. Some believe that this will negatively affect women in developing countries, who often undertake much of the heavy labor.

Experts believe that GM food has yet to make an impact on securing the global food supply because it is not practiced on the crops that matter most to people in developing countries: potatoes, cassava, rice, wheat, millet, and sorghum.[52] Ignoring these in favor of frost-resistant strawberries and stay-ripe bananas leaves GM food in the realm of a boutique industry rather than a marketplace necessity. There is no economic incentive for private companies to invest in research and development into the crops grown by subsistence farmers in the developing world, a phenomenon known as the "molecular divide."[53] Technology typically originates in the developed world but without economic incentives does not transfer to areas where it could help others most, particularly sub-Saharan Africa. That may be changing. In 2008, Monsanto announced plans to develop seeds that would double corn, soybean, and cotton yields by 2030 using less land and water. The effort is directed at "improv[ing] the lives of small and poor farmers by sharing [Monsanto's] technology" without charging royalties.[54] As the journalist Andrew Pollack explained, the plan is "aimed at least in part at winning acceptance of genetically modified crops by showing that they can play a major role in feeding the world."[55]

Pest Control

RESISTANCE TO GM CROPS

Charles Darwin called it survival of the fittest. In order for organisms to evolve and continue to thrive, they must adapt to their environment. In the past, adaptation took place on an evolutionary timescale over millions of years. But everything happens faster in the modern world, including evolutionary change. Although GMOs have existed for a mere 20 years, scientists are already seeing signs that microorganisms can adapt to them in order to ensure their survival.

Bt cotton developed to ward off the destructive bollworm appears to be the first casualty. Bollworms that have developed a genetic mutation that enables them to withstand the Bt toxin have been found in GM cotton crops in Mississippi and Arkansas.[56] This development occurred in less than a decade and was fastest in areas where monoculture practices eliminated nearby groves, weeds, and trees where the bollworms could have survived without destroying the Bt cotton.

HYDROPONICS AND INTEGRATED PEST MANAGEMENT

Hydroponics is the practice of growing plants without soil. Removing soil from the equation means bypassing many of the problems associated with traditional agriculture. Hydroponic biotechnology involves growing crops in a water-based solution containing nutrients. It has many advantages over soil farming. For example, farmers can grow plants indoors year-round using organic techniques, all while conserving water and eliminating the need for pesticides and herbicides.

Hydroponics was practiced by the ancient Babylonians and Aztecs, and in modern times it is common in Israel and other countries where the soil and climate are too harsh for widespread agriculture. In 1627, Sir Francis Bacon's posthumously published manuscript *Sylva Sylvarum* was the first written work to outline the best techniques for growing plants without soil. Botanists improved techniques in the 1860s, and in 1937 the term *hydroponics* was coined by the University of California, Berkeley, scientist William Frederick Gericke, author of *The Complete Guide to Soilless Gardening.* As of 2009, the largest commercial hydroponic operation is Eurofresh Farms in Willcox, Arizona, which produces more than 100 million pounds of tomatoes annually. Anything described as "hothouse grown" typically refers to some variation of hydroponics.

Integrated pest management is an umbrella term for a combination of tactics used to grow agricultural crops without the commercial use of pesticides and herbicides. The main idea is to limit pests to an acceptable level—not eliminate them altogether, which is costly, dangerous to the environment, and often unnecessary. Instead, farmers choose the right crops for the existing land conditions and maintain them by hand-pulling weeds, tilling the soil, and setting manual bug traps. The next step is to introduce natural pest control by releasing bugs or roundworms that eat pests or species of fungi that actively kill insects. Pesticides are the final step, used sparingly and only as necessary. This lessens the chance that bugs will become resistant to pesticides and ensures that agriculture will remain sustainable in a given environment.

Perry Adkisson and Ray F. Smith received the 1997 World Food Prize for their work in developing integrated pest management. Smith spent his career as a professor of entomology at the University of California, Berkeley, and helped form the FAO's Panel of Experts on Integrated Pest Control and the Consortium for International Crop Protection, both of which brought integrated pest control techniques to farmers throughout the world.

Transgenic Contamination

GENE WANDERING AND GENE STACKING

Gene wandering refers to the comingling of GM crops and non–GM crops through unintentional natural processes, such as seed blowing across a highway from a GM field to a non–GM one. Once the GM seed is released into the environment, it is nearly impossible to contain. Its modified genome will mingle with non-modified genomes, thereby altering numerous species within a given ecosystem. Farmers will no longer be able to control exactly what they are growing. "The biotechnology companies attempt to place the responsibility on the farmers whose crops have been contaminated, with a patent infringement suit as their ultimate weapon," wrote Jane Matthews Glenn in the *Washburn Law Journal*.[57] This would hurt the organic farming industry the most because once GM seed is deposited in an organic crop field, the crop can no longer be considered organic, and a farmer's entire livelihood could be lost. Gene wandering is a problem for countries that regulate their seed pools and require an approval process for plants with new traits. Organic farmers in Canada have filed a lawsuit against biotechnology companies because of this issue, and it is likely that more cases will arise in an attempt to "salvage traditional and alternative agriculture from the stranglehold of agricultural biotechnology."[58] Another concern is gene stacking, where two or more GM plants crossbreed in the wild, resulting in a variety that requires more toxic herbicides than currently used to control rogue plants.[59]

THE QUIST-CHAPELA REPORT

In 2001, Ignacio Chapela and David Quist, professors at the University of California, Berkeley, published a paper in *Nature* claiming that GM corn grown in the United States had contaminated indigenous varieties of maize (corn) in a remote area of Oaxaca, Mexico. Maize is indigenous to Mexico, where it has been cultivated for more than 7,000 years; Mexico is also home to dozens of varieties that are not grown in the United States. In an effort to protect its biodiversity, Mexico banned GM corn from being grown in the country in 1998, although GM corn harvested in other countries is allowed to be sold there. Chapela and Quist's claims were met with a firestorm of controversy; *Nature* soon issued a retraction—something the journal had never done in its 133-year history—and Chapela himself was initially denied tenure at his university, which had ties to the biotech industry.[60] "The Quist-Chapela report not only suggested that transgenic corn had been widely planted, it also reported that the foreign DNA appeared in diverse locations within the maize genome—in other words, the transgenes that were spliced into corn plants were able to 'jump around' the chromosomes. Such movement would pose the risk of disrupting the functioning of other genes."[61] Further dangers could arise, such as superweeds or pests that are resistant to GM corn but are capable of destroying neighboring crops.

Chapela and Quist's detractors believed the scientists' findings were the result of false positives obtained through the process of polymerase chain reaction. The *Nature* editors asked Quist and Chapela to provide further documentation of their claims, so they conducted different tests using a process known as dot blotting. These tests appeared to substantiate their original claims. Detractors were still not convinced, even though they agreed with Quist and Chapela that transgenic corn existed in Mexico. "In consequence, the dispute is less over the likely presence of transgenic maize than whether Chapela and Quist actually demonstrated it, and whether foreign DNA is as widespread and unstable as they claim."[62] Controversy rages because the stakes are high. The public backlash against GM food would increase if it were found to be uncontrollable within an ecosystem.

Bioterrorism and Biosecurity

Weaponized anthrax (*Bacillus anthracis*) mailed to members of the U.S. government and news media killed five people and infected 22 others in September and October 2001. It was one of the most disruptive acts of bioterrorism in history, requiring a years-long $200 million cleanup of U.S. postal facilities and a $41.7 million cleanup of Capitol Hill. Early suspicions that the anthrax came from al-Qaeda terrorists were mostly dismissed when it was discovered that it possessed the genetic fingerprint of the Ames strain that is held at Fort Detrick, Maryland. After a seven-year investigation, the FBI presented evidence that the crime was perpetrated by Bruce E. Ivins, a U.S. Army microbiologist, who committed suicide before criminal charges were filed against him. The episode prompted the passage of the Project Bioshield Act in 2004, which provided $5 billion for vaccines in case of a bioterror event, and the Biodefense and Pandemic Vaccine and Drug Development Act of

2005 ("Bioshield 2"), which cuts approval time for new drugs to hit the market in the case of a pandemic. Both laws provide immunity to pharmaceutical companies if their vaccines are found to be harmful to humans.

Anthrax is a highly lethal bacteria that in its natural form affects mainly grazing mammals. It caused deadly plagues in animal herds in ancient times, and it often infected humans who came into contact with diseased animals. Louis Pasteur developed the first vaccine for anthrax in 1881, and today it can be treated with potent antibiotics. Anthrax was first used as a biological weapon against the Russian army in World War I, and during World War II the British army conducted extensive weapons testing with anthrax bombs on Gruinard Island in Scotland, although these were never used in battle. The island remained uninhabited until it was decontaminated in 1990. The strain of anthrax used on Gruinard was more lethal than the Ames strain.

Bioterrorism and biowarfare refer to the practices of altering or using naturally occurring bacteria, viruses, or toxins to kill or harm individuals. The ancient Romans threw rotting animal carcasses into wells to pollute an enemy's water supply, while in medieval times marauding invaders used the bubonic plague as an inducement to get people to flee a city. In the Americas, the smallpox virus carried by Europeans wiped out entire indigenous populations, and during the American Revolution smallpox spread by British forces decimated the colonies, especially during the siege of Boston, leading some to believe the British were using the virus as a weapon. Mustard gas used as a weapon killed thousands during World War I, but not nearly as many as the unintentional Spanish flu epidemic that washed over the globe at the war's end. As weapons, biological agents have many advantages; the raw materials needed to develop them are often readily available, and they are hard to trace and easily disseminated.

Biological agents were outlawed as weapons of war by the Geneva Protocol of 1925, but the protocol did not prohibit the development of weapons and did not contain an enforcement clause. The United States continued to develop bioweapons until President Nixon halted the program in 1969. Although episodes of biowarfare and bioterrorism have been few since World War I, many countries continue to develop the weaponry through secret programs. In Iraq, mustard gas was used on Kurdish citizens in the 1980s, killing thousands. Such biological agents were part of the weapons of mass destruction reasoning behind the U.S. invasion of the country in 2003.

Bioethics on the International Stage

Bioethics came into its own with the publication of Van Rensselaer Potter's *Bioethics: Bridge to the World* in 1971. Potter was a professor of oncology at the University of Wisconsin–Madison and coined the term *bioethics* to denote a philosophy that takes biology, medicine, ecology, and human values into consideration. He did not equate the term with biomedical ethics, which takes specific medical research subjects into consideration, although many do. Despite the more recent coining of the term, most trace the beginning of bioethics to the drafting of the Nuremburg Code, when countries outraged by Nazi atrocities during World War II agreed that certain forms of science have no place in civilized society. Most of the atrocities in question were committed by Josef Mengele, a physician and SS officer at the Auschwitz-Birkenau concentration camp. He was interested particularly in heredity and twins, and he conducted many experiments on the camp's inmates, usually without anesthesia. These included dissecting live infants, castration, amputation, sterilization, organ removal, and injecting chemicals into peoples' eyes.[63]

THE NUREMBERG CODE

When Mengele's atrocities came to light following the war, the Nuremberg Code of 1947 was established to limit experiments involving human beings. The code is nonbinding, meaning violations cannot be prosecuted, but along with the 1964 Declaration of Helsinki developed by the World Medical Association to guard against unethical human experimentation, the Nuremberg Code was incorporated into many countries' legal documents, including the Code of Federal Regulations established by the U.S. Department of Health and Human Services.

The Nuremberg Code contains 10 points, and its cornerstone is informed voluntary consent of all participants in any research study and their right to back out of the experiment at any time for any reason. Coercion of any kind—economic, physical, or mental—is prohibited. Additionally, the true nature and extent of the experiment must be made known to the subject. The study should be designed well, contribute to the public good, and not be feasible without human testing. No test should take place if suffering or pain is likely to be a known result. Many of these points have become standard in the decades since; all doctors in the United States are required to have patients sign informed consent policies.

CODEX ALIMENTARIUS

The Codex Alimentarius ("food code" in Latin) contains the standards and practices developed by the Codex Alimentarius Commission, established in 1963 by the FAO and the World Health Organization (WHO). The commission's priorities are to protect the health of consumers and ensure fair international trade. By 2000, it was clear that the Codex needed to address issues pertaining to biotechnology, so the Codex Alimentarius ad hoc Intergovernmental Task Force on Foods Derived from Modern Biotechnology met in Japan. The task force acknowledged that GMOs may help many countries attain food security but also recognized that GMOs may present new risks to human and animal health and the environment. It therefore advocated a stringent case-by-case evaluation of the risks and benefits of each new GMO.[64] The task force also noted that most research and development takes place in the private sector in wealthy countries and encourages efforts to transfer technology to developing countries through public funding. The FAO has also established a Commission on Genetic Resources for Food and Agriculture, which is open to all UN members and acts as an honest broker between nations involved in issues of plant and animal biotechnology.[65]

Since then the United Nations has created other committees designed to guide discussions between countries on biotechnology issues. The UNESCO International Bioethics Committee was created in 1993 to monitor advances in life sciences. The panel of 36 independent experts does not make judgments about particular technologies, but it encourages education. In 1997, the UN General Assembly unanimously adopted the Universal Declaration on the Human Genome and Human Rights as a tool to "safeguard respect for human rights, fundamental freedoms and human dignity and to protect public health."[66]

INTERNATIONAL HISTORY

The Fertile Crescent

Agriculture originated about 10,000 B.C.E. in the Fertile Crescent, a moon-shaped area spanning Egypt's lower Nile Valley, the top of the Arabian Peninsula, and the wedge of land between the Tigris and Euphrates Rivers in modern-day Iraq. The Fertile Crescent was unique; it boasted rich soil and the eight indigenous Neolithic founder crops (including emmer, einkorn, barley, flax, chick peas, and lentils) that allowed Neolithic peoples to transition from hunting and gathering to agriculture. Seeds were selected, stored over winter, and planted in spring. The process of choosing certain seeds for their expressed characteristics and learning how to irrigate and rotate crops was the genesis of biotechnology. Indigenous animals such as cows, goats, sheep, and pigs aided these early farmers with labor and were themselves sources of food.

Bread and Beer

Fermentation followed closely on the heels of agriculture, even though the scientific properties of enzymes and yeast were not understood. The Egyptians were probably the first to brew beer on a wide scale almost 9,000 years ago, and other cultures independently evolved their own brewing processes. Bread baking probably followed shortly thereafter, and people also learned to domesticate animals and keep bees to make honey. Animal husbandry arose in which specific animals were bred to be docile, easily herded, and suited to labor.

The Dawn of Medicine

The first antibiotic was moldy soybean curd, used by the Chinese 2,500 years ago to treat skin infections. The Chinese also practiced variolation, a primitive type of vaccination in which people were purposely and mildly infected with smallpox to create immunity against future infections. The eighth-century Indian physician Madhav described a similar process in his medical textbook, *Nidāna*. The Sudanese-Nubian civilization of Africa used a form of what is now known as tetracycline as an antibiotic as early as 350 C.E. In the Middle Ages in Europe, tinctures made from plant extracts or cheese curds were used to ward off infection.

The smallpox vaccine was improved over the centuries. Lady Mary Wortley Montagu witnessed the vaccination process in Turkey in 1717 and brought it to England, but it failed to catch on. Within a few decades, a vaccine using cowpox as an immunization was developed by Edward Jenner, which became standard. Even without knowledge of how viruses work, early scientists were able to administer effective vaccines.

The Greeks Explain It All

Around 500 B.C.E., the Greek mathematician Pythagoras surmised that heredity was the dominion of males. Females provided nourishment and safety for the unborn, but the father provided all the traits—physical and otherwise—that a child possessed. A few years later, the philosopher Empedocles, noting obvious similarities between children and their mothers, added to Pythagoras's idea by theorizing that the man's semen mixed with fluids inside the woman's body, resulting in a child with characteristics of both parents. Aristotle, building on his predecessors' theories on heredity, formulated his own around 350 B.C.E. He believed that both the mother and father contributed to the physical makeup of their offspring, but he attributed heredity to a mixture of semen and menstrual blood. This idea was overturned only in the 17th century by the English physician William Harvey, who was the first to suggest the process of fertilization took place inside the woman's body. Harvey's ideas were verified by the Dutch scientist Antoni van Leeuwenhoek, the father of microbiology, who in 1677 developed a microscope powerful enough to view sperm for the first time and witnessed the fertilization of an egg.

Introduction

The Age of Science

GREGOR MENDEL: THE FATHER OF GENETICS

Gregor Mendel was an Austrian monk who studied the inherited characteristics of pea plants, resulting in the first documented understanding of dominant and recessive phenotypes. A phenotype is any observable physical characteristic of an organism. (An organism's genotype is its inherited set of instructions, which can be influenced by the environment.) Mendel crossbred pea plants with various characteristics and meticulously noted which physical traits were passed on to the offspring. For example, a pea plant with wrinkled seeds crossed with a pea plant with smooth seeds created a plant with smooth seeds. From this, Mendel concluded that smooth seeds were a dominant phenotype, or physical trait, passed on by alleles—the coding sequence portion of a gene—obtained from the parent plant. Mendel investigated other traits too, such as blossom color, pod color, and pod shape. His in-depth experiments led him to understand which phenotypes predominated when various combinations of plants were bred. Sadly for Mendel, his landmark 1866 paper, "Experiments on Plant Hybridization," was ignored by the scientific establishment upon its publication. Recognition came posthumously in the early 1900s, when scientists rediscovered his work and were able to replicate his findings.

LOUIS PASTEUR AND GERM THEORY

Louis Pasteur was a French microbiologist and chemist whose work led to greater understanding of how diseases are transmitted. While much of his early work involved the chemical properties of crystals, he is perhaps most well known for his vaccines for rabies, cholera, and anthrax. He discovered the anthrax vaccine in 1877, when he infected sheep with anthrax and saprophytic bacteria (bacteria that feeds on dead tissue) at the same time, and the sheep remained healthy. Pasteur also developed pasteurization, the process of heating milk to kill bacteria that can cause illness. The Pasteur Institute in France, named for him, continues to be a leading institution in the field of bioengineering.

The cornerstone of Pasteur's contribution to science was his experiments that demonstrated germ theory. Simply stated, germ theory is the idea that microorganisms cause disease. The prevailing theory before Pasteur's time was that living organisms arose from decaying matter; that is, mice were generated by haystacks, and maggots were generated by rotting meat. Pasteur, building on his predecessors' work, especially that of van Leeuwenhoek, believed that fermentation of decaying matter allowed microorganisms to reproduce. Germs were microorganisms, and if allowed to reproduce, they would cause disease. Pasteur's ideas led to the development of basic hygiene practices in the medical profession. Doctors began washing their hands before operating on patients, resulting in a much lower mortality rate from infection. People finally began to realize that epidemics could be traced to polluted water or food.

CHARLES DARWIN AND *ON THE ORIGIN OF SPECIES*

Charles Darwin's first experience with the giant Galápagos tortoise was a culinary one. During his five-year voyage on the HMS *Beagle,* the ship dropped anchor on the remote islands off the coast of Ecuador, and the crew feasted on the large amphibians. Darwin was a clergy student and amateur geologist when he joined the survey expedition as a companion to the ship's captain; he took voluminous notes on his observations of South America, Tierra del Fuego, Australia, and the ship's other ports of call and sent home many exquisite fossils of extinct animals. Only in hindsight did the diversity of species among the tiny Galápagos Islands spark the eureka moment that led to Darwin's theory of natural selection. The publication of his ideas in *On the Origin of Species* in 1859 was a seminal moment in the history of science.

The theory of natural selection states that all species of plants and animals have evolved over time from common ancestors—useful traits are preserved in a species, while detrimental ones are extinguished. Evolution takes place as members of a species survive and pass their traits on to their offspring. In this way, species change. If a bird needs a long beak to nab worms from its rocky environment, birds born with short beaks will die out, leaving only long-beaked birds to reproduce. Much of Darwin's theory has proven useful to the study of heredity and environment. One misstep was his theory of pangenesis, which stated that gene-carrying gemmules that moved through the bloodstream were responsible for inherited traits.

Darwin adapted his ideas about natural selection to human evolution in *The Descent of Man, and Selection in Relation to Sex,* published 12 years after *On the Origin of Species.* Much of this book concerns the idea of polygenism, as opposed to monogenism. Polygenism was the prevailing opinion among anthropologists at the time; it stated that each race of people was a distinct species. The idea was used to justify slavery, a practice Darwin found abhorrent. Darwin railed against polygenism and promoted monogenism, or the idea that all humans belong to the same species. According to monogenism, differences in skin color are only superficial and mainly the result of a people's physical environment. The idea of sex selection explained differences between males and females of a given species: males who displayed sought-after characteristics were more likely to attract and mate with females and pass on their genes. Sex selection explains the resplendent plumes of the male peacock, for instance, whose length and colors would at first appear to be a hindrance to it. But the male's colorful display is intended to attract females, so the longer and more colorful the feathers, the better. Darwin used the same idea to explain aesthetic instincts in human beings and how women and men are attracted to each other and select their mates.

FRANCIS GALTON: THE FATHER OF EUGENICS

Charles Darwin's cousin, Sir Francis Galton, was captivated by *On the Origin of Species*. Galton was a renowned intellectual himself, and inspired by Darwin's book, he embarked on his own exploration of human variation. Such was the birth of eugenics, a word of Galton's own devising, which is defined as the improvement of humankind through selective breeding, sterilization, and other forms of intervention. Galton's seminal work, *Hereditary Genius*, was published in 1869 and launched the fierce modern debate of nature versus nurture. Galton believed that abilities were inherited and cited his own eminent, privileged family as evidence. His theory was the culmination of meticulous surveys in which he collected information from a wide variety of individuals about their race, heritage, birth order, and occupation. While it was more remarkable as a sociological study than as an argument for eugenics, it was influential in many countries for many years.

Several years later, Galton turned to twins in order to explore the nature versus nurture concept further. His research culminated in the publication of *The History of Twins*, in which he outlined the similarities and differences between twins reared in similar environments and those raised in divergent environments. Further research into heredity led Galton to conduct experiments that ultimately refuted Darwin's notion of pangenesis. In the end, Galton's research veered close to Mendel's, and his mathematical approach to studying human characteristics laid the groundwork for modern biometrics.

EUGENICS AROUND THE GLOBE

In the mid-19th century, Galton's ideas about eugenics became the basis of a social movement that spread around the world. In their quest to improve humankind, advocates divided eugenics into two categories: positive eugenics, in which procreation to pass along "desirable" qualities was encouraged; and negative eugenics, in which procreation among those possessing "undesirable" traits was discouraged or prohibited. Many countries institutionalized some form of eugenics, spurred on by three international conferences on the subject. The First International Eugenics Conference took place in 1912 and was inspired by Francis Galton and presided over by Leonard Darwin, Charles Darwin's son, who believed that eugenics was a practical application of evolutionary principles. Winston Churchill attended the conference, as did the ambassadors of Norway, Greece, and France. Those who addressed the delegates promoted compulsory sterilization and the idea of better breeding principles for humans.

The inventor Alexander Graham Bell was the honorary president at the second conference held at the American Museum of Natural History in New York in 1921. Much of the discussion was led by American scientists, whose work had not suffered the interruptions that plagued European researchers during World War I. The third conference also took place at the Museum of Natural History in 1932, and as tensions rose between Axis and Allied powers, it proved to be the last.

Beyond the conferences, eugenics became integrated into the fabric of American culture. For example, President Theodore Roosevelt formed a "heredity commission" to encourage "the increase of families of good blood and discourage the vicious elements in the crossbred American civilization."[67] Charles Davenport and Harry H. Laughlin formed the Eugenics Record Office at Cold Spring Harbor, New York, in 1910, at a laboratory that still conducts much research today, although genetics has replaced eugenics as their main interest. Laughlin tirelessly campaigned for compulsory sterilization laws and immigration limits before Congress, citing high levels of insanity suffered by southern and eastern European immigrants. His high-profile work provided the basis for Germany's 1933 Law for the Prevention of Hereditarily Diseased Offspring, which resulted in the sterilization of 350,000 people. For his contributions to the science of racial cleansing, Laughlin was awarded an honorary degree from the University of Heidelberg in 1936.

Twentieth-Century Advances
GOOD NEWS, BAD NEWS

The 20th century was a period of unprecedented scientific discovery. Diseases that once killed millions were eradicated, and antibiotics including penicillin put an end to high mortality rates. People lived longer on average than at any other time in human history. More foods of greater variety were produced to feed the world's surging population. Scientists unlocked the secrets of DNA, and couples once unable to have children had the option of in-vitro fertilization. Medically speaking, it was a good century.

Yet the 20th century also saw the worst pandemic in human history. At the close of World War I, when millions of troops were arriving home from the battlefields, a flu epidemic ignited worldwide. The Spanish flu, as it was called, killed between 20 and 100 million people between 1918 and 1920, more than twice the number who died in the war, and an estimated 2.5 to 5 percent of the planet's population. It was an unusually virulent strain of influenza. While many flu viruses strike those with compromised immune systems—namely the young, the sick, and the elderly—this one struck healthy young men and women in the prime of life. It hijacked a person's immune system to unleash a storm of harmful proteins called cytokines, which created a feedback loop in which healthy immune cells facilitated uncontrollable reproduction of cytokines. The Spanish flu could kill a person within a few hours; there were reports of asymptomatic people dropping dead on city streets. Though the Spanish flu epidemic was an unusual event, influenza itself is not. Every year the flu kills roughly 36,000 people in the United States and upwards of 500,000 worldwide.[68] Ironically, the H1N1 pandemic of 2009 and 2010, the first pandemic designated in over 40 years by the WHO, had a relatively low death rate. Of the 1 million Americans infected in the early months of the outbreak, 302 died—making the virus a moderate one.

Viruses are sneaky organisms that mutate in order to survive. Epidemiologists therefore keep tabs on various influenza viruses, and because of this the Influenza Genome Sequencing Project (IGSP) was born. Run by the Institute for Genomic Research (TIGR) and funded by the U.S. National Institutes of Health (NIH), the IGSP sequenced more than 1,800 influenza genomes in its first few years of existence and placed them in the public domain. This information will prove invaluable to researchers developing new vaccines, which it is hoped will lessen the severity of the annual flu season and any widespread outbreaks in the years ahead.

BIOTECHNOLOGIE

The term *biotechnology* was coined in 1917 by a Hungarian inventor named Karl Ereky. In his book *Biotechnologie,* he described how technology could be used to transform plants and animals into products more useful than in their natural state. These products, Ereky theorized, could benefit society, for example by solving food and energy problems. Thus, the Anglicized term *biotechnology* became a post–World War I buzzword. When the Eighteenth Amendment to the U.S. Constitution was ratified and Prohibition went into effect in 1920, prosperous American breweries were forced to transform into biotech companies that conducted fermentation to make products other than alcoholic beverages. One of the first large-scale endeavors was to ferment agricultural waste, an idea central to chemurgy, to produce munitions stockpiles when rubber and petroleum resources were scarce. After World War II, fermentation moved into the pharmaceutical realm. Penicillin was produced in mass quantities using the process of deep fermentation. Steroids and cortisone soon followed. Fermentation created a revolution in health care, with many debilitating diseases becoming no more dangerous than the common cold.

BIOTECHNOLOGY AND GENETIC ENGINEERING

THE ROAD TO THE DOUBLE HELIX

In 1869, the Swiss physician Friedrich Miescher observed a substance in the nuclei of cells that he called nuclein. It was actually DNA. Then, in 1919, the Russian-born biochemist Phoebus Levene, working at the Rockefeller Institute of Medical Research in New York City, theorized that nuclein (also known as nucleic acid) was a string of nucleotides—molecules containing a sugar and a base held together with phosphates. He also correctly identified the nucleotides in question as adenine, guanine, thymine, and cytosine. He had previously identified ribose and deoxyribose. But Levene believed that each molecule contained only four nucleotides and that nucleotides were much too simple to contain the genetic code. The genetic code, it was thought at the time, resided in the protein of a cell. Oswald Avery, also at the Rockefeller Institute, took the biggest leap in 1944 when he claimed that DNA was the key to genes and chromosomes. His work was verified in 1952 by geneticists Alfred Hershey and Martha Chase in the famous Hershey-Chase experiment, which garnered Hershey (but not Chase) a Nobel Prize.

Finally, on April 25, 1953, James D. Watson and Francis Crick published "A Structure for Deoxyribose Nucleic Acid" in *Nature*. It was a landmark moment in the history of science, on par with Mendel's phenotype experiments and Darwin's *On the Origin of Species*. It was the world premiere of the indelible, elegant image of a double-helix strand of DNA in a sketch by Crick's wife, which was later transformed into a three-dimensional model. Watson and Crick, along with their colleague Maurice Wilkins, received the 1962 Nobel Prize in physiology or medicine for "their discoveries concerning the molecular structure of nucleic acids and its significance for information transfer in living material." Not included in the award was X-ray crystallographer Rosalind Franklin, who generated the legendary "photograph 51" X-ray diffraction photo that revealed DNA's double-helix structure. Her death two years earlier had made her ineligible for the Nobel Prize, and subsequently her role has been overshadowed by the long, illustrious careers of her colleagues. What did the discovery of the double-helix structure mean? It meant that molecular biologists had a framework with which to understand how genetic information for all living organisms is coded.

ALEXANDER FLEMING

Thanks to the research of the Scottish biologist Alexander Fleming, the world has penicillin, the first widely manufactured antibiotic, which is developed from the mold of the *Penicillium* genus. Initially, it was effective against a multitude of diseases, including gonorrhea, scarlet fever, pneumonia, meningitis, and diphtheria. It was the miracle drug of the late 1940s and 1950s, drastically changing public health rules and saving countless lives in the process. Along with the elimination of smallpox and polio, the development of penicillin was one of the major medical accomplishments of the 20th century. Although Fleming isolated *Penicillium* and recorded its antiseptic properties, he was unable to transform it into a medically viable treatment. Ernst Chain and Howard Florey at the Dunn School of Pathology at Oxford University in Britain developed the means to produce enough stable penicillin to treat disease on a large scale.

JONAS SALK

Just a couple of generations ago, poliomyelitis was a deadly scourge, killing thousands around the world each year and paralyzing many more. No one who was around in the 1950s can forget the sight of children immobilized for life in iron lungs, unable to breathe on their own. A fast-moving disease that begins with flulike symptoms and often leads to paralysis of the legs or entire body, polio killed 6,000 people in the United States in 1916 alone. In 1952, the year Jonas Salk's vaccine was first tested, 57,628 polio cases were reported. Within two years of the vaccine's introduction, reported cases of polio had dropped by 90 percent. As of 2010, many countries have been deemed polio free by the WHO, with the remaining outbreaks concentrated in India and Nigeria. Although Salk's vaccine, an injection that required several booster shots, was replaced in the early 1960s with the less invasive Sabin oral vaccine, Salk's research, particularly his use of the double-blind test, changed preventative health care forever.

As an epidemiologist at the University of Michigan School of Public Health, Salk worked on developing a flu vaccine for the U.S. Army during World War II. A few years later, he moved to the University of Pittsburgh, where he headed the virus research lab and began developing a polio vaccine by injecting dead polio cells into the body. His first methodical studies were conducted on schoolchildren in neighboring areas; this success led to the largest medical experiment in history, a double-blind study involving 1.8 million children in 44 states, known as the Francis Field Trial. Results were announced in 1955, and mass inoculation began almost immediately. In the mid-1960s, Salk moved to La Jolla, California, where he founded and directed the Salk Institute for Biological Studies. Researchers there focused on molecular biology and genetics. In Salk's later years, he worked toward creating a vaccine for the AIDS virus.

THE DEATH OF SMALLPOX

Smallpox was a scourge on human history for thousands of years. The virus first struck humans around 10,000 B.C.E. and killed roughly 30 percent of those who contracted it. The insidious virus first manifested as small blisters, often concentrated on the face and throat. Those who survived the infestation of pustules might suffer scarring, blindness, or deformities. In the 20th century alone, smallpox killed some 300 to 500 million people.[69] In 1967, as a massive vaccination program sponsored by the WHO got underway, 2 million people around the world died of the disease, but by 1979 it was pronounced extinct—the first disease to be eradicated in human history. The elimination of smallpox stands as one of the greatest medical triumphs of all time.

MAD COW DISEASE

In 1984, a cow on a farm in West Sussex, Britain, became sick and died. Laboratory tests on the carcass failed to identify the animal's disease, which was bovine spongiform encephalopathy (BSE), or mad cow disease (MCD), a degenerative neurological condition that slowly destroys the brain. The disease can manifest in human beings as variant Creutzfeldt-Jakob Disease (vCJD), a fatal condition in which a person's brain is infected by a prion—a viruslike protein—that destroys the tissue. Creutzfeldt-Jakob has a long incubation period—years, in most cases—which means that many people could be infected without knowing it. Patients die roughly a year after the first symptoms appear, and there is no cure.

That first dead cow in Britain represented a missed opportunity. The Ministry of Agriculture did not realize the cow had BSE until two years after its death. By 2001, some 179,000 cattle had died of BSE and another 4.4 million were destroyed as a precautionary measure, costing billions and nearly destroying the country's beef industry. It is thought that roughly 80 people in the country have died after developing vCJD from eating beef derived from infected cows. Because the disease has no cure and a long incubation period, this toll is expected to rise.

What caused the precipitous rise in BSE? It was discovered that the culprit was the practice of using animal remains as feed for existing livestock—cannibalism, in effect, which had been a common practice in Britain's cattle industry since at least 1926.[70] The first sick cow possibly contracted BSE from a naturally occurring mutant gene. Once the mutant gene entered the food chain, there was no way to eliminate it other than to kill everything in that chain. The mad cow scare underscores the dangers of a contaminated food supply, especially in this era of industrialized feedlots.

UNLOCKING THE GENOME

The next step for scientists after the discovery of the double-helix structure of DNA was to map the genes that comprise the structure. Frederick Sanger was the first person to map the genome of a living organism, using the chain termination method, now known as the Sanger method. He sequenced the genome of bacteriophage (a virus that infects bacteria) Phi X 174 in 1977—by hand. The bacteriophage is single-stranded and contains only 11 genes and 5,386 bases. This paved the way for the sequencing of other organisms, culminating in the Human Genome Project. Sanger has received two Nobel Prizes in chemistry for his work on amino acids, proteins, and insulin.

CLONING

The 1982 film *Blade Runner* presents a dystopian view of the United States in the year 2019. Clones called replicants are created to work in space colonies; some escape and return to Earth, where they are hunted down and killed by blade runners, elite law enforcement officers. The replicants have human feelings and memories and look identical to the general population; their uprising is caused by their desire for freedom to live on Earth. The movie capitalizes on people's fear of clones—genetically identical copies of an organism.

While humankind is a long way from populating space colonies with clones, we have entered the era in which organisms are cloned for a number of purposes. Plants have been cloned and genetically manipulated ever since Mendel began his pea plant experiments. Different types of animals have been cloned in the laboratory, with varying levels of success. No one has yet cloned a human being, but many believe it is only a matter of time.

Animal Cloning

Animals were cloned for the first time in 1958, when a frog from the species *Xenopus laevis* was reproduced from an embryonic cell.[71] And on July 5, 1996, Ian Wilmut and Keith Campbell of the Roslin Institute in Edinburgh, Scotland, announced the birth of Dolly, a Finn-Dorset ewe, cloned from the mammary cell of a full-grown female through the process of nuclear transfer. The experiment proved that an entire organism could be created from a mature somatic cell or any cell other than a sperm or an egg. The process of nuclear transfer entails removing the nucleus from an egg cell and replacing it with the nucleus of a mammary cell. The new cell is given an electric shock to encourage it to divide. In Dolly's case, the resulting blastocyst was implanted into a surrogate mother, and from there gestation proceeded normally. The process was not easy; Dolly was the only surviving sheep out of 277 attempts.[72]

Other than her unusual conception, Dolly was normal. She mothered six lambs but died at the age of six from a type of lung cancer that is common to her breed. Her stuffed and mounted remains are on display at the Royal Museum of Scotland. Some scientists speculated that Dolly was in effect 12 when she died and thus at the end of her normal life expectancy because the mammary cell she was created from was from a six-year-old sheep. However, the scientists at the Roslin Institute dispute this notion.

Since Dolly's birth, which was heralded as a milestone in modern biotechnology, other animals including horses, deer, and bulls have been cloned using the same process of nuclear transfer. With the refinement of the process, some believe that in the near future cloning may become a feasible way to prevent endangered species from becoming extinct.[73]

Human Cloning

Artificially created human clones, as opposed to natural-born identical twins (who are essentially clones), are what the fuss is all about. Such clones could be produced via somatic cell nuclear transfer—the same process that resulted in Dolly the sheep—or parthenogenesis, an asexual process in which female egg cells divide without fertilization. Parthenogenesis occurs naturally in many plant species and some animal species such as bees, wasps, reptiles, and fish, and it almost always results in the propagation of females. As of 2009, no verified artificial human clones have been born, but urban legends abound. In 1978, the well-regarded science writer David Rorvik published *In His Image: The Cloning of a Man*, in which he claimed to have personal knowledge of a living human clone. The book was a best seller but was debunked as a hoax, though Rorvik has continued to defend it.[74]

The South Korean researcher Hwang Woo-Suk published a paper in 2005 claiming to have cloned 11 human embryonic stem cells. The claims were later found to be false, and Hwang was indicted on bioethics violations and embezzlement charges.[75] Hwang went on to create the world's first verified cloned dog, an Afghan hound named Snuppy, in 2005, and in June 2008 he claimed to have created 17 clones of Tibetan mastiff dogs at the request of the Chinese Academy of Sciences[76] before his research license was revoked. In August 2008, RNL Bio, a South Korean firm headed by one of Hwang's former colleagues, delivered the world's first cloned puppies for commercial purposes. An American woman paid $50,000 and received five pit bull puppies cloned from DNA obtained from her dog, which had died of cancer two years earlier. The company plans to clone several hundred dogs per year and would like to branch out into camel cloning for wealthy Middle Eastern clients.[77]

Advanced Cell Technology (ACT), a California-based company, has cloned human embryos up to the six-cell stage using the somatic cell nuclear transfer (SCNT) method.[78] Efforts to clone cells using parthenogenesis have so far failed. In January 2008, Andrew French and Samuel Wood of California-based Stemagen announced they had successfully cloned human embryos from adult skin cells. Furthermore, the embryos survived to the stage that would have been appropriate for transfer into a surrogate but were instead destroyed. This step was taken because the researchers were not interested in creating a human clone; rather, their goal was to prove it is possible to create personalized stem cells that could be used to generate replacement tissues for patients suffering from disease. Such tissue could not be rejected by the body's immune system—a common problem—because it would be from that same body.[79]

ETHICS GUIDELINES

By the late 20th century, after the misdeeds of pre–World War II eugenicists and morally questionable research such as the Tuskegee Study had raised public awareness, many countries developed legal guidelines to protect patients' rights. In the United States, Congress established the National Commission for the Protection of Human Subjects of Biomedical and Behavioral Research in 1974.

In 1975, the International Food Policy Research Institute (IFPRI) was established, which now houses the Consultative Group on International Agricultural Research (CGIAR). CGIAR has many programs that deal with agricultural biotechnology and genetic engineering, particularly as they pertain to international trade. The most basic issue is that countries that grow GM crops want to sell them on the international market, but the lower production costs and increased yields are seen as an unfair advantage in countries that do not have the economic power to compete. "As a result," according to Peter W. B. Phillips of IFPRI, "disadvantaged farmers may join with consumers in importing countries concerned about the safety of these products in calling for increased controls on these products."[80] Phillips notes that the divide normally pits the United States, Canada, Japan, and Mexico against the EU, New Zealand, and Australia. The United States, Argentina, and Canada have adopted policies of voluntary labeling of GMOs, whereas the EU, China, Japan, Australia, and New Zealand have implemented mandatory labeling procedures.

THE HUMAN GENOME PROJECT

The term *genome* refers to an organism's genetic material. The human genome is comprised of a DNA sequence of four different nucleotides—adenine, thymine, guanine, and cytosine—that make up the 20,000 to 25,000 genes in each person's 23 chromosomes (or 46, if you count each diploid chromosome twice). A nucleotide is a compound that consists of a nitrogenous ring (composed of atoms of nitrogen, carbon, and oxygen) joined by a sugar molecule and a phosphate molecule. The nucleotides are arranged in roughly 3 billion base pairs, with each base pair consisting of either adenine and thyminde or guanine and cytosine. These nucleotides determine which proteins are manufactured in the body and in what amount, leading to each person's individual looks, traits, and characteristics.

The Human Genome Project is one of the largest undertakings in scientific history—a $3 billion international research endeavor initiated by the U.S. Department of Energy (DOE) in 1990 and initially headed by Nobel laureate James D. Watson at the newly created National Center for Human Genome Research at the NIH. Geneticists from Great Britain, China, France, Germany, and Japan were integral to the project. Watson resigned in 1992, and the project continued under geneticist Francis Collins. In 1997, the center was renamed the National Human Genome Research Institute. Unique to the Human Genome Project is its effort to address the ethical, legal, and social issues (abbreviated as ELSI) that may arise from the data and the government's effort to transfer the data to the private sector (in a process called technology transfer).

The completion of a rough draft of the human genome was announced jointly on June 26, 2000, by U.S. president Bill Clinton and British prime minister Tony Blair. The project continued until April 2003, two years before its projected end date, because of the number of people working on it and the somewhat friendly rivalry between government and private initiatives. In September 2007, Celera Genomics published founder Craig Venter's complete genome: a 6-billion-nucleotide chain. Work continues on the initiative as of 2009 because although the genome sequence is complete, many details remain to be explored in highly repetitive areas of the sequence that do not contain DNA. The ever-increasing body of public domain data has spawned other long-term projects. For example, the International HapMap Project uses genomic data supplied by people from a variety of cultures to identify patterns of genetic variation common to people around the world.

The goal of the Human Genome Project was to map the sequence of chemical base pairs that comprise the 25,000 genes in the human genome. Even before the complete map was published, researchers had already developed genetic tests that could evaluate whether a person is genetically predisposed to developing breast cancer, cystic fibrosis, and liver disease. Tests for many other types of cancer and Alzheimer's disease are expected to be developed in the near future. In addition to the human genome, the genomic sequences of other species have been mapped. The bacterium *E. coli,* the fruit fly (*Drosophila*), and the laboratory mouse have had their genomes mapped, and work continues on the genomes of zebrafish, yeast, nematodes, and many microscopic organisms. All organisms' genomes are composed of adenine, thymine, guanine, and cytosine—the only thing that differentiates a person from a grain of rice, genetically speaking, is the order of the nucleotides.

Work at the NIH was augmented by researchers around the world, particularly from Great Britain and Canada through the Wellcome Trust Sanger Institute. Nevertheless, Celera Genomics began a simultaneous, privately funded human genome project with the goal of developing commercial health care products from its findings. Celera used a different technique than the Human Genome Project, the "whole genome shotgun sequencing process," and finished sooner and at a fraction of the cost. However, critics note that Celera was able to capitalize on data already generated from the Human Genome Project and made available to the public through the National Center for Biotechnology Information's Gen-Bank, an online database that contains all known, published nucleotide bases. Celera also sought legal protection of its findings, unlike the government initiative, but was denied this when President Clinton announced in 2000 that the genome sequence could not be patented. Celera's stock, and indeed the whole biotechnology sector, suffered heavy losses on the NASDAQ.

The importance of the Human Genome Project has been underscored on the international political scene. The Universal Declaration on Human Genome and Human Rights was developed by the United Nations Educational, Scientific and Cultural Organization (UNESCO) in 1997 to provide an ethical framework for research that might arise from the project. The declaration holds that reproductive cloning is contrary to human dignity and will distort family relationships, limit human genetic diversity, promote the view that certain people have the right to determine the purpose of another person's existence, and risk turning humans into manufactured objects. The World Medical Association's (WMA) Resolution on Cloning in 1997 came to many of the same conclusions.

CONVENTION ON BIOLOGICAL DIVERSITY AND THE CARTAGENA PROTOCOL

The Convention on Biological Diversity was opened for signature at the Earth Summit in Rio de Janeiro in June 1992. It promotes sustainable agriculture, conservation, and biosafety to ensure that the planet retains the greatest number of plant and animal species possible. Biological diversity, researchers agree, is a key indicator of an environment's health. At the convention, biotechnology was defined as "any technological application that uses biological systems, living organisms, or derivatives thereof, to make or modify products or processes for specific use."[81]

A major component of the treaty is that it regulates access to genetic resources in an effort to protect cultures from bioprospecting, or the act of exploiting a natural resource for financial gain without proper compensation to those who rely on that resource for their livelihood. It also requires a transfer of technology to those who can benefit from it, without regard to financial considerations.

The Cartagena Protocol on Biosafety is a later treaty, adopted in 2000 and coming into force in 2003. It expanded on some of the features of the Convention on Biological Diversity. Its goal is to protect existing biodiversity from organisms modified through modern biotechnology. It allows countries to ban GMOs if they believe such technology poses a threat to their well-being, and it requires other countries to label GMOs in the international marketplace so people remain informed about the goods that cross their borders.

THE INTERNATIONAL SEED BANK

The Svalbard Global Seed Vault opened in 2008 in the Arctic Circle in Norway, just 700 miles from the North Pole. The underground vault holds seeds from 1.5 million plant varieties; it is an insurance policy against worldwide ecological devastation and will help ensure continued genetic diversity. The Svalbard Vault is not the first—more than 1,000 seed vaults exist throughout the world under various levels of security and containing various numbers of seeds—but it was designed to be the most secure, most complete facility of its kind. It is managed by the Norwegian government, the Global Crop Diversity Trust, and the Nordic Genetic Resource Center, and it has the capacity to house more than 4.5 million sealed seed packets in conditions that will maintain their usability for hundreds to thousands of years.

CONCLUSION

The very nature of biotechnology ensures that it will continue to raise controversial issues for years to come. The scientific ability to manipulate the building blocks of human life brings with it a host of decisions that require open and informed debate among all members of society and good governance on the part of its leaders. What should we eat, and how will it affect our health as well as the environment? Should we use genetics to cure disease, to prevent disease, to change our physical appearance, or to give our offspring traits that we wish we had ourselves? How can we ensure that genetic testing does not lead to discrimination? What are the moral guidelines for cloning, and what happens when countries adopt policies that are at odds with the world community? Understanding that biotechnology has been around as long as society itself lends the issue perspective. Knowledge of the past, whether scientific, political, or cultural, can help guide us in the future.

Introduction

[1] Martin W. Bauer. "Controversial Medical and Agri-Food Biotechnology: A Cultivation Analysis." *Public Understanding of Science* 11 (2002): 93–111.

[2] Clive James. "Table 1: Global Area of Biotech Crops in 2007: By Country." *International Service for the Acquisition of Agri-Biotech Applications Brief 37—2007: Executive Summary.* Available online. URL: http://www.isaaa.org/Resources/Publications/briefs/37/executive summary/default.html. Accessed May 13, 2009.

[3] Andrew Pollack. "Narrow Path for New Biotech Food Crops." *New York Times* (5/20/04). Available online. URL: http://www.nytimes.com/2004/05/20/business/narrow-path-for-new-biotech-food-crops.html. Accessed September 9, 2009.

[4] *King Corn* (2007), directed by Aaron Woolf. Balcony Releasing.

[5] I. R. Dohoo et al. "A Meta-analysis Review of the Effects of Recombinant Bovine Somatotropin." *Canadian Journal of Veterinary Medicine* 67.4 (October 2003): 241–264.

[6] The information on rBGH use in these countries is from Laura Sayre. "Protecting Milk from Monsanto." *Mother Earth News* (June–July 2008): 27.

[7] "Facts about Antibiotic Resistance." U.S. Food and Drug Administration. Available online. URL: http://www.fda.gov/oc/opacom/hottopics/antiresist_facts.html. Accessed May 13, 2009.

[8] Peter Smith. "'Hormone-free' Milk Spurs Labeling Debate." *Christian Science Monitor* (4/21/08). Available online. URL: http://features.csmonitor.com/environment/2008/04/21/hormone-free-milk-spurs-labeling-debate/. Accessed May 13, 2009.

[9] Agricultural Marketing Resource Center. "Organic Dairy Profile." Available online. URL: http://www.agmrc.org/commodities__products/livestock/dairy/organic_dairy_profile.cfm. Accessed May 13, 2009.

[10] Sayre. "Protecting Milk from Monsanto." *Mother Earth News* (June–July 2008): 27.

[11] Union of Concerned Scientists. "Antibiotics and Food" (6/23/08). Available online. URL: http://ucsusa.wsm.ga3.org/food_and_environment/antibiotics_and_food/. Accessed May 13, 2009.

[12] "Maine Says No to Drugs." *Current Science* 92.7 (December 1, 2006): 12.

[13] Ken Midkiff. *The Meat You Eat: How Corporate Farming Has Endangered America's Food Supply.* New York: St. Martin's, 2004, p. 41.

[14] Lee Silver. "Why GM Food Is Good for Us." *Newsweek International* (3/20/06): 57–58.

[15] James D. Watson and Francis H. C. Crick. "Molecular Structure of Nucleic Acids." *Nature* 171 (1953): 737–738.

[16] M. B. Gerstein et al. "What Is a Gene, Post-ENCODE? History and Updated Definition." *Genome Research* 17.6 (2007): 669–681.

[17] Michael Kaback. "Screening and Prevention in Tay-Sachs Disease: Origins, Update, and Impact." *Advances in Genetics* 44 (2001): 253–265.

[18] Ramez Naam. "More Than Human." *New York Times* (7/3/05). Available online. URL: http://www.nytimes.com/2005/07/03/books/chapters/0703-1st-naam.html. Accessed September 9, 2009.

[19] Maggie Fox. "Gene Therapy Shows Promise in Rare Brain Disease." *Reuters* (5/16/08). Available online. URL: http://www.reuters.com/article/scienceNewsidUSN1338828620080513. Accessed September 9, 2009.

[20] "World's first IVF baby marks 30th birthday." Agence France-Presse (7/23/08). Available online. URL: http://afp.google.com/article/ALeqM5iqs7hQfKFma-avGEqVGUGbLJ5xXQ. Accessed May 13, 2009.

[21] Kirsty Horsey. "Three Million IVF Babies Born Worldwide." June 28, 2006. Available online. URL: http://www.ivf.net/ivf/three_million_ivf_babies_born_worldwide-o2105-en.html. Accessed May 13, 2009.

[22] Stephen L. Baird. "Designer Babies: Eugenics Repackages or Consumer Options?" *Technology Teacher* 66.7 (April 2007): 12.

[23] Marilynn Marchione and Lindsey Tanner. "More Couples Use Embryo Tests for 'Designer Babies'; Trend Alarming to Medical Ethicist Who Says the Risks Are Not Justifiable." *Houston Chronicle* (10/1/06): p. 3.

[24] "Designer Baby Transplant Success." BBC News (7/27/04). Available online. URL: http://news.bbc.co.uk/2/hi/health/3930927.stm. Accessed May 13, 2009.

[25] Baird. "Designer Babies: Eugenics Repackaged or Consumer Options?" *Technology Teacher* 66.7 (April 2007): 12.

[26] Julian Ma et al. "Molecular Farming for New Drugs and Vaccines: Current Perspectives on the Production of Pharmaceuticals in Transgenic Plants." *EMBO Reports* 6.7 (2005): 593–599.

[27] Andrew Pollack. "New Ventures Aim to Put Farms in Vanguard of Drug Production." *New York Times* (5/14/00). Available online. URL: http://www.nytimes.com/2000/05/14/business/new-ventures-aim-to-put-farms-in-vanguard-of-drug-production.html. Accessed September 9, 2009.

[28] "Go-ahead for 'Pharmed' Goat Drug." BBC News (6/2/06). Available online. URL: http://news.bbc.co.uk/1/hi/sci/tech/5041298.stm. Accessed May 13, 2009.

[29] "ATryn—Recombinant Human Antithrombin." *GTC Biotherapeutics* (2006). Available online. URL: http://www.gtc-bio.com/products/atryn.html. Accessed May 13, 2009.

[30] Tom Murphy. "Dow AgroSciences Seeks Better Vaccine: Plant-Based Preventative Measure Loaded with Potential." *Indianapolis Business Journal* (7/17/06): 21.

[31] Medicago, Inc. "Medicago's H5N1 Pandemic Flu Vaccine Effective in Key Ferret Animal Model with Single Dose." Press Release (6/3/08). Available online. URL: http://www.medicago.com/upload/MDG%20ferret%20study%20release%20FINAL%20EN.pdf. Accessed May 13, 2009.

[32] Jeffrey M. Smith. *Seeds of Deception: Exposing Industry and Government Lies about the Safety of the Genetically Altered Foods You're Eating.* Fairfield, Iowa: Yes! Books, 2003, p. 143.

[33] Michael Pollan. "Playing God in the Garden." *New York Times Magazine* (10/25/98). Available online. URL: http://www.nytimes.com/1998/10/25/magazine/playing-god-in-the-garden.html. Accessed September 9, 2009.

[34] Ron Smith. "GMO Peanuts Could Improve Health, Production Efficiency." *Delta Farm Press* (4/7/08).

[35] Daniel Charles. "Notes on Sources." *Lords of the Harvest: Biotech, Big Money, and the Future of Food.* New York: Perseus, 2001, p. 320.

[36] Philip Kitcher. "Manipulating Genes: How Much Is Too Much?" Available online. URL: http://www.pbs.org/wgbh/nova/genome/manipulate.html. Accessed May 13, 2009.

Introduction

[37] Jonas Salk. Interview footage in *Sicko*, directed by Michael Moore. Dog Eat Dog Films, 2007.

[38] Kitcher. "Manipulating Genes: How Much Is Too Much?" Available online. URL: http://www.pbs.org/wgbh/nova/genome/manipulate.html. Accessed May 13, 2009.

[39] Caroline Fredrickson. "ACLU Letter to the Senate Health, Education, Labor and Pensions Committee Regarding the Genetic Information Nondiscrimination Act for 2007" (1/31/07). Available online. URL: http://www.aclu.org/privacy/gen/34993leg20080423.html. Accessed May 13, 2009.

[40] Philip Pullella. "Vatican Lists 'New Sins,' Including Pollution." Reuters (2008). Available online. URL: http://www.reuters.com/article.topNews/idUSL109602320080310. Accessed September 9, 2009.

[41] Southern Baptist Convention. "Resolution #7: On Human Embryonic and Stem Cell Research, Adopted from the SBC Convention, 16 June 1999." Available online. URL: http://www.johnstonsarchive.net/baptist/sbcabres.html. Accessed May 13, 2009.

[42] C. L. Richard. "Why Biotech Foods Are Kosher." Available online. URL: http://www.agbioworld.org/biotechinfo/religion/kosher.html. Accessed May 13, 2009.

[43] Information on Islamic law, aside from cloning, is from Bushra Mirza. "Islamic Perspectives on Biotechnology." In Michael C. Brannigan, ed., *Cross-cultural Biotechnology*. Oxford, England: Rowman & Littlefield, 2004, p. 112.

[44] John Tierney. "Are Scientists Playing God? It Depends on Your Religion." *New York Times* (11/20/07). Available online. URL: http://www.nytimes.com/2007/11/20/science/20tier.html. Accessed September 9, 2009.

[45] Tierney. "Are Scientists Playing God? It Depends on Your Religion."

[46] *Fed Up! Genetic Engineering, Industrial Agriculture and Sustainable Alternatives*. Directed by Angelo Sacerdote. Microcinema International, 2002.

[47] Erwin Northoff. "The Gene Revolution: Great Potential for the Poor, but No Panacea." FAO Newsroom (5/17/04). Available online. URL: http://www.fao.org/newsroom/en/news/2004/41714/index.html. Accessed on May 13, 2009.

[48] Northoff. "The Gene Revolution: Great Potential for the Poor, but No Panacea."

[49] C. S. Prakash. "Gene Revolution and Food Security." *AgBioWorld* (3/2/00). Available online. URL: http://www.agbioworld.org/biotech-info/topics/dev-world/revolution.html. Accessed May 13, 2009.

[50] Alexei Barrionuevo. "Can Gene-Altered Rice Rescue the Farm Belt?" *New York Times* (8/16/05). Available online. URL: http://www.nytimes.com/2005/08/16/business/16biorice.html. Accessed September 9, 2009.

[51] William J. Broad. "Food Revolution That Starts with Rice." *New York Times* (6/17/08). Available online. URL: http://www.nytimes.com/2008/06/17/health/17iht-17rice.13764265.html. Accessed September 9, 2009.

[52] Pollack. "Narrow Path for New Biotech Food Crops." *New York Times* (5/20/04).

[53] Lousie O. Fresco. "A New Social Contract on Biotechnology." *FAO* magazine (May 2003). Available online. URL: http://www.fao.org/ag/magazine/0305sp1.htm. Accessed May 13, 2009.

[54] Pollack. "Monsanto Plans to Boost Food Supply." *New York Times* (6/5/08). Available online. URL: http://www.nytimes.com/2008/06/05/business/worldbusiness05crop.html. Accessed September 9, 2009.

55 Pollack. "Monsanto Plans to Boost Food Supply."

56 "First Evidence Emerges of Pest Resistance to GM Crops: Scientists." Agence France-Presse (2/8/08). Available online. URL: http://afp.google.com/article/ALeqM5jInSHTJCaw7ZebsL-SoTa35Ot1HQ. Accessed September 9, 2009.

57 Jane Matthews Glenn. "Footloose: Civil Responsibility for GMO Gene Wandering in Canada." *Washburn Law Journal* 43.3 (7/12/04): 547–573.

58 Chidi Oguamanam. "Tension on the Farm Fields: The Death of Traditional Agriculture?" *Bulletin of Science, Technology, and Society* 27.4 (August 2007): 265.

59 Glenn. "Footloose: Civil Responsibility for GMO Gene Wandering in Canada." *Washburn Law Journal* 43.3 (7/12/04): 547–573.

60 Charles C. Mann. "Transgenic Data Deemed Unconvincing." *Science* 296.5566 (4/12/02): 236.

61 Mann. "Transgenic Data Deemed Unconvincing."

62 Mann. "Transgenic Data Deemed Unconvincing."

63 Robert Jay Lifton. *The Nazi Doctors: Medical Killing and the Psychology of Genocide*. New York: Basic Books, 1986, 2000.

64 Food and Agriculture Organization of the United Nations. "FAO Statement on Biotechnology." Available online. URL: http://www.fao.org/Biotech/stat.asp. Accessed May 13, 2009.

65 Commission on Genetic Resources for Food and Agriculture. "Functional Statement." Available online. URL: http://www.fao.org/ag/cgrfa/fsC.htm. Accessed May 13, 2009.

66 Federico Mayor. "The Universal Declaration on the Human Genome and Human Rights." UNESCO (12/3/97). Available online. URL: http://portal.unesco.org/en/ev.php-URL_ID=13177&URL_DO=DO_TOPIC&URL_SECTION=201.html. Accessed May 13, 2009.

67 Harry Bruinius. *Better for All the World. The Secret History of Forced Sterilization and America's Quest for Racial Purity*. New York: Knopf, 2006.

68 W. Thompson et al. "Mortality Associated with Influenza and Respiratory Syncytial Virus in the United States." *JAMA* 289.2 (2003): 179–186. Available online. URL: http://jama.ama-assn.org/cgi/content/full/289/2/179. Accessed May 13, 2009.

69 David A. Koplow. *Smallpox: The Fight to Eradicate a Global Scourge*. Berkeley: University of California Press, 2003.

70 Information on England's cattle industry and the numbers of cows infected and/or destroyed is from David Brown. "The 'Recipe for Disaster' that Killed 80 and Left a £5bn Bill." *Telegraph News* (6/19/01). Available online. URL: http://www.telegraph.co.uk/news/uknews/1371964/The-recipe-for-disaster-that-killed-80-and-left-a-5bn-bill.html. Accessed May 13, 2009.

71 J. B. Gurdon, T. R. Elsdale, and M. Fischberg. "Sexually Mature Individuals of *Xenopus laevis* from the Transplantation of Single Somatic Nuclei." *Nature* 182 (7/5/58): 64–65.

72 Ian Wilmut et al. "Viable Offspring Derived from Fetal and Adult Mammalian Cells." *Nature* 385.6619 (1997): 810–813.

73 Alan O. Trounson. "Future and Applications of Cloning." *Methods of Molecular Biology* 348 (2006): 319–332.

[74] William J. Broad. "Publisher Settles Suit, Says Clone Book Is a Fake." *Science* 216 (4/23/82): 391.

[75] "Disgraced Korean Cloning Scientist Indicted." *New York Times* (5/12/06). Available online. URL: http://www.nytimes.com/2006/05/12/world/asia/12korea.html. Accessed September 9, 2009.

[76] Kwang-Tae Kim. "South Korean Ex-Professor Claims Dog Clones." Associated Press (6/19/08). Available online. URL: http://www.usatoday.com/tech/science/2008-06-19-2670862260_x.htm. Accessed September 9, 2009.

[77] Hyung-Jin Kim. "S. Korean Firm Delivers First Commercial Dog Clones." Associated Press (8/5/08). Available online. URL: http://www.usatoday/news/offbeat/2008-08-05-dog_N.htm. Accessed September 9, 2009.

[78] Jose B. Cibelli, Robert P. Lanza, and Michael D. West. "The First Human Cloned Embryo." *Scientific American* (11/24/01). Available online. URL: http://www.sciam.com/article.cfm?id=0008B8F9-AC62-1C75-9B81809EC588EF21.

[79] Rick Weiss. "Mature Human Embryos Created from Adult Skin Cells." *Washington Post* (1/18/08). Available online. URL: http://www.washingtonpost.com/wp-dyn/content/article/2008/01/17/AR2008011700324.html? hpid=topnews. Accessed May 13, 2009.

[80] Peter W. B. Phillips. "Policy, National Regulation, and International Standards for GM Foods." In Philip G. Pardey and Bonwoo Koo, eds. *Biotechnology and Genetic Resource Policies*. Washington, D.C.: International Food Policy Research Institute, 2003, brief 1, p. 2.

[81] *Convention on Biological Diversity.* Available online. URL: http://www.cbd.int/convention/convention.shtml. Accessed May 13, 2009.

2

Focus on the United States

CURRENT SITUATION
Steroids, Hormones, and Antibiotics

Consider these scenarios: a bodybuilder takes steroids for muscle enhancement; a 12-year-old boy takes a growth hormone in hopes that he will grow taller than his current height of four feet; a transgender man takes female hormones to develop breasts and lose facial hair. All these situations involve the consumption of supplements designed to change the way people look. Is it wrong? What about cosmetic surgery or injecting permanent dye into the skin in the form of a tattoo?

Much of the discussion surrounding biotechnology involves drawing a line between what is acceptable and what is not. Would it be wrong to modify germ line cells (those that can be passed on to future generations) to ensure that a child has blue eyes? Is that more wrong than aborting a fetus with grave medical defects? Obviously, people have differing ideas about where the line should be drawn. And in practice a double standard is sometimes evident. For example, growth hormones are illegal in the United States for bodybuilders and athletes, but the U.S. meat supply is heavily dependent on them to boost production and meet consumer demand.

The United States is a world leader in biotechnology and genetic engineering. From government-funded institutions such as the National Institutes for Health (NIH) to private-sector companies such as Genentech and Monsanto to world-class research universities such as Johns Hopkins and MIT, the United States attracts the best and brightest in the fields of medicine, genetics, agronomy, and bioinformatics. U.S. scientists were integral to the international Human Genome Project, one of the major scientific undertakings of the late 20th and early 21st-first centuries, and U.S. agriculture companies plant more crops from genetically engineered (GM) seed than any other country. Politically, debates over GM food, cloning, stem cell research, and the like characterize the national discussion, while medical breakthroughs pertaining to genetics and pharmaceuticals are commonplace.

Agriculture

LIVESTOCK

In the United States, small family farms are largely a thing of the past. Agribusiness is the new paradigm, in which farming is concentrated in the hands of a few large corporations. Arkansas-based Tyson Foods, for instance, is the world's largest meat producer, with 123 processing plants, 107,000 employees, and $25.6 billion in annual revenues. It supplies chicken, beef, and pork products to McDonald's, Burger King, Wendy's, KFC, Taco Bell, and Wal-Mart. For its chicken, Tyson contracts with local chicken farmers, who raise the broilers in climate-controlled sheds that house around 24,000 birds at one time. Tyson owns the chickens but outsources their growth.

When it comes to livestock, the western United States initially proved ideal as a range. For centuries, wild bison roamed the prairies and plains, grazing lightly on vast tracts of land. But with settlement in the 19th century, the bison became a nuisance, especially when railroads were built and the thousands of miles of track needed to be kept clear. The American solution was the methodical slaughter of millions of wild bison, which were replaced by domesticated cattle that grazed across millions of acres of public land until it was depleted of nutrients.

In the late 20th century, ranching became mechanized under a system called Concentrated Animal Feeding Operations (CAFOs). On reaching 650 pounds, cattle are transferred to feedlots, and for the next three to four months they are fed a diet of mainly corn until they gain another 400 pounds. The economies of scale that CAFOs enable have allowed the U.S. beef and cattle industry to become the world's largest, with 34.3 million animals slaughtered and retail sales of $74 billion in 2007.[1] The desire to raise cattle to finishing weight as quickly as possible has led to reliance on growth hormones, and the crowded conditions of feedlots make them fertile ground for disease, so animals are also given antibiotics. Even the cattle's diet, corn instead of grass, causes acidosis, which likewise requires treatment with antibiotics. The animals' diet also results in meat with significantly higher levels of saturated fat than that from cattle that has been allowed to graze freely on the range. Finally, consumers who eat too much saturated fat have a greater risk of heart disease.

THE FLAVR SAVR TOMATO

In 1994 the Calgene Corporation of California debuted its Flavr Savr tomato in supermarkets across the country. It was the first commercially grown GM food to be licensed by the Food and Drug Administration (FDA) for human consumption. The tomato was modified by the insertion of a gene that stalled production of enzymes that ripen the fruit. The Flavr Savr was allowed to ripen on the vine, whereas most commercially harvested tomatoes have to be picked while they are green to prevent damage during transportation. The Flavr Savr would not be damaged during transport, resulting in a shipment of ripe, firm, and tasty produce—or so it was hoped. In reality, the Flavr Savr was plagued with problems from the beginning in terms of regulations, public relations, and quality. The product was not popular with consumers, who saw no advantage to the higher-priced Flavr Savr over traditional varieties, and it was pulled from store shelves within a year.

It turned out that Calgene had chosen a variety of tomato for genetic modification that was more suitable for processing than for eating; it bruised easily, and the inserted gene did not change that. The tomato, which had been developed in California, was grown for production in Florida and proved unsuited to the soil conditions there. Finally, its high production cost collided with the low price of tomatoes on the world market. Monsanto accused the small Calgene company of patent infringement and quickly bought it out.[2]

While consumer backlash against the tomatoes may have had more to do with quality than genetic modification, some people were concerned with the process by which the gene was added to the fruit. This genetic marker caused a secondary genetic modification that made the tomatoes resistant to the antibiotic kanamycin. If the Flavr Savr had been commercially successful, kanamycin-resistant bacteria could have developed, thus negating the effectiveness of that antibiotic.

ROUNDUP READY SOYBEANS

Roundup is a commercial herbicide that has been manufactured by agriculture giant Monsanto since 1973. Its active ingredient, glyphosate, kills weeds and plants that without intervention will choke agricultural crops. Around 90 million pounds of Roundup are used annually on U.S. crops; it is also popular as a weed treatment for lawns in the consumer market. In 1996, Monsanto created its first Roundup Ready soybean seed. The seeds were genetically engineered to be resistant to glyphosate, meaning that a farmer who plants Roundup Ready soybeans can spray an entire field with Roundup herbicide and be confident that the crops will be safe and that only the weeds will die. But if Roundup is sprayed on crops planted with non–Roundup Ready seeds, those crops will die. The seed proved popular with farmers attracted by its labor-saving convenience.

Because Monsanto owns the patent on Roundup Ready seed (although its patent on Roundup expired in 2000), farmers who save seed for the next year's planting are in violation of the corporation's patent. This upsets a process that began with the dawn of agriculture: Each year's crops were sowed with seed reaped from the previous year's crops. But farmers who use GM seed are required to destroy the excess and buy new seed each year, adding to their expenses. In addition, if GM seed lands in the field of a farmer who did not buy it—due to wind, birds, or some other inadvertent process—that farmer can be sued for patent infringement. This is what happened to Percy Schmeiser.

Percy Schmeiser had been farming canola in Saskatchewan, Canada, for decades when a private investigator hired by Monsanto to enforce its "technology use agreement" obtained samples of canola plants from Schmeiser's property. Those plants were found to have grown from Monsanto's Roundup Ready canola seeds, and in 1998 the company sued Mr. Schmeiser for patent infringement. Schmeiser claimed that the seed, which had been found near public right-of-ways, had blown off trucks driving past his fields. He had never planted it. A bitter, protracted legal battle ensued that pitted Monsanto, the large corporation, against Schmeiser, a small family farmer. Though the legal issue was patent infringement, the subtext was transgenic contamination: If GM seed could germinate seemingly randomly, what would that mean for the farmers who did not want it on their property, and what would it mean for the patent holder? Furthermore, Schmeiser claimed the investigator had obtained the canola samples illegally by trespassing on his property.

The case made it to the Supreme Court of Canada, and on May 21, 2004, the court ruled 5–4 that Monsanto's patent had been infringed but that Schmeiser was not required to pay any damages. While this was seen as a win for Monsanto, it failed to address the issue of gene wandering, or transgenic contamination, or even Schmeiser's allegation that the seed had blown onto his property, most likely from a passing truck.

BT CORN: THE STARLINK CONTROVERSY

Bacillus thuringiensis (Bt) is a naturally occurring bacteria that is toxic to many insects. It is available as an insecticide under the trade names Dipel and Thuricide. In 2000, the pharmaceutical company Aventis CropScience created corn that was genetically modified with the Bt bacteria. StarLink Bt corn, as it was called, was marketed to farmers as a weapon against the European corn borer moth, a common and nasty pest. Because the Bt insecticide was genetically built into the seed, farmers did not need to spray their fields with either Dipel or Thuricide. Additionally, because of the slim possibility that the Bt bacteria could cause an allergic reaction in humans, the StarLink seed was approved only for crops grown as animal feed, not for human consumption.

BIOTECHNOLOGY AND GENETIC ENGINEERING

All was well until the StarLink corn was found in taco shells manufactured by Kraft Foods. The corn was traced back to farmers who said they had not been told that StarLink corn was not approved for human consumption. Though there was no evidence that the taco shells caused any illness, Kraft recalled 2.5 million boxes of them, Aventis pulled StarLink seed off the market and bought the remaining crop to keep it out of circulation, and the Environmental Protection Agency (EPA) revoked Aventis's license to sell it. However, when the issue went to court, a federal judge upheld the FDA's policy on GM foods. The FDA's 1992 statement said that GM foods are safe and are not subject to monitoring or testing as food additives and that they do not require labeling.[3] Nevertheless, "the episode showed that seeds planted on less than 1 percent of America's corn acreage could easily spread from farm to farm, contaminate the nation's grain handling system and seep into global food supplies."[4]

Proponents of Bt corn, also called transgenic corn, believe it is superior to traditional corn because it allows farmers to reduce their use of pesticides. This saves farmers time and money and allows them to produce more food on the same amount of land. Opponents of Bt corn caution that it has not been adequately tested to ascertain its effects on humans who eat it. They believe that FDA standards are not stringent enough and that to proclaim transgenic corn safe without conducting long-term research is irresponsible, especially because once it is planted outside the laboratory, it cannot be controlled. Wind and birds will spread the seed, so containing the genetically altered corn may prove impossible.

Could Bt corn kill monarch caterpillars? That was the concern in 1999 when researchers at Cornell University found that pollen from Bt corn is toxic to monarch caterpillars and can settle on the milkweed that is crucial to the caterpillars' survival. Subsequent studies found that the level of pollen needed to affect the monarchs was unlikely to accumulate naturally in the environment, but the public outcry overshadowed this finding. In fact, monarch populations have increased in the United States since the introduction of Bt corn, mainly because the Bt corn requires smaller amounts of pesticides than non–GM corn, which on balance leaves the monarchs' environment healthier. Nevertheless, if Bt corn becomes common, the theory goes, the corn borer will develop a resistance to it, and farmers will not just be back to square one but will need to use more toxic pesticides than ever before. Other agricultural experts believe the chances of this are slim; by law farmers must plant non–Bt corn in areas adjacent to Bt corn in order to allow the pests a "safe harbor" where they can pass on their non–Bt tolerant genes to subsequent generations, providing a gene pool that is varied enough to ensure that Bt corn remains effective.

THE DANGERS OF MONOCULTURE

The family farm had seen better days by the time Earl Butz was named Secretary of Agriculture by President Nixon in 1971. Butz was a farmer by trade, a former Dean of Agriculture at Purdue University, and the chair of the U.S. delegation to the United Nations (UN) Food and Agriculture Organization (FAO) when he took the reins of the United States Department of Agriculture (USDA). He sought to transform the agricultural industry by reforming many New Deal era policies (that were designed to help small farmers) by replacing them with policies that exemplified his mantra, "Get big, or get out." Butz grew up during the Great Depression, and his goal was to make food affordable enough so that no one in the country ever went hungry again.

Butz encouraged farmers to plant commodity crops that could be traded on the world market. Corn became the country's staple crop as family farms failed and agriculture companies consolidated land holdings to thousands of acres. Thus came a new era in American agriculture—the era of monoculture. Monoculture is the practice of planting a single crop over a wide area; in the United States this means mainly corn and wheat, although any large-scale agricultural operation can be considered monoculture, including raising cattle in vast feedlots. There are some good reasons to do this. Farmers can get greater yields from a finite piece of land. Maintenance is easier; farmers can practice economies of scale with planting, spraying, irrigation, and harvesting. Plants tend to be more uniform as a result of consistent soil and climate conditions. Most important, monoculture results in surplus crops and reduced food prices—two essentials when it comes to feeding a hungry world.

On the other hand, monoculture can turn a small problem into a major catastrophe. In 1845, a blight struck Ireland's potato crop, wiping out the harvest on which one-third of the population relied for its survival. In the following three years, approximately 1 million people died of starvation and disease, and another million emigrated to other countries. In total, Ireland lost about 25 percent of its population, permanently affecting its social and political landscape.

Though many factors contributed to the Great Irish Famine, monoculture was partly to blame. Because Ireland's only crop was the potato, the blight spread quickly throughout the fields, destroying everything in its path. Without a steady supply of food, people could no longer function. Death and social mayhem swept through the countryside and reverberated through the towns.

While no one believes the United States will suffer a disaster on par with the Great Irish Famine any time soon, there is no doubt that monoculture has caused the country to rely disproportionately on certain crops (particularly corn) and that soil degradation may have a negative impact on biodiversity. Monoculture crops often require high doses of pesticides and herbicides, as natural predators are extinguished from the environment. Disease becomes commonplace in cattle, and large doses of antibiotics are necessary to maintain the safety of the meat supply. Pesticide runoff and animal waste gathers in waterways and becomes concentrated at major delta regions, such as the Mississippi Delta, resulting in miles-wide algal blooms that choke out oxygen and suffocate marine life. Traces of pesticides and antibiotics filter through the commercial food system, along with their attendant consequences on human health.

THE RISE OF ORGANIC FOOD

The organic food movement is in part a reaction against monoculture. Organic foods are grown without chemical fertilizers, pesticides, radiation, food additives, contaminated water, or antibiotics. Organic food by definition precludes genetically modified organisms (GMOs). The movement began in the 1980s with small family farmers recognizing the opportunity to sell organic produce to health-conscious customers. Initially, organic products were expensive and found mainly at local farmers' markets. Besides fruit and vegetables, the organic market includes free-range chickens and their eggs, grass-fed beef, milk and cheese derived from cows that are not injected with hormones and antibiotics, and other meats from animals raised outside the CAFO model.

The organic market has been expanding around 20 percent per year since 1990, although it is still small and prices are still quite high. The National Organic Program was created by the USDA in 2002 to grant certification to organic growers. Their standards document runs more than 500 pages and outlines all conditions that must be met before a product can be labeled certified organic. Other countries have their own certification standards and governing bodies. Industry groups have developed their own standards to ensure consumers receive what organic products offer and no more. In response to cross-contamination of organically grown crops with GM crops, the Non-GMO Project was initiated in 2009. The group's goal is to make sure that manufacturers of organic foods follow procedures that allow for no more than 0.9 percent biotech ingredients in their products—the same level allowed in Europe.[5]

Biotechnology and Medicine

DIETARY SUPPLEMENTS: FOODS OR DRUGS?

On February 17, 2003, Steve Bechler, a 23-year-old pitcher for the Baltimore Orioles, died of heat stroke during spring training. He had been taking the herbal supplement ephedra in order to lose weight. A Florida coroner ruled that the substance was a contributing factor in Bechler's death.[6] Far from an isolated event, Bechler's was only one of a number of deaths to which ephedra contributed. Citing safety and health concerns, the FDA banned the substance in early 2004.

Ephedra sinica, the plant from which ephedra is derived, has been used for more than 5,000 years in Chinese medicine as a treatment for asthma and hay fever. Available over the counter in the United States in several forms before it was banned and as an ingredient in Sudafed and in the weight-loss pill Metabolife, ephedra's stimulant properties were well known. It increases blood pressure and heart rate. It was both a cold medication and a weight-loss supplement. Who was responsible for regulating it?

The ephedra scenario has been played out several times in the consumer marketplace, from the controversy over the fenfluramine-phentermine cocktail known as fen-phen that was prescribed as a weight-loss aid until it was found to cause heart damage, to the decades-old controversy over whether saccharine and aspartame—two sugar substitutes—cause cancer or other health problems.

The FDA regulates dietary supplements as food, not drugs. The Dietary Supplement Health and Education Act of 1994 was enacted to make a few of these distinctions. According to the legislation, supplements include vitamins, minerals, herbs, botanicals, and amino acids.

PERSONALIZED MEDICINE

Traditionally, a doctor diagnoses a patient with a condition and then prescribes an antidote. High blood pressure, high cholesterol, depression, and chronic pulmonary heart disease are all common conditions that are treatable with a multitude of pharmaceuticals. Often it takes a bit of fine-tuning between patient and physician to find the right medication and the right dosage for the maximum intended effect. Side effects vary from person to person, as do allergic reactions and the way medications interact with any others the patient may be taking. All in all, treatment is an inexact science in which the risks and benefits must be weighed on an individual basis.

Personalized medicine seeks to change all that. With the completion of the Human Genome Project, scientists are beginning to understand how a person's genotype could guide treatment for his or her ailments. Simply taking stock of which diseases a person is genetically predisposed to constitutes personalized medicine in its most basic form. But personalized medicine could lead to discrimination on the part of health insurers. They may be unwilling to cover individuals who carry the gene for a particular disease, such as breast cancer or cystic fibrosis. In the eyes of health insurers, these people would be a bad risk.

THE SPROUTING BRANCHES OF MEDICINE

The more scientists know, the more specialized their fields of study become. The term *biotechnology* encompasses a myriad of disciplines, many of which overlap, and all of which are at the forefront of current research and development. A few of these are:

- Pharmacogenetics: the study of how a person's inherited genetic tendencies interact with specific medicines. Scientists who work in this field hope to create safer, more effective drugs and vaccines that are tailored to a person's proteins, RNA, and DNA. This field is also called pharmacogenomics.

- Proteomics: the study of proteins, enzymes, and protein modification for medicinal purposes. Proteomics is analogous to but more complex than genomics because proteins change from cell to cell, whereas a person's genome remains constant.

- Epigenetics: the study of inherited traits that are not the result of DNA.

- Nuclear medicine: the science of medical imaging to reveal biological processes at the subcellular level. It often relies on radiopharmaceuticals and medical isotopes.

- Nanobiotechnology: the study of chemical elements for the purpose of creating technical or medical devices. It is also called bionanotechnology.

RISING INEQUALITY

In 2007, health care expenditures in the United States totaled $2.3 trillion. That comes to $7,600 for every person in the country, or 16 percent of the nation's gross domestic product (GDP).[7] At the same time, 45.7 million Americans had no health insurance at all.[8] These are the stark facts that point to the disparity in health care in the United States. The most modern advances of biotechnology carry a hefty price tag and are generally available only to those with insurance. Diagnostic tests, lifesaving medications, chemotherapy for cancer patients, organ transplants, long-term care, etc., can quickly bankrupt a person without proper coverage. This is an ethical and political issue that overshadows much of the tremendous gains that scientists have made in recent decades. Health care reform was a major issue during President Obama's first year in office, with contentious partisan debate over how to bring down the cost of health care, cover those who are uninsured, while maintaining benefits for those who are, and how to fund such an initiative during a deep recession. Attempts by previous administrations to streamline the system and cover more people have ended in failure because of the enormous influence of lobbyists from the insurance and for-profit medical industries. It is important to remember that despite the promises of biotechnology in terms of medicine and health care, not everyone has access to its benefits.

HISTORY

From its earliest days, the United States has been an agrarian society. Immigrants were attracted by the country's wide open spaces, the ability to start over, and the plentiful resources that made it possible. Settlers headed west to stake their claim on land that was fertile and abundant. Families were large; children were a cheap source of labor for the endless tasks the family farm required. In the South, the economy grew up around cotton and tobacco farming; in the Midwest corn and wheat dominated. Self-sufficiency was key; communities were held together by their shared interest in providing what they needed to survive and grow.

In 1900, about 50 percent of the U.S. population lived on farms or was otherwise involved in agriculture. By 2003, that figure had dwindled to 0.7 percent.[9] The intervening century saw agriculture become concentrated in the hands of a few large corporations. Gone was the tradition of raising many different breeds of chickens, hogs, beef, and fish alongside each other in a proliferation of small farms from coast to coast. The change began during the Great Depression, when farms began to fail and families sold their land and moved to urban areas. It was compounded by the storms of the dust bowl, which turned once-fertile fields barren. Then in the 1980s it reached its apex, as many family farms were bought out by agribusiness, which streamlined farming by cutting back to only one breed of animal or one crop. Profit was forged on economies of scale. Monoculture brought about an inadvertent decline in regional biodiversity. People knew less and less about how their food was made; it no longer came from the backyard or the neighborhood farmer. It was packaged, canned, frozen, or processed. Instead of sustenance, the buzzword was convenience.

Immigration, Poverty, Discrimination, and Eugenics: The Nineteenth Century

As railroads united the east and west coasts, the frontier closed, and the nation's urban areas began to grow exponentially. In just 10 years, between 1890 and 1900, the population of New York City more than doubled from 1.5 million to 3.4 million.[10] Immigrants from Europe poured into the shoddy tenement buildings of the Lower East Side of Manhattan, where violence, poverty, and a lack of sanitation created dismal living conditions. Similar situations arose in Chicago, San Francisco, and Philadelphia. These immigrants often had large families and short life expectancies. Birth-control activist Margaret Sanger understood that one of the most effective ways to combat poverty in these urban environments was to encourage people to stop having so many children.

As in so many other countries, the eugenics movement in 19th-century America aimed to reduce the numbers of impoverished city dwellers, who were believed to be a burden on society. Poverty was considered to be the result of "feeblemindedness" rather than a lack of economic opportunities and education. The poor were often blamed for their own situation; thus, reducing the number of poor people by limiting their fertility rates became a favored strategy among social theorists.

Of all the social critics, Sanger was one of the most vocal. She is remembered today for pioneering the family planning movement, promoting birth control, and founding the American Birth Control League, the forerunner of today's Planned Parenthood. In Sanger's day, birth control was uniformly illegal, whether a woman was married or not. But Sanger recognized that multiple pregnancies were dangerous for a woman's health, especially in the absence of adequate sanitation, health care, and nutrition. Moreover, women who died as a result of childbirth would not be there to care for their children, which tended to exacerbate the cycle of poverty and crime. Birth control, therefore, represented the best way to improve women's lives and give their existing children the chance for a better future.

To say that Sanger faced stiff opposition is putting it mildly. She was harassed, jailed, and even deported, yet her beliefs never wavered. In a society that made women the property of their husbands, giving women control over their reproductive systems was considered outrageous and immoral. Nevertheless, when it came to eugenics—including forced sterility—Sanger faced less censure. That was because she was a proponent of negative eugenics; that is, she wanted to limit the fertility of the country's most undesirable citizens. These included the economically disadvantaged, the illiterate, the handicapped, the criminally minded, and those deemed to be lazy or stupid. Toward this end, she rallied for sterilization as a major weapon in her crusade to improve living conditions on the Lower East Side of New York. She waged her campaign through her newspaper column "What Every Girl Should Know," and she distributed pamphlets on family planning throughout New York City, a practice that flagrantly violated the 1873 Comstock Laws, which forbade dissemination of "obscene material." She coined the term *birth control* in her newsletter, "The Woman Rebel," and was soon charged with violating obscenity laws.

In 1927, Sanger organized the first World Population Conference in Geneva, Switzerland. She lived long enough to see the introduction of the birth control pill—itself a culture-shifting product of biotechnology—in 1966, and the Supreme Court decision in *Griswold v. Connecticut* in 1965, which made it legal for married couples to use birth control. Sanger's books include *Woman and the New Race* (1920), *Happiness in Marriage* (1926), and *My Fight for Birth Control* (1931). Eugenics, as espoused by Sanger and others, represented the intersection of biotechnology and public policy during the first half of the 20th century.

"THREE GENERATIONS OF IMBECILES IS ENOUGH"

Carrie Buck was born in 1906 to a mentally impaired woman with a record of delinquency and prostitution. She was placed as a foster child with the Dobbs family in Virginia, where she attended school until the sixth grade. After that she remained with the family as a helper. When she was 17, Carrie was raped by a relative of Mrs. Dobbs and became pregnant. In an effort to protect the family's reputation, the Dobbses had Carrie committed to the Virginia State Colony for Epileptics and Feebleminded, citing promiscuous behavior and their belief that she had the mental ability of a nine-year-old. At the State Colony she gave birth to a daughter, Vivian, who was also raised by the Dobbses and was assumed to be feebleminded. At the time, Virginia state law mandated compulsory sterilization for the mentally impaired, so the superintendent of the State Colony authorized Carrie Buck's sterilization. Carrie herself, through a legal guardian, opposed the operation, and the resulting lawsuit reached the United States Supreme Court in 1927. On May 2, 1927, the decision against Buck was handed down by Justice Oliver Wendell Holmes, who wrote:

> We have seen more than once that the public welfare may call upon the best citizens for their lives. It would be strange if it could not call upon those who already sap the strength of the State for these lesser sacrifices, often not felt to be such by those concerned, in order to prevent our being swamped with incompetence. It is better for all the world, if instead of waiting to execute degenerate offspring for crime, or to let them starve for their imbecility, society can prevent those who are manifestly unfit from continuing their kind. The principle that sustains compulsory vaccination is broad enough to cover cutting the Fallopian tubes. . . . Three generations of imbeciles are enough.[11]

Buck v. Bell was a landmark moment in negative eugenics, when even an American hero such as Holmes hailed an idea that within a few short years would be associated with Nazi atrocities. With the Supreme Court on their side, many states joined Virginia in enforcing eugenics statutes. Many of these statutes were based on the model law written by Harry H. Laughlin, the pioneering eugenicist of the Eugenics Records Office in Cold Spring Harbor, New York, and these laws remained on the books until the 1970s.

Laughlin was instrumental in the founding of the American Eugenics Society (AES) in 1926, a professional organization that had more than 1,200 members, organized academic conferences, and promoted research. The society also sponsored "Fittest Family" contests at state fairs throughout the country, in which families competed to be deemed the most genetically talented, good-looking, and physically adept. One of the AES's favorite pronouncements was that a child was born every 16 seconds in America, a feebleminded child every 48 seconds, and a criminal every 50 seconds. All these feebleminded, criminal children, the AES argued, cost taxpayers $100 every 15 seconds (in 1930s Great Depression dollars).

In the end, Carrie Buck was sterilized and released from the State Colony. Her daughter Vivian died of an illness at age eight, after having completed two years of school in which she proved to be a decent student, even qualifying for the honor roll one semester. Buck eventually married a man named William Eagle, and they lived together until his death 25 years later; their only regret was not being able to have children. In her later years, Buck was an avid reader, and it became obvious that she—like her daughter—was not mentally impaired in any significant way.[12]

Patent Medicines, Jake Leg, and *The Jungle*

Food and drugs were not regulated in the early 1900s. This provided a perfect opportunity for con artists to travel from town to town, posing as doctors and selling patent medicines that were made of secret ingredients, the foremost of which was usually alcohol. In an era that frowned upon excess consumption of alcohol—especially after Prohibition was enacted in 1920—patent medicines were an easy way to circumvent the law. Since no agency existed to verify claims made by these snake oil salesmen, they guaranteed their potions would cure everything from allergies and insomnia to impotence and depression. Patent medicines enjoyed wide popularity among alcoholics looking for booze and gullible customers longing for a cure for their illnesses. By the time the latter realized the medicine was phony, the good doctor was long gone.

Coca-Cola, as created by the druggist John Pemberton in 1885, was initially a patent medicine that contained nine milligrams of cocaine per serving.

Pemberton claimed it cured morphine addiction, headaches, and impotence, among other things. (Pemberton himself suffered from morphine addiction.) Cocaine was removed from Coca-Cola in 1903, but caffeine remained. By 1911, citing the harmful effects of caffeine, the Food and Drug Act required labeling of the ingredient. Thus, food additives and labeling laws vary according to the moral winds of the time. Even in the 21st century, noncarbonated Coca-Cola syrup is still used as a remedy for upset stomach.

One patent medicine in particular, Jamaican Ginger Extract, or "Jake," crippled tens of thousands of people in 1930. As Prohibition dragged on, Jake gained a loyal clientele, given that it was 70 to 80 percent alcohol. To discourage its use as an illegal form of drink, government officials required Jake to contain a high concentration of ginger solids, which made it too bitter for most taste buds. A pair of bootleggers attempting to circumvent the requirement replaced the ginger solids with tri-o-tolyl phosphate (TOCP); they hoped their Jake would pass government inspection but still be palatable to consumers. It turned out that TOCP was a neurotoxin that damaged nerve cells in the spinal cord. Those who drank the batch of Jake doctored with TOCP began to lose the use of their hands and feet, then their arms and legs. Some victims were permanently and completely paralyzed, while others were partially paralyzed. Over time, a few regained feeling in their limbs, but many others suffered permanent muscle atrophy and were disabled for the rest of their lives. Most of the victims were poor or black, and none had any recourse, legal or otherwise, to obtain social services or medical treatment for their disabilities. Though existing laws allowed federal officials to prosecute the bootleggers, no law existed to help the victims.[13]

At the turn of the century, the Chicago meatpacking industry was rife with corruption and violence and completely lacking in sanitation. The Union Stockyards were the nation's largest, employing 25,000 people and butchering 12 million head of cattle and hogs each year. The vast compound of kill houses and processing plants was strategically placed outside of town in order to shield the citizens of Chicago from the odor, smoke, and mounds of animal waste they produced. Blood, bones, and entrails were dumped by the ton into the water supply, and workers' families lived in squalid conditions amid the livestock and animal carcasses. Because the workforce was composed mainly of impoverished immigrants—many of whom knew little English—the situation remained off the radar of public awareness. Then Upton Sinclair, later a Pulitzer Prize–winning author, arrived on the scene and immersed himself in the immigrant culture to research his muckraking novel *The Jungle*. The book was a fictionalized account of oppressed workers, including women and children, and dangerous working conditions that often led to chronic illness, disabilities, or death.

The Jungle caused a firestorm of protest when it was published in 1906, just as Sinclair had hoped. It caused people across the nation to protest the horrific conditions of the stockyards and slaughterhouses of Chicago and elsewhere. The uproar led directly to the Meat Inspection Act and the Pure Food and Drug Act, which called for the formation of the FDA. The law was strengthened in 1938, and the FDA has long been one of the most important government regulatory agencies.

Land-Grant Universities

The Morrill Acts of 1862 and 1890 gave federal land to each state to establish colleges that would teach agricultural science and animal husbandry in the belief that graduates would help ensure the survival of a prosperous farming middle class. The archetype for the colleges was the Iowa Agricultural College and Model Farm, established in 1858. The Hatch Act, passed in 1887, created agricultural experiment stations at each school to focus on crop and livestock issues. In 1914, the Smith-Lever Act was passed; it enabled the colleges to establish cooperative extensions for the purpose of assisting community members with matters relating to agriculture and horticulture. Many extension centers still operate today as a resource for people interested in gardening, horticulture, agriculture, and environmental issues.

Government funding of these land-grant colleges helped the country become a world leader in scientific research. In 1877, for example, the botany professor William J. Beal of Michigan State University produced the first genetic hybrid corn variety, and in the 1930s, also at Michigan State University, the food science professor G. Malcolm Trout invented the process of milk homogenization. The botanist George Washington Carver attended Iowa State University; he helped the South recover from soil depletion and a boll weevil plague caused by its cotton monoculture. He initiated crop rotation with peanuts, sweet potatoes, soybeans, pecans, and cowpeas to diversify the economy and reinvigorate the land. Texas A&M, an agricultural and mechanical land-grant college, is today a world leader in animal cloning. Its College of Veterinary Medicine has cloned the first cat, cattle, Boer goat, pig, deer, and horse.[14] Researchers at MIT have made many biogenetic discoveries, including gene splicing, protein synthesis, and reverse transcriptase. At the University of Wisconsin–Madison, vitamins A and B were discovered between 1913 and 1916, and in 1923 researchers there developed a way to add vitamin D to milk. Still a leader in biotechnology, the University of Wisconsin–Madison's James Thomson was the first to isolate and use human embryonic stem cells in 1998.

Today, every state still has as least one agricultural land-grant school. Many of these are major state universities that have excellent reputations for all of their programs, such as Tuskegee University, California Institute of Technology, Purdue, Cornell, and Virginia Tech. U.S. territories, including American Samoa, Guam, and Puerto Rico, also have land-grant universities. Of course, many other non–land-grant universities are home to world-class biotechnology research departments, including Johns Hopkins University in Baltimore, which has the largest research and development budget of any university in the country.

Thomas Hunt Morgan and the Mutant Fruit Fly

Thomas Hunt Morgan received his Ph.D. from Johns Hopkins University in 1890 and became the first geneticist to win the Nobel Prize in physiology or medicine, which he received in 1933 "for his discoveries concerning the role played by the chromosome in heredity."[15] His work on *Drosophila melanogaster*, the common fruit fly, at Columbia University led to much groundbreaking work in genetics. Although Morgan admired Charles Darwin, he objected to his description of natural selection as a slow-moving force in which minute changes accumulate over numerous generations. Instead, Morgan was inspired by Gregor Mendel, whose work had just been rediscovered and indicated that careful manipulation of a species can instantly introduce major genetic changes. Morgan's goal was to replicate Mendel's pea plant experiments with fruit flies.

To do this, he would find a mutant fruit fly and cross-breed it with other fruit flies in order to observe the heritability of the mutant traits. Hundreds of thousands of flies and several years later, Morgan found a male mutant *Drosophila* with white eyes. He bred the fly with the normal red-eyed flies and noted that their progeny were all red-eyed. Those red-eyed progeny were bred and produced white-eyed males. From this, Morgan determined that some traits, such as eye color, are sex-linked. Further research isolated more traits, which Morgan and his students observed through subsequent generations of cross-breeding. In 1915 he and his laboratory partners published *The Mechanism of Mendelian Heredity*, a landmark book that paved the way for future work on chromosomes and genetics and made *Drosophila melanogaster* the model organism for much of that research. One of the book's coauthors, Alfred Sturtevant, in 1913 became the first person to create a genetic map.

"Bad Blood": The Tuskegee Study of Syphilis, 1932–1972

"The Tuskegee Study of Untreated Syphilis in the Negro Male" is a sinister chapter in the history of American medicine. The long-term study was conducted by the U.S. Public Health Service at the all-black Tuskegee Institute. It analyzed 400 poor, mostly illiterate black males in Macon County, Alabama, who suffered from syphilis, a debilitating venereal disease. A control group of 200 syphilis-free males was also followed. In 1932, with manufacture of antibiotics still in the future, existing treatments for the disease were dangerous, toxic (for example, mercury ointment), or ineffective. The doctors conducting the study wanted to see if withholding treatment would be a better option. The experiment's initial design called for observation of untreated syphilis for a period of six to eight months, followed by a treatment phase. Some involved with the study disagreed with that plan, and a power struggle ensued. The head doctor resigned from the study, and Oliver C. Wenger assumed control, changing the experiment into a long-term observational study with no treatment component.

The patients were not told they had syphilis, only that they had "bad blood." Wenger and the other doctors gained the patients' participation by promising them transportation to the clinic and funeral costs in case of their death. The study was unethical, racist, and violated participants' human rights, ultimately costing many of them their lives. Worst of all, after penicillin became the standard treatment for syphilis, the men were prevented from receiving it, even as public health officials visited towns throughout the South and distributed it free of charge.[16] In fact, after 250 of the subjects were diagnosed with syphilis during their World War II draft physicals and ordered by the military to take penicillin, the Tuskegee Study's doctors forbade them to do so.

The prohibition against treatment for the Tuskegee test subjects remained in effect even after the Nuremberg Code was established to protect individual rights after Nazi experiments during World War II came to light. Dr. John R. Heller, who presided over the study in its later years, saw no reason to inform the participants of their true health situation. "The men's status did not warrant ethical debate," Heller said. "They were subjects, not patients; clinical material, not sick people."[17] The study was scheduled to conclude only after the last patient died and was autopsied. By 1972, when the study finally shut down, only 74 of the original 400 men were still alive. More than 100 had died as a result of untreated syphilis, 40 of their wives had been infected, and 19 of their children were born with congenital syphilis.

Over the years, several individuals raised questions about the ethics of the study, but they were repudiated by the study's supporters, which included the American Medical Association (AMA), the National Medical Association (NMA), and many African-American medical doctors (including some who assisted with the Tuskegee study). In 1966, Peter Buxtun, a Public Health Service (PHS) investigator, having voiced his concerns to a deaf medical establishment, finally went to the press. The story became front-page news in the *New York Times* on July 26, 1972. A congressional hearing and a class action lawsuit by the National Association for the Advancement of Colored People (NAACP) followed. The tragedy of the Tuskegee Study led to the 1974 National Research Act and the 1979 Belmont Report, a landmark document outlining ethical procedures for experiments on human subjects and establishing institutional review boards to oversee experiments carried out on people.

Many opposed to latter-day biotechnology experiments cite the Tuskegee Study as a prime example of what can happen when scientists operate in secrecy, without proper oversight. Indeed, one legacy of the study is that some African Americans mistrust public health initiatives, especially those concerning human immunodeficiency virus, or HIV.

Chemurgy

Chemurgy is the practice of using agricultural substances in consumer products. The term was coined in 1934 by the chemist William J. Hale, whose book *The Farm Chemurgic* served as a how-to guide. Even before its publication, many products, such as celluloid, linoleum, and printers' ink, were chemurgic in nature. Celluloid, for example, was created from plant matter (cellulose) and nitric acid and was used to make film, toys, and pens. Hale and the fellow farm journalist Wheeler McMillen sought to organize chemurgists into a national organization. In 1935, the Farm Chemurgic Council was created to advocate the use of renewable raw materials in industry. But the council soon faced opposition from Franklin Roosevelt, who believed that the chemurgists hindered the policies of the USDA and the petroleum industry, which promoted the use of oil over plant-based substances.

Henry Ford was an early proponent of soy as a raw material and made many parts of his automobiles from soy products, including buttons, knobs, switches, gears, glues, and paints. At one point, each Model T manufactured by the Ford Motor Company contained 60 pounds of soybeans in its paint and molded parts. George Washington Carver, a chemurgist even if he did not acknowlege it, developed glue, soaps, and paints from peanuts and sweet potatoes. Chemurgy became useful during World War II when rubber supplies were interrupted. Corn was used to make synthetic rubber, and even milkweed found new life as a material used in constructing military life jackets.

When inexpensive plastics and petrochemical products flooded the market in the 1950s, chemurgy lapsed into obscurity. Only with the spike in oil prices in the first decade of the 21st century did interest in the field revive. The Ford Motor Company's green technology initiative harkened back to Henry Ford's use of soybeans by outfitting its 2008 Mustang seats in soy-based foam.[18]

Even the push toward biofuels embodies elements of chemurgy. In Brazil, 80 percent of all cars run on ethanol, which is made primarily from sugarcane. Ethanol in the United States is made mostly from corn—5.4 billion gallons as of 2006.[19] Farmers are running into problems with corn-based biofuels; land is scarce, and crops that are bought for fuel have reduced the amount available for food. The result has been a rise in corn commodity prices. But interest in chemurgy is growing, even if the term itself has fallen out of favor. The New Uses Council, for example, was formed in 1990 to carry on the work of the defunct Farm Chemurgic Council, and the Association for the Advancement of Industrial Crops is composed of agriculture scientists, chemists, and geneticists who promote commercial use of natural crops.

Irradiated Food: The 1960s

Food irradiation became a common practice in the 1960s as a way to safeguard meat, poultry, fruits, vegetables, and even spices from harmful viruses and bacteria such as *E. coli* and *Salmonella*. The process entails exposing food briefly to gamma rays, electron beams, or X-ray accelerators, in a manner industry officials compare to the pasteurization of dairy products. Irradiation kills bacteria and fungi, slows the ripening of fruits and sprouting of vegetables, increases juice yield, improves hydration, and kills insects and parasites. The FDA promotes food irradiation as a safe procedure; indeed, it is practiced in more than 40 countries and is approved by the World Health Organization (WHO). The UN FAO requires nations to irradiate food shipped internationally as aid, and the USDA quarantines imported food if it is not irradiated.

Irradiation does not make food radioactive, yet some consumers remain leery of the practice. As a concession, irradiated food is labeled with the international symbol for irradiation, a graphic emblem known as the radura. Different foods undergo different levels of irradiation. Fruits receive a low dose, and prepackaged meats receive the highest dose. It is possible that irradiation results in a loss of a food's nutrients, but experts say that any loss is no more than what occurs during normal cooking. In the end, most agree that the benefits far outweigh the deficits; irradiated food is often used in hospitals for those whose immune systems are compromised.[20]

Many other household goods are subjected to irradiation at higher levels than food, yet the public remains mostly unaware of the fact. Medical equipment is routinely irradiated, as are automobile parts, plastics, hardware supplies, and precious gemstones. The former head of the FDA's Office of Biotechnology, Henry I. Miller, believes that irradiation offers the best hope for ensuring the safety of the country's food supply. Food contaminated with microorganisms causes 76 million cases of illness each year. "The only way to make a cultivated field completely safe from microbial contamination is to pave it over. But you can't eat asphalt," Miller once wrote.[21] Instead, he has advocated increased irradiation to kill bacteria and viruses, as well as rDNA technology—gene-splicing—to eliminate toxins such as staph and botulinum that are impervious to radiation. In addition, rDNA technology can also invest food with antibodies and proteins that will make it healthier than in its natural form.

The People behind the Science

RACHEL CARSON AND *SILENT SPRING*

Rachel Carson was a marine biologist and the author of the seminal 1962 book *Silent Spring*. It revealed the environmental dangers posed by DDT, a pesticide that was sprayed generously, repeatedly, and unquestioningly over populated areas from coast to coast after World War II to eliminate mosquitoes and other insects. DDT was hugely popular; its inventor, the Swiss chemist Paul Hermann Müller, even received the Nobel Prize in physiology or medicine for developing it. Carson was the first to raise safety concerns. She maintained that DDT caused reproductive problems and thinning egg shells in birds, which contributed to high death rates for many avian species. The silent spring of her title refers to a future in which there are no birds left to fill the spring air with their songs.

Carson went on to explain how DDT also harmed species other than birds, by working itself through the food chain via the process of bioaccumulation. Humans, being at the top of the food chain, are at risk of consuming the highest amounts of DDT. Carson believed that chemical companies that promoted DDT had failed to research its adverse affects and in some cases even knowingly disseminated misinformation, leading the public to believe it was safe. In 1962, many government and industry leaders railed against Carson and her book, calling her histrionic and trying to discredit her science. One such critic was Robert White-Stevens, a biochemist in charge of agricultural research for the chemical giant American Cyanamid, who wrote that "if man were to follow the teachings of Miss Carson, we would return to the Dark Ages, and the insects and diseases and vermin would once again inherit the earth."[22]

Yet Carson never advocated a complete ban on all pesticides—only caution and restrictions on their use, and independent research studies ultimately verified many of her claims. Her book is credited for launching the modern environmental movement and prompting the formation of the U.S. EPA. By 1972, DDT was banned in the United States. Four decades later, it remains in use in many areas of the world still plagued by high rates of malaria. In some cases, public health experts have shown that the risk to human health from DDT is much less than the risk of illness and death from malaria, a claim with which Carson readily agreed.

PAUL BERG AND THE ASILOMAR CONFERENCE

Paul Berg is a Stanford University biochemist who won the 1980 Nobel Prize in chemistry with Walter Gilbert and Frederick Sanger for his research into nucleic acids. At first, Berg and his colleagues had no qualms about conducting biochemical research involving bacteria and viral vectors. Then in 1972 Berg was working on a project in which he combined DNA fragments from the monkey virus SV40 with the bacteriophage lambda. His goal was to insert the new combined genetic material into a bacterium of *E. coli*. His fellow scientists urged him not to complete the experiment because SV40 was known to cause cancer in mice, and *E. coli* was known to inhabit the human intestinal tract. What if the new genetic material inadvertently infected the lab workers? Could they develop cancer? Berg agreed that his experiment should not proceed until he and his fellow biochemists came to a consensus about how to conduct the research safely. The National Academy of Sciences appointed a committee to look into the matter, and it recommended an international conference. Berg organized the 1975 Asilomar Conference on Recombinant DNA for the purpose of establishing guidelines for biotechnology research in which an organism's genome is altered by the insertion of genes. It represented an early example of the precautionary principle.

Two main principles arose from the Asilomar Conference: 1) Containment of recombinant DNA material is an essential element in the design of a research experiment; and 2) the effectiveness of the containment effort should match the estimated risk. Along with these principles, conference participants determined that biological barriers should be used to limit the spread of recombinant DNA, such as bacteria and vectors that cannot survive outside a laboratory environment.[23] Conference attendees forbade experiments that would clone recombinant DNA material made from pathogens and toxin genes or that would result in large amounts of recombinant DNA material that could be harmful to plant, animal, and human life. In the ensuing decades, researchers have acknowledged the importance of the Asilomar Conference on the burgeoning biotechnology industry, particularly for its mission to keep scientific knowledge in the public eye, lest scientists working in the area be accused of conspiracy and secrecy.

E. B. WILSON AND NETTIE STEVENS

Edmund Beecher Wilson was one of the first cell biologists in the United States and a proto-geneticist whose work in zoology led to insights in 1905 regarding the X and Y chromosomes, which determine an organism's gender. Wilson spent his career at Columbia University; he laid the groundwork for Thomas Hunt Morgan and his fruit fly experiments, which led to the understanding of genetic mutations and phenotype (observable characteristics). Both Wilson and Morgan owe a debt to Nettie Stevens, one of America's first prominent female scientists. Stevens graduated from Stanford in 1900 with a master's degree and went on to study cytology at Bryn Mawr. In 1905, she discovered that females of a species have two X chromosomes and that males have an X and a Y chromosome. This was a breakthrough—Stevens was the first scientist to link chromosomes with physical traits. Her work laid the foundation for modern embryology and cytogenetics, or the study of chromosomes and cell division. She is also credited with introducing fruit flies to Morgan as an ideal species for his experiments. Like Rosalind Franklin after her, Stevens died young, and her accomplishments were downplayed by her male colleagues. Nevertheless, she remains a key figure in the establishment of cytogenetics.

JOSHUA LEDERBERG: BIOTECH PRODIGY

Joshua Lederberg, the son of an Orthodox rabbi, graduated from high school at 15 and received a Ph.D. from Yale University at the age of 22. He won the Nobel Prize in physiology or medicine at 33 for discovering that bacteria engage in sexual reproduction and exchange genes. He founded the field of molecular biology in the 1950s and advised nine U.S. presidents. He also studied extraterrestrial life, artificial intelligence, and biological warfare.

The breadth of Lederberg's knowledge resulted in a sharper focus on the role of microbes in evolution and human history. The relationship between humans and germs, he understood, is symbiotic. Though germs are much more efficient at adapting to their environment than humans and are also capable of unleashing global pandemics that kill millions, they need us to survive. If germs led to the death of all humans, they would bring about their own extinction:

> *Biologically speaking, the reason we are still here is because microbes need live hosts for their own survival. This reality allows us to establish some of the ground rules of evolutionary success in the microbial world. It is as if they have read the Bible and know Genesis: they go forth and disseminate as their first rule. They multiply. Next, according to Malthusian and Darwinian doctrine, they have to be the fittest in order to survive so that they can produce the largest number of offspring they can.*[24]

Lederberg conducted much of his research at the University of Wisconsin beginning in 1947, where he founded the department of medical genetics.

Along with his wife, Esther Zimmer, and a few dedicated graduate students, Lederberg's work with bacteria, especially *E. coli* and *Salmonella,* shed light on how they become resistant to antibiotics and overturned the notion that they were primitive organisms incapable of evolution. He later founded the department of genetics at Stanford University and became the president of Rockefeller University.

Infrastructure: The FDA, NIH, CDC, and More

The FDA is an agency of the U.S. Department of Health and Human Services (HHS). It was formed in 1906 by President Theodore Roosevelt to ensure the safety of the country's food, dietary supplements, drugs, cosmetics, and vaccines. In 2008, the FDA and its subdivisions had a budget of $2.1 billion. Subdivisions include the Center for Biologics Evaluation and Research, which ensures the safety of blood products, vaccines, cell and tissue-based products, and gene therapy products; the Center for Food Safety and Applied Nutrition, which regulates diet supplements, cosmetics, health claims of various products, and bottled water (but not public drinking water; that is the job of the EPA); and the Center for Drug Evaluation and Research, which is entrusted with making sure all prescription and over-the-counter drugs are safe.

The NIH is another agency of the HHS. Its precursor was the Laboratory of Hygiene, established by the government in 1887, but it has since grown into one of the largest public biomedical research organizations in the world, with 27 separate institutes, including the National Human Genome Research Institute, the National Institute of Biomedical Imaging and Bioengineering, and the Bioinformatics Resource Center.

The HHS is also the oversight agency for the Centers for Disease Control and Prevention (CDC) in Atlanta, Georgia. The CDC is the latter-day incarnation of the Communicable Disease Center, established by the government in 1946 with the goal of eradicating malaria through the liberal use of DDT. The CDC is home to one of only two official repositories of the smallpox virus in the world (the other is in Russia), and it is a leader in research on bioterrorism, public health genomics, and biotechnology that may one day make babies' shots obsolete.[25] Along with the WHO, the CDC is the leading organization for public health and disease prevention in the world.

The Rise of the Pharmaceutical Industry

What is the most profitable industry in the United States today? Not manufacturing, or automobiles, real estate, finance, or communications. It is pharmaceuticals. In 2006, the industry generated $286 billion in profits.[26] It grew rapidly after World War II, when biomedical research led to a string of cures and treatments for diseases. The FDA became its regulatory agency, and soon the divide between prescription and over-the-counter drugs became entrenched, along with the arduous testing process required before new drugs could go on the market.

Prior to World War II, much medical research and development took place in Europe, especially Germany, Switzerland, the Netherlands, Belgium, and Italy. In the 1950s, research and development on the U.S. side of the Atlantic surged. Oral contraceptives, cortisone, blood pressure drugs, and heart medications changed people's expectations of health care, and many such substances became part of everyday life. In the early 1960s, antipsychotic medications and tranquilizers became popular; Valium was marketed in 1963 and quickly became the most prescribed drug of all time. As pharmaceuticals became more entrenched in popular culture, patent legislation paved the way for the distinction between name brand and generic drugs, which in turn influenced how drug companies chose to invest their research and development funds. By the 1980s, small biotechnology companies were struggling to survive, and many of them were bought by pharmaceutical companies. The 1990s saw the introduction of antidepressants and a growing concern that doctors were overprescribing medication or overdiagnosing certain conditions. In the 2000s, the divide between biotechnology, genetic engineering, and pharmacology is increasingly blurred.

COUNTERSTRATEGIES
Congressional Committees

Should embryonic stem cells be used for research? Should animals be cloned? Should people be cloned? Who should decide if a person can receive gene therapy? In the United States, the public dialogue takes place between citizens, religious organizations, scientists, and government institutions. Government, more often than not, makes the final decision, usually because it holds the purse strings. And yet those purse strings are funded with taxpayer dollars, and government officials are elected by the people. More than ever, an informed citizenry is necessary for a reasoned national debate on biotechnology.

The House of Representatives and the Senate have established a number of committees to research biotechnology issues and suggest policies. They hold hearings and sponsor legislation. Getting legislation passed by Congress and signed into law by the president, however, is another matter entirely. The most notable committee is the House Committee on Science and Technology, which was formed in 1958. It has jurisdiction over scientific research and development that is not defense related. Some of its recent projects include preparations for future pandemics and legislation to strengthen science education and retain brainpower within the country. The House Committee on Agriculture has jurisdiction over federal agricultural policy and nutrition guidelines. It was founded in 1820 and regulates the livestock industry, the dairy industry, plants, and seeds.

The U.S. Senate has two standing committees that intersect with biotechnology: the Committee on Agriculture, Nutrition, and Forestry, formed in 1825, is responsible for the farm bills that shape the country's agriculture industry, and the Committee on Commerce, Science, and Transportation regulates toxic materials.

President's Council on Bioethics

President George W. Bush established the President's Council on Bioethics (PCBE) on November 28, 2001, in order to "advise the President on bioethical issues that emerge as a consequence of advances in biomedical science and technology." The members of the council are appointed by the president but cannot be current members of the U.S. government. In its first eight years, the council published several papers on stem cell research and human cloning.

In his essay *Human Dignity and Respect for Persons: A Historical Perspective on Public Bioethics,* committee member F. Daniel Davis summarized the issues central to the President's Council's mission:

> *Today, more than ever before, we seem poised for mastery over many aspects of human life, including those that unite us with nonhuman animals and those that separate us from them. For some, these achievements of the ongoing revolution in biomedicine and biotechnology testify to the triumph of human ingenuity and to the efficacy of the human will to fashion our environment—and ourselves—as we wish. For others, the claim that all these impressive achievements make positive contributions to human flourishing is misguided and even dangerous, neglecting the sober lesson of Tuskegee: that the quest for new knowledge, and for new applications of that knowledge, can be perverted so as to inflict egregious harms on our fellow human beings—harms that go far beyond the failure to secure their voluntary informed consent.*[27]

Health Insurance Legislation

One of the dangers of genetic testing is that health insurance companies may use the results of such tests to deny coverage to patients. Even if patients alter their lifestyle to try to avoid a disease for which they are genetically predisposed, health insurers could still claim they are a high risk. To counteract this possibility, the U.S. Congress passed the Genetic Information Nondiscrimination Act of 2007, which prohibits the improper use of genetic information by health insurers and employers. The bill was signed into law by President George W. Bush on May 21, 2008. The legislation was supported by the NIH National Human Genome Research Institute as necessary for the advancement of biomedical research. Opponents of the bill claimed it was overly broad and may contradict state laws.

Political Climate Leads to Scientific Discoveries

In 1995, the U.S. Congress passed the Dickey Amendment, which was signed into law by President Clinton. The bill prohibits the HHS (which funds the NIH) from appropriating funds for research in which human embryos are destroyed. Three years later, a private company developed technology that allowed researchers to conduct studies on human embryonic stem cells, which can be obtained from human embryos. President Clinton recommended that the Dickey Amendment be changed to allow stem cell research on embryos that were scheduled to be destroyed, typically after they were no longer needed for IVF purposes.

President George W. Bush changed the law to allow federal funding for stem cell research only on embryos previously created for reproduction that were no longer needed and already scheduled for research use prior to August 10, 2001.[28] There were no laws barring embryonic stem cell research conducted privately or on adult stem cells. (President Obama, in one of his first acts in office, lifted the strict limitations on stem cell research.) Bush's ban prompted many scientists to move to places such as Singapore, which lured them with state-of-the-art research facilities; however, it also forced scientists who remained in the United States to think outside the box. In November 2007, induced pluripotent stem cells (iPSCs) were created independently by two researchers, James Thomson at University of Wisconsin–Madison and Yamanaka Shinya at Kyoto University in Japan. Both scientists' teams, working independently, devised a way to turn adult human skin cells into the equivalent of embryonic stem cells by adding four genes, all of which act as master regulator genes that turn other genes on and off.

"Pluripotent" means a cell has the ability to become any kind of the 220 different types of cells in the human body; it can become a liver cell, a skin cell, a kidney cell, etc. Embryonic stem cells are pluripotent, which is why they have been so crucial to stem cell research. Now that researchers have found a way to turn adult skin cells into pluripotent cells (that is, induced pluripotent cells), it is possible that embryonic stem cells will someday no longer be needed for research. As of 2010, iPSCs present several research problems: 1) The processes used to create them are not yet efficient enough for large-scale research; 2) there is some concern that one of the genes involved in the process is a cancer gene and that another may cause mutations that result in cancer; and 3) there is concern that the viral vectors and retroviruses used to carry out the procedure may have unintended consequences.[29] These hurdles will likely be overcome, and when they are, the advantages will be huge. Not only would iPSCs bypass the ethics involved in destroying human embryos, but they would also make moot concerns about donor eggs and cloning of eggs. Cells generated for therapy would not be rejected by a patient and would be a boon particularly to Alzheimer's research, which has previously required cloned cells.

States Take Matters into Their Own Hands

Silicon Valley, the area south of San Francisco, California, was a hotbed of activity in the information technology revolution of the 1990s and 2000s. Home to several major research universities and boasting a high per capita rate of scientists and venture capitalists to fund new research, the area was poised to take advantage of the biotech advances of the 21st century. Governor Arnold Schwarzenegger promoted the state to biotech start-up companies, offering incentives for them to relocate and take advantage of the intellectual capital of the area. In 2004, California voters passed legislation that would fund the California Institute for Regenerative Research with $3 billion.[30]

Massachusetts, particularly the Boston area, is also home to several major research universities, including the Massachusetts Institute of Technology (MIT) and Harvard University. The Greater Boston Chamber of Commerce has made life sciences one of its key areas of focus; it sponsors forums and speakers on biotechnology and genetics issues from both the academic and business realms. It markets itself as having the resources for companies that want to develop new drugs and processes. When President George W. Bush denied federal funds to institutions that conduct research using human embryonic stem cells, Massachusetts governor Deval Patrick stepped in and dedicated $1 billion in state funds to continue such research.[31] The goal is to provide a favorable climate in which to conduct private-public research that takes advantage of the area's wealth of top scientists and business leaders. Some of the money will be granted to universities for fellowships, grants, and training; some will be dispensed in the form of tax incentives to encourage companies to expand in the area; and $500 million will go toward building a stem cell bank at the University of Massachusetts Medical School.

State by state, governments have found a way to respond to the wishes of citizens who may not agree with federal policy. Government officials in

Maine adopted a plan to halt antibiotic resistance by discouraging public entities from buying meat from farms that put antibiotics in animal feed. The goal is to become a leader in taking a stance against overconsumption of antibiotics by animals to the detriment of human health.[32]

Medical Philanthropy

Research and development of biotech drugs have grown increasingly expensive and also more divisive as some protest animal research, others voice religious objections, and yet others seek treatments for rare diseases that affect few people (so-called orphan diseases). Some individuals are taking matters into their own hands by funding research projects that appeal to them by donating significant amounts of money to institutions that will conduct research on their behalf. This practice is called medical philanthropy, and its roots go back to before World War II, when the Rockefeller Institute funded the development of a vaccine for yellow fever and worked to eliminate hookworm.[33] The Wall Street financier Michael Milkin has given millions to research prostate cancer from which he is in remission, and the actor Michael J. Fox has created a foundation to research Parkinson's disease, from which he suffers.

The world's largest philanthropic organization is the Bill and Melinda Gates Foundation, with an endowment of $34.6 billion, of which at least $1.5 billion is granted each year to initiatives worldwide. In 2008, the Gates Foundation gave $164.5 million to the Alliance for a Green Revolution in Africa, which will be used to improve the soil of 4 million African farmers. The same year, the foundation gave $3 million to the International Centre for Genetic Engineering and Biotechnology in Trieste, Italy, for a program to expand biotechnology programs in Africa.[34] Nearly $38 million has been pledged by the group for research that seeks to genetically modify mosquitoes to be immune to malaria and dengue fever so they cannot infect humans.[35]

The Wellcome Trust in Great Britain has provided funding for the Human Genome Project and pledged £340 million to the Wellcome Trust Sanger Institute from 2006 to 2011 for further genetic research.[36] The Howard Hughes Medical Institute, founded by idiosyncratic millionaire aviator Howard Hughes in 1953, grants $780 million each year for research and education on biotech topics at 64 laboratories at U.S. universities.[37] It was founded specifically for biomedical research, and after Hughes' death in 1976, it transitioned to genetics. James and Virginia Stowers, financial investors who have both survived cancer, created the Stowers Institute for Medical Research in 2000 to undertake gene research they felt was being ignored by other organizations. The Cystic Fibrosis Foundation funds studies that test new drugs and supports budding biotechnology firms—both things the NIH are reluctant to do.

Philanthropists have stepped in where they see a lack of resources committed to necessary programs by governments and private industry. Some are forming public-private trusts to develop drugs for so-called neglected diseases. In particular, some charitable organizations have pursued embryonic stem cell research in light of the government's prohibitions on using federal funds for it. Business executives in California, for example, have pledged $14 million to a state stem cell program that is no longer funded with state money.[38]

A different form of philanthropy, in the form of prize money, is the idea behind the Archon X Prize for Genomics, which is offering a reward of $10 million for the first team to sequence 100 human genomes, in 10 days, at a cost of less than $10,000 per genome.[39] The idea behind the X Prize for Genomics is to advance the technology of genome sequencing to the point that it becomes commonplace, affordable, and accessible. If a person's genome can be sequenced in a few days, personalized medicine will be off to a running start. Patients could receive treatment tailored to their DNA, and researchers will be able to pinpoint the genetics of common killers, such as heart disease, cancer, Alzheimer's disease, and asthma. Part of the prize money will be given by J. Craig Venter, leader of the first private company to sequence a human genome (which happened to be his own). The X Prize Foundation believes that offering prize money is an efficient way to fund technology breakthroughs that benefit humanity. It requires people to work together for a common goal and creates an irresistible sense of competition that is generally lacking in traditional research grants. Prize money taps into the entrepreneurial spirit that is often absent at large research institutions and encourages people to think outside the box.

---

[1] USDA Economic Research Service. "U.S. Beef and Cattle Industry: Background Statistics and Information." Updated April 28, 2008. Available online. URL: http://www.ers.usda.gov/news/BSECoverage.htm. Accessed May 13, 2009.

[2] Soil Association Information Sheet. "Flavr Savr Tomato and GM Tomato Puree: Problems with the First GM Foods." July 29, 2005. Available online. URL: http://www.soilassociation. org/web/sa/saweb.nsf/librarytitles/1E946.HTMl/$file/Flavr%20savr.pdf. Accessed May 13, 2009.

[3] Andrew Pollack. "Judge Upholds F.D.A. Policy on Genetically Altered Foods." *New York Times* (10/4/00). Available online. URL: http://www.nytimes.com/2000/10/09/business/judge-upholds-fda-policy-on-genetically-altered-foods.html. Accessed September 9, 2009.

[4] David Barboza. "As Biotech Crops Multiply, Consumers Get Little Choice." *New York Times* (6/10/01). Available online. URL: http://www.nytimes.com/2001/06/10/us/as-biotech-crops-multiply-consumers-get-little-choice.html. Accessed September 9, 2009.

[5] William Neuman. "Bio-tech Free, Mostly." *New York Times* (8/28/09). Available online. URL: http://www.nytimes.com/2009/08/29/business/29gmo.html. Accessed September 9, 2009.

[6] Hal Bodly. "Medical Examiner: Ephedra a Factor in Bechler Death." *USA Today* (5/13/03). Available online. URL: http://www.usatoday.com/sports/baseball/al/orioles/2003-03-13-bechler-exam_x.htm. Accessed May 13, 2009.

[7] J. A. Poisal et al. "Health Spending Projections Through 2016: Modest Changes Obscure Part D's Impact." *Health Affairs* (2/21/07): W242–253.

[8] U.S. Census Bureau. "Health Insurance Coverage: 2007." Updated August 26, 2008. Available online. URL: http://www.census.gov/hhes/www/hlthins/hlthin07/hlth07asc.html. Accessed May 13, 2009.

[9] Ken Midkiff. *The Meat You Eat.* New York: St. Martin's, 2005.

[10] U.S. Census Bureau. "Population of the 100 Largest Cities and Other Urban Places in the United States: 1790 to 1990" (June 1998). Available online. URL: http://www.census.gov/population/www/documentation/twps0027.html. Accessed May 13, 2009.

[11] U.S. Supreme Court. *Buck v. Bell,* 274 U.S. 200 (1927). Available online. URL: http://laws.findlaw.com/us/274/200.html. Accessed May 13, 2009.

[12] Stephen Jay Gould. "Carrie Buck's Daughter." *Natural History* (July 1984). Available online. URL: http://findarticles.com/p/articles/mi_m1134/is_6_111/ai_87854861/print. Accessed May 13, 2009.

[13] Dan Baum. "Jake Leg." *New Yorker* (9/15/03): 50–57.

[14] Juan A. Lozano. "Texas A&M Cloning Project Raises Questions Still." *Bryan-College Station Eagle* (6/27/05).

[15] "The Nobel Prize in Physiology or Medicine 1933." Available online. URL: http://nobelprize.org/nobel_prizes/medicine/laureates/1933/index.html. Accessed May 13, 2009.

[16] Centers for Disease Control. "U.S. Public Health and Service Syphilis Study at Tuskegee" (1/9/08). Available online. URL: http://www.cdc.gov/tuskegee/timeline.htm. Accessed May 13, 2009.

[17] J. Jones. *Bad Blood: The Tuskegee Syphilis Experiment: A Tragedy of Race and Medicine.* New York: Free Press, 1981.

[18] Information on the use of soy-based products in Ford's cars is from "Ford, Lear to Launch Industry's First Soy Based Seat Foam in 2008 Ford Mustang." Lear MediaRoom (7/12/07). Available online. URL: http://lear.mediaroom.com/index.php?s=press_releases&year=2007. Accessed May 13, 2009.

[19] Energy Information Administration. "Biofuels in the U.S. Transportation Sector" (February 2007). Available online. URL: http://www.eia.doe.gov/oiaf/analysispaper/biomass.html. Accessed May 13, 2009.

[20] U.S. EPA. "Food Irradiation" (7/31/08). Available online. URL: http://www.epa.gov/radiation/sources/food_irrad.html#irradiation_bacteria. Accessed May 13, 2009.

[21] Henry I. Miller. "It's Frankenfood v. the Killer Tomatoes." *New York Post* (6/11/08).

[22] Dorothy McLaughlin. "Fooling with Nature: Silent Spring Revisited." *PBS Frontline.* Available online. URL: http://www.pbs.org/wgbh/pages/frontline/shows/nature/disrupt/sspring.html. Accessed May 13, 2009.

[23] Paul Berg et al. "Summary Statement of the Asilomar Conference on Recombinant DNA Molecules." *Proceedings of the National Academy of Sciences* 72.6 (June 1975): 1,981–1,984.

[24] Joshua Lederberg. "Getting in Tune with the Enemy—Microbes." *Scientist* 17.11 (6/2/03): 20.

[25] "Biotechnology Is Working to Make Baby Shots Obsolete." Centers for Disease Control Press Release (May 1997). Available online. URL: http://www.cdc.gov/od/oc/media/pressrel/newvac1.htm. Accessed May 13, 2009.

[26] Donald L. Barlett. "Why We Pay So Much." *Time* (2/2/04). Available online. URL: http://www.time.com/time/magazine/article/0,9171,993223,00.html. Accessed September 9, 2009.

[27] F. Daniel Davis. "Human Dignity and Respect for Persons: A Historical Perspective on Public Bioethics." In *Human Dignity and Bioethics: Essays Commissioned by the President's Council on Bioethics.* Washington, D.C., 2008. Available online. URL: http://bioethics.gov/reports/human_dignity/human_dignity_and_bioethics.pdf. Accessed May 13, 2009.

[28] National Institutes of Health. "Human Embryonic Stem Cell Policy Under Former President Bush." (3/10/09). Available online. URL: http://stemcells.nih.gov/policy/2001policy.htm. Accessed May 14, 2009.

[29] Gina Kolata. "Scientists Bypass Need for Embryo to Get Stem Cells." *New York Times* (11/21/07). Available online. URL: http://www.nytimes.com/2007/11/21/science/21stem.html. Accessed September 9, 2009.

[30] Jason Szep. "Massachusetts to Spend $1 Billion on Biotechnology." Reuters (6/16/08). Available online. URL: http://www.reuters.com/article/technologyNews/idUSN1626132920080616. Accessed September 9, 2009.

[31] Szep. "Massachusetts to Spend $1 Billion on Biotechnology."

[32] "Maine Says No to Drugs." *Current Science* 92.7 (12/1/06): 12.

[33] Pollack. "Fighting Diseases with Checkbooks." *New York Times* (7/8/06). Available online. URL: http://query.nytimes.com/gst/fullpage.html?res=9E0DE6DB1030F93BA35754COA9609C8B63. Accessed September 9, 2009.

[34] Bill and Melinda Gates Foundation. "ICGEB Receives Grant to Strengthen and Expand Biosafety Systems in Subsaharan Africa" (6/27/08). Available online. URL: http://www.gatesfoundation.org. Accessed May 13, 2009.

[35] Maria Cheng. "Genetically Modified Mosquitos May Combat Malaria." Associated Press (6/19/08).

[36] Mark Walport. "The Wellcome Trust and the Human Genome Project" (5/1/06). Available online. URL: http://genome.wellcome.ac.uk/doc_WTX031720.html. Accessed May 13, 2009.

[37] Howard Hughes Medical Institute. "Discovering New Knowledge" (7/10/08). Available online. URL: http://www.hhmi.org/about. Accessed May 13, 2009.

[38] Pollack. "Fighting Diseases with Checkbooks." *New York Times* (7/8/06).

[39] Archon X Prize for Genomics Web site. Available online. URL: http://genomics.xprize.org. Accessed May 13, 2009.

3

Global Perspectives

INTRODUCTION

Each country's approach to biotechnology and genetic engineering depends on its unique history, government, economy, and culture. Detailed analyses of Japan, Germany, India, and South Africa are offered here. But first, here is a word about key developments in a few other countries with unique circumstances. Iceland, Singapore, and Great Britain have all made headlines for the way they have tackled issues dealing with genetic technology that have arisen within their borders.

ICELAND

Iceland's demographics lend themselves well to genetic research. This frozen landmass just south of the Arctic Circle is home to roughly 300,000 people; its isolation means it has one of the most homogenous populations in the world. Descendants of the Nordic Vikings who settled the coast more than 1,000 years ago, Icelanders have a gene pool that has evolved with few outside influences. The country's genealogy records date back 1,000 years, and the government began collecting detailed medical information on its citizens in 1915. By the 1950s, scientists had begun compiling an exhaustive tissue bank that contains the genetic material of many residents. All of these factors have proven to be a bonanza for scientists seeking to map the population's shared genetic history.[1]

Kari Stefansson, a native of Iceland and a former Harvard professor of neurology, founded deCODE Genetics, Inc., Iceland's first biotechnology firm, in 1996 to capitalize on the wealth of government data at his disposal. Backed by U.S. venture capital money, the Reykjavik-based company worked to track down the genes responsible for diseases that tend to run in families, such as heart disease, schizophrenia, and asthma. The goal was to find the genes linked to the diseases in order to expedite research into new treatments and cures. By 2003, however, due to widespread concerns about privacy and informed consent, the Icelandic Supreme Court barred deCODE Genetics from implementing its Icelandic Health Sector Database, which was designed to facilitate the company's research. "The Icelandic court decision shows clearly that there are limits as to how far genetic research can intrude into the private lives of participants and how far the government can pass legislation on the scientists' behalf," according to Renate Gertz.[2] The legal debate underscores the divide between what is scientifically possible and what is ethical, a situation that will most likely become more common as genetic advances accrue.

When it comes to agricultural biotechnology, Iceland's fragile ecosystem presents challenges unfamiliar to other countries. Its terrain has few trees, a harsh climate, and rocky soil that make it unsuitable for large-scale farming. Grassland is scarce but of high quality, and it must be grazed sparingly. Because of these conditions, until the mid-20th century most Icelanders lived on small farms run by one or two families that raised cattle, sheep, and perhaps chickens. Today, many market-bound vegetables are grown in greenhouses, and livestock is still the domain of the small, family-run farm. "The country's rugged cows and hardy sheep are virtually unchanged genetically from the Vikings' first imports in the ninth century. All the animals graze on Iceland's rich grasslands; there are no intensive animal confinement systems in the country," according to the environmental reporter Jim Motavalli.[3]

Many farms have been owned by the same family for centuries; they have on average 18 dairy cows and 152 sheep.[4] Lamb is a dietary staple and is of a high quality. Most grain is grown as feed for sheep. At one point the government, recognizing the need to prevent degrading the land through monoculture, recommended that the country's 5,000 farmers universally adopt organic agricultural practices. Even before this, the industry had typically used few artificial fertilizers, hormones, and antibiotics. Iceland's approach to agriculture differs from that of many other developed countries in that it does not rely on genetically modified (GM) seed or widespread monoculture.

SINGAPORE

Singapore is a small Southeast Asian island nation nestled between Malaysia and Indonesia with a population of roughly 4.6 million.[5] It was a British colony until it gained independence in 1965, and it is one of the wealthiest countries in the world due to its key position on international trade routes. Its highly educated, English-speaking business class is a major asset in this era of globalization, and its success in the electronic, petrochemical, and financial services industries has given it an edge in attracting emerging high-tech companies. After losing some major businesses to fast-developing neighboring cities such as Kuala Lumpur, by the turn of the 21st century Singapore had repositioned itself as a leading destination for scientists whose work was hampered by regulations in their home countries. The government invested $3 billion in biotechnology infrastructure and developed the Agency for Science, Technology and Research (A*STAR) to attract world-class talent and oversee groundbreaking research and development.

Singapore welcomed David and Birgitte Lane, a cancer researcher and skin cell expert, respectively, from Great Britain in 2004. David Lane became the chair of Singapore's Biomedical Research Council and the executive director of its Institute of Molecular and Cell Biology. Alan Colman, the cocreator of Dolly the cloned sheep, relocated from England in 1996, and Neal G. Copeland and Nancy A. Jenkins, formerly of the National Cancer Institute in Maryland, arrived not long after that. To capitalize on its existing assets, the government built a $300 million research park—the Biopolis—to attract nascent companies, a deal sweetened by the park's first occupants, the Institute of Bioengineering and Nanotechnology and the Genome Institute of Singapore, both of which actively develop technologies they seek to license to start-up companies.[6] Another National Cancer Institute researcher, Edison Tak-Bun Liu, moved from the United States to become the head of the Genome Institute, and Jackie Y. Ying of the Massachusetts Institute of Technology (MIT) became head of the Institute of Bioengineering and Nanotechnology.

Singapore has concentrated on attracting scientists who work with stem cells, erasing the regulatory red tape that hampers such research in many western countries. Both growing stem cells and therapeutic cloning are legal in Singapore. In 2006, ES Cell International became the first company to produce and sell human embryonic stem cells on the open market, under the direction of CEO Colman. A vial of stem cells could be obtained via the Internet for USD$6,000,[7] a price that Colman hoped would raise enough funds to allow the company to continue research in diabetes and heart disease. Pharmaceutical companies, following the research and the banking industries, are also moving to Singapore, aided by the government's generosity in building state-of-the-art infrastructure.

This push toward biotechnology hit a snag in 2007 when some of these research stars accepted positions back in their home countries. A Reuters reporter noted that "a World Bank report said Singapore had only a 50-50 chance of succeeding in its biomed drive, and warned a big part of the biotech sector is made up of 'footloose' star researchers who could leave the city-state at short notice."[8] David and Birgitte Lane went to Scotland to head up the division of molecular medicine at Dundee's College of Life Sciences, and Colman returned to King's College in London to head the Stem Cell Centre.

Nevertheless, with the infrastructure in place and Singapore's history of business-friendly practices, it seems likely that it will remain a hotbed of activity for decades to come.

GREAT BRITAIN

When it comes to dissent over GM food, Great Britain is one of the most vocal nations in the European Union (EU). Overall, the EU is cautious; surveys reveal that a majority of the population would prefer not to eat GM food, citing the lack of studies on its safety and long-term effects. This hesitancy began in the 1990s, when GM food became a prominent issue in England and when the country placed tougher restrictions on GM grains and products than did the United States. Regulation of genetically modified organisms (GMOs) in Great Britain is entrusted to the Department of Environment, Food, and Rural Affairs (DEFRA). DEFRA is pro-consumer and not anti–GM; it believes that each type of GM seed needs to be evaluated independently. It seeks to minimize cross-contamination between GM and non–GM seed so that individuals can decide whether or not to consume GM food.

Prince Charles has long been one of the country's most high-profile opponents of GM food. Widespread adoption of GM agriculture, he said in 2008, would be "the biggest disaster, environmentally, of all time."[9] He believes it will destroy the livelihoods of small farmers, driving them from their land into urban areas, leaving agriculture solely in the hands of large corporations that practice unsustainable monoculture. His beliefs are echoed by many citizens of Great Britain, which has resisted GM agriculture for years and whose meat industry suffered the mad cow disease catastrophe in the 1990s.

The Pusztai Affair

The Pusztai affair is the most well-known GM scandal in Great Britain. It began when Árpád Pusztai, a high-ranking, well-respected scientist at the Rowett Research Institute in Aberdeen, Scotland, discussed his research on GM potatoes in a 1998 television interview. His test evaluated the safety of potatoes modified with a lectin (protein) obtained from the snowdrop plant. The snowdrop lectin was engineered to exhibit characteristics of a pesticide; it was toxic to insects but not mammals. The idea was that potatoes genetically altered with lectin would not need to be treated with pesticide in the field. Pusztai, who was a proponent of GM food at the time, also thought the lectin would be a good candidate for inserting into other agricultural crops.

Pusztai tested the potatoes by comparing their effects on three groups of rats. The first group was fed potatoes containing genetically altered lectin; the control group was fed non-modified potatoes; the third group was fed potatoes containing lectin, but it was not genetically altered. What Pusztai found surprised him. Rats in the first group experienced some intestinal damage as well as a weakened immune system. The other two groups exhibited no changes at all. Specifically, the research showed that the modified gene itself was not the problem; it was the way it was inserted into the potato. The inserted protein altered the entire potato genome.[10] Although the findings were controversial even among his fellow scientists, the resulting research paper was published in the highly respected peer-reviewed journal the *Lancet*.

The real trouble started when Pusztai appeared on a British television show on August 12, 1998, and announced in response to the interviewer's question that he would not be willing to eat potatoes containing genetically altered lectin. He also said that the lack of other studies like his was a concern to him and that marketing GM foods without such studies essentially turned the public into guinea pigs.[11] Pusztai was immediately fired from the Rowett Research Institute, where he had worked for more than 30 years. His papers were confiscated, and he was legally prohibited from talking about his work with anyone. The orders had come from the highest echelons of the British government and filtered down to the parliamentary Science and Technology Committee, which was concerned about the effect Pusztai's work might have on Great Britain's agricultural industry. The head of the Rowett Research Institute denounced Pusztai's research in the media and accused the scientist of misconduct; the Royal Society—the country's national academy of science—also denounced Pusztai.

Pusztai's opponents believed that the idea of GM crops being guilty until proven innocent was faulty. "Contrary to all the hype and nonsense, what we are considering with GM crops is ultimately just another kind of plant breeding," wrote John Gatehouse, a molecular biologist and transgenic plant specialist at the University of Durham. "GM crops have the potential to bring great benefits to agriculture throughout the world, which could not be achieved by other means. Logically, the testing procedures required for GM crops should be similar to those already used for new varieties of crops produced by 'conventional' techniques."[12] While this view may have predominated among researchers, it did not echo the opinion of the general public, which remained much more leery of genetic modification.

The Pusztai affair illuminated how public opinion can influence scientific research, which is supposed to be impartial, and thereby compromise objectivity. Many in the scientific community came to Pusztai's defense, arguing that their work should be immune to public opinion and that Pusztai was fired unfairly. Twenty-two scientists from 13 countries signed a statement of support and railed against the limits on academic freedom. Ultimately, Pusztai was exonerated; however, he was not allowed to resume his research.

JAPAN

History

SOY SAUCE, SAKE, AND TOFU

Biotechnology in Japan dates back to the dawn of its cuisine, which has long relied on the staples of soy sauce, sake, and tofu. Sake is a wine made from fermented rice in a process similar to the brewing of beer. History has obscured its origins, but it is thought to have been invented in China or Japan as early as the third century C.E. In its initial form, sake had a porridgelike consistency. Rice or other grains were chewed and spat into a pot. Enzymes from the saliva initiated fermentation, and the low-alcohol result was mixed with unchewed grain and then served. By the eighth century C.E., the process had been refined significantly, and Kyoto's Imperial Palace had its own sake brewmasters in residence.

Tofu is made by coagulating proteins and oil in soy milk, with salt (calcium chloride or calcium sulfate) or acid (glucono delta-lactone). It likely originated in China, probably in the second century B.C.E., and made its way to Japan in the eighth century C.E. Its popularity spread with the rise of Buddhism, as it provided valuable protein in a vegetarian diet. Soy sauce is made from soybeans fermented with yeast. Like tofu, it originated in China more than 2,500 years ago and was introduced to Japan in the seventh century C.E. by Buddhist monks.

A more recent addition to Japanese cuisine is monosodium glutamate (MSG), a food additive used as a flavor enhancer, which was isolated in 1907 and patented by the Ajinomoto Corporation. In 1956, the Kyowa Hakko Kogyo company discovered microorganisms that could be used to make MSG.[13] MSG is produced by fermenting starch, sugar beets, sugarcane, or molasses; it is found in many foods today, especially fast-food items, snack foods, processed foods, and powdered mixes. Some believe that glutamate interferes with the body's ability to remove toxins such as mercury and should therefore be avoided.

Sake, soy sauce, and tofu are still staples of the Japanese diet, and their production has always been taken seriously, combining equal measures of art and science to create quality products that appeal to the Japanese palate. Knowledge was passed down through societies affiliated with specific religious temples, whose monks improved processes and passed their knowledge on to subsequent generations, creating the foundation for Japan's modern biotechnology industry. The Osaka Brewing Society, for example, was founded in 1923 and became the Society for Biotechnology in 2003. The government set up the Fermentation Research Institute in 1940. In the 1950s, the Kyowa Hakko Kogyo Company, a brewery, adapted its fermentation processes to produce pharmaceuticals such as streptomycin, an antibiotic used to treat tuberculosis.

For most of the 20th century, Japanese biotechnology focused on fermentation, almost completely excluding genetics. Not until the 1980s did the country realize the need to invest in genetics in order to remain competitive on the world stage. Initial endeavors involved partnering with established international companies and licensing existing technology. Homegrown research and development came later and, when it did, the companies involved were those associated with food. Suntory, Ltd., a liquor company, used a synthetic gene to produce gamma-interferon for the treatment of cancer. In 1983, the company transferred the technology to the U.S. pharmaceutical company Schering-Plough, Japan's first successful transfer of genetic intellectual property.[14] However, for the next 20 years or so, U.S.-Japanese partnerships with U.S. companies providing the licensed technology remained the norm.[15]

EUGENICS, HYGIENE, AND UNIT 731

Like many countries that participated in the first International Eugenics Congress in 1912, Japan embraced eugenics and discouraged miscegenation in order to maintain "wholesome" bloodlines throughout the island nation. The modern Japanese push toward eugenics began in 1884 with the publication of Takahashi Yoshio's *A Treatise on the Improvement of the Japanese Race* and continued in subsequent decades through alliances with many German scientists, such as Erwin von Baelz, who lived in Japan for 30 years, was a doctor to the imperial family, and promoted the idea of an unsullied Yamato (pure Japanese) race (despite the fact that he married a Japanese woman and had two children with her). Unlike later proponents, Yoshio Takahashi promoted miscegenation between whites and "yellows" because he thought it would result in a race of taller, more beautiful Japanese people. Similarly, in 1939 Susumu Ijichi advocated miscegenation between Japanese and Manchurians. Both Takahashi and Ijichi were refuted by those such as Hiroyuki Katô, who preached that Japanese blood must remain pure.

Yûseigaku was the term used to denote the science of superior birth, and those of pure Japanese ancestry were considered members of the Yamato race. It was imperative among Japanese eugenicists that Yamato blood not be compromised. "Blood (*chi, ketsu*) began to be invoked by the mid-19th century in Japan as a metaphor for both hereditary material and racial essence," wrote Jennifer Robertson.[16] Toward this end, eugenics focused on the hygiene and health of girls and women. The goal was to increase the number of "ideal," healthy citizens and prevent reproduction of mentally and physically "inferior" citizens who would be a burden on society.

Opposition to these ideas came from those who believed in the Shinto religious tradition of divine origin and argued that if the Japanese people were descendants of a divine being, they should not be treated as animals requiring eugenic intervention.[17] As in Germany and the United States around the same time, eugenics was inspired by nationalism and involved compulsory sterilization of people deemed "undesirable," including those suffering from leprosy, the mentally ill, criminals, and alcoholics. Leprosy, in particular, was subject to the legal spotlight in Japan because it was believed to be inherited. Leprosy prevention laws were passed in 1907, 1931, and 1953; they outlined a policy of isolation and sterilization of those suffering from the disease. The laws stayed on the books until 1996.[18]

Shigenori Ikeda, a journalist who had spent time in Germany, advocated positive eugenics that took the form of reproductive education. His magazine *Eugenics* popularized the idea of a "eugenic marriage," in which partners of equal social standing and good breeding were encouraged to marry and submit their information to the government to create a genealogy database that would further the cause of eugenicists. Ikeda sponsored "Blood Purity Day" on December 21, 1928, in which people could receive a free blood test at the Tokyo Hygiene Laboratory. Negative eugenics was espoused by Hisomi Nagai, the director of the Japanese Society of Health and Human Ecology, established in 1930.

In 1940, the National Eugenic Law was passed under the administration of Prime Minister Fumimaro Konoe. It limited sterilization to the mentally defective, promoted genetic screening, and limited the availability of birth control. Then, as is still somewhat true in the 21st century, marriage between a Japanese citizen and a member of another culture, race, or nationality was frowned upon. Prior to World War II, this primarily meant Koreans, who had immigrated in significant numbers to Japan as laborers. The idea of maintaining Japanese racial purity held sway over the country. The Program on Population Policy was enacted to increase numbers of "healthy" people. After World War II, the National Eugenic Law was replaced by the 1948 Eugenic Protection Law, which allowed sterilization and abortion with the consent of a woman and her spouse. This law is typically regarded as the advent of legal abortion in Japan, a policy that holds to the present day. Ironically, the birth control pill became widely available only in 1999.

During World War II, eugenics in Japan took a sinister detour into biological and chemical warfare, just as it did in the United States and Germany. The Imperial Japanese Army's Unit 731 performed experiments on roughly 10,000 civilians and soldiers from China, Korea, Mongolia, and Russia. Officially deemed the "Epidemic Prevention and Water Purification Department of the Kwantung Army," Unit 731 used human subjects to conduct experiments such as vivisection without anesthesia, infection of victims with disease or removal of organs while subjects were still alive, impregnating women against their will and then removing the fetus for examination, and amputating limbs and even reattaching them to other parts of the body. Other subjects were used as human targets to test grenades, flamethrowers, and chemical weapons. Medical experiments focused on infecting people with diseases such as syphilis and gonorrhea, cholera, anthrax, and bubonic plague. Most of the large-scale experiments were carried out on Chinese nationals, with an estimated 200,000 civilians killed in the process.[19]

NASCENT BIOMEDICINE

Interferons (IFNs) are proteins that are produced by the immune system to combat viruses and parasites. In the 1950s, pioneering work on interferon began in Japan. Although Alick Isaacs and Jean Lindenmann, British and Swiss researchers respectively, were credited for discovering and naming interferon in 1957 after the publication of their work on the influenza virus, it is now known that two researchers at the University of Tokyo, Yasuichi Nagano and Yasuhiko Kojima, discovered the role of interferon through their rabbit-skin tests three years earlier while working on a vaccine for smallpox. Interferon therapy is crucial in treating hepatitis C, multiple sclerosis, and some types of cancer.

Apart from this, biotechnology progress from the 1950s through the end of the century was slow in Japan. Even after the Human Genome Project put human DNA sequences into the public domain, Japanese researchers were slow to move into proteomics, or the study of proteins, which many believe is the next frontier in biomedicine. Developing these industries in Japan is the responsibility of the Science and Technology Agency.

SHIFTING ATTITUDES

Japan has a reputation for paternalistic attitudes when it comes to informed consent. For years, doctors used patient data freely and without permission for what they believed was the greater good. The trend was established hundreds of years ago with the rise of the medical profession, which was centered in the privileged classes. The cloistered brotherhood of medical professionals kept medical knowledge close to their vests. Patients trusted the experts and did not question details of their treatment out of respect for doctors.

As recently as 2001, a group of colon cancer patients were attended to by doctors who forged consent documents and used the patients' tissue samples to obtain genetic information for use in research.[20] While this event was thought to be isolated, it brought more attention to the need for privacy measures in medicine, and the idea of informed consent, or "truth-telling," has gained currency in recent years.[21]

Current Situation

HIGH TECH; LOW BIRTH RATE

Japan has long had an economically competitive, highly educated population. Science is heavily valued in the culture, a fact that led to the country's post–World War II growth and transformation into a world leader in electronics and automobiles. By 2006, Japan's pharmaceutical market was the world's second largest, even though the Japanese pay the highest drug prices in the world. This trend will eventually collide with the fact that Japan's low birth rate and aging population will result in a society that requires more health care than its shrinking workforce can pay for or its government can provide—a problem that is plaguing countries as varied as Germany and China. On the other end of the medical spectrum are younger people who would like to receive genetic tests and genetic counseling to guide their family planning, but because the government provides universal health care to all its citizens, such nonessential services can be hard to come by.[22]

BIOETHICS: *SEIMEI RINRI*

According to Takao Takahashi, author of *Taking Life and Death Seriously: Bioethics from Japan,* bioethics (*seimei rinri*) can be divided into three distinct periods. The first, roughly 1965 to 1980, corresponds with Koichi Bai's writings on informed consent, which were inspired by German notions of the idea and resulted in the formation of the Japanese Association of Medical Law in 1969. In 1974, the Japan Brain Wave Society drew up guidelines for diagnosing brain death, which were based on the 1968 Declaration of Sydney and outline the requirements for organ transplants from a brain-dead person. The second period of bioethics began in earnest around 1980, roughly 10 years after the topic emerged in the United States. It evolved from the four principles outlined in the book *Principles of Biomedical Ethics,* published in 1979: respect for autonomy, nonmaleficience, beneficence, and justice. The era was characterized by the evolution of institutions and research procedures that took bioethics into consideration. The third period began in 1990 and continues to the present day. The focus of this period is on genetics research conducted under well-defined bioethics policies.[23]

Bioethics in Japan is based heavily on translations of Western publications, and since 1986 scholars at Chiba University have been engaged in a long-term translation project that brings Western documents on bioethics into Japanese collections. In 2000, the Bioethics Committee Council for Science and Technology (BCCST) published *Fundamental Principles of Research on the Human Genome,* and it has since published papers on embryonic stem cell research, gene therapy, epidemiology, and clinical research. Additionally, the Center for Biomedical Ethics and Law (CBEL), established at the University of Tokyo in October 2003, conducts research and develops policy recommendations.

When it comes to bioethics, the discussion in Japan seldom focuses on religion. Though both Shinto and Buddhism are widespread, most people do not practice their faith regularly. In terms of peoples' attitudes toward biotechnology and genetic engineering, there is a low level of concern over the rights of the fetus and/or negative eugenics. For instance, only 1 to 2 percent of Japanese surveyed said a fetus has a right to life. Nevertheless, when it comes to children born with a genetic disease, Japanese parents suffer more shame and guilt than their Western counterparts. This indicates that genetic testing, if widely available, would become a valuable resource for those of childbearing age. In vitro fertilization (IVF) is widely practiced in Japan, with few public complaints; more than 11,000 children were born from IVF in 1999, but the practice of surrogate parenting remains unlawful.

Unlike in neighboring China, 76 percent of respondents in one survey said that if they could have only one child, they would prefer a girl. This indicates that if prenatal sex selection became common, abortions of male fetuses might rise. For this reason, in 1986 the Japan Society of Obstetrics and Gynecology and the Japan Medical Association recommended that sex selection genetic testing should be available only to those from families with sex-linked genetic disorders (for example, muscular dystrophy and hemophilia).[24]

BIOAGRICULTURE

Japan was slow to accept GM food, but surging grain prices in recent years have forced people to reconsider their stance.[25] As of 2008, Japanese farmers did not plant GM seed, but manufacturers had turned to buying corn starch and corn syrup made from GM corn as ingredients for processed food.[26] Only 15 percent of all land in Japan is arable, and combined with the country's high population density, conventional wisdom would suggest that the higher yields offered by GM crops would be popular in the nation. Rice is a staple crop and is one of the grains that is often subject to genetic modification.

By the 21st century, Japan had a trade imbalance when it came to food. The country grew only 60 percent of what its inhabitants consumed and imported the remainder. Because much of the food grown elsewhere in the world and available for export (particularly U.S. soybean and corn crops) is genetically modified, Japan came under increasing pressure to accept GM food. Consumers were wary, however; in a 2006 survey, 61 percent of those questioned were reluctant to eat it.[27] Though still a majority, this figure represents a decline over a 2003 survey, in which 80 percent of respondents were unwilling to eat GM food. Consumers cited fears of mad cow disease (bovine spongiform encephalopathy, or BSE) and avian flu as their main concerns, despite tenuous, if any, connection between those diseases and GM food. Laws require some GM foods to be labeled but exempt others such as soy sauce, oil, and animal feed.

Livestock is not a big industry in Japan. Historically, the Japanese have eaten little meat, subsisting mostly on seafood, vegetables, and grains. But by the 1980s, increased global demand for Japanese Kobe beef, derived from cattle raised in luxurious conditions, resulted in more farmers raising livestock. Most Kobe beef is exported; meat for domestic consumption is largely imported, although pork—the country's most popular meat—and dairy farms provide substantial supplies for the Japanese.

Counterstrategies

Even though there are few limits on embryonic stem cell research in Japan, Yamanaka Shinya, at the Institute for Integrated Cell-Material Sciences at Kyoto University, pioneered a process that may eliminate the need for embryonic stem cells in future research. The process would have wide ramifications in many nations. Yamanaka created a way to turn adult skin cells into the equivalent of embryonic stem cells—cells that can be used for all the same functions as stem cells but do not require the destruction of an embryo.[28] To do this, he added "master regulator" genes to chromosomes on the skin cells. The master regulator genes can turn other genes on and off, giving the cell the same flexibility as a stem cell. This discovery, although duplicated almost simultaneously by researchers at the University of Wisconsin, was applauded in Japan as an important step toward the country becoming relevant in the field of genetic engineering. Researchers caution that this development is still not feasible for gene therapy on human beings; it is for basic research only because of an increased risk of the gene turning cells cancerous.

GOVERNMENT'S BIOSTRATEGY 2002

In December 2002, Japan's Biotechnology Strategy Council drafted the Biotechnology Strategy Guidelines, which aimed to stimulate the economy while addressing the issue of Japan's aging population and depletion of natural resources.[29] Noting that the life sciences will be to the 21st century what electronics were to the 20th, council members acknowledged that Japan was behind the Western curve in developing a robust biotech industry. By way of example, it was noted that the budget for the U.S. National Institutes of Health (NIH) was seven times the amount spent by Japan's government on research in the life sciences.[30] Thus, the council focused its goals on the country's strengths: genomics research, pharmaceuticals, microbial and bioprocess engineering, and functional foods. (Functional foods are those that have health or disease-prevention benefits beyond their basic nutritional content. Fermented foods are often considered functional because of their ability to reestablish gut flora and balance the digestive system.) The council believes that genome-based pharmaceuticals will make people healthier and that microbial engineering may revolutionize the food supply. The three main strategies of the Biotechnology Strategy Guidelines are 1) revamp research and development, 2) enhance the process of industrialization, and 3) impart to the public an understanding of biotech's importance.

Japanese research strengths include advances in rice and silkworm technology. One experiment used the silkworm (*Bombyx mori*) to produce a large volume of recombinant protein (that is, protein derived combining DNA sequences that do not normally occur together) in the form of a cocoon. The process would provide a partial solution to the problem of generating sufficient quantities of medically useful proteins.[31]

INDIA

India, with more than 1 billion people, is the world's most populous democracy, and it will soon surpass China as the world's most populous country. It is still considered a developing country; vast slums and extreme poverty are pervasive in many areas. Nevertheless, India also boasts a large, growing middle class and a strong educational system that has made it competitive with the world's most developed regions. The southern part of the country around the city of Bangalore has been developed into a biotech friendly region, with universities and research laboratories that attract both Indian and international scientists.

History

OF MYTHS AND SMALLPOX

India traces its origins back to the ancient civilizations of the Indus River Valley, which formed around 3,000 B.C.E. in an area that today straddles Pakistan and northern India. The fertile land was perfect for growing wheat, legumes, cotton, sugarcane, and rice. The same crops are still cultivated in that area today, but the surging population has brought the challenge of feeding more people on less arable land, with dwindling water resources and poor soil conditions.

The major text of Hinduism, the *Mahabharata*, takes the form of an epic poem. It is one of the longest ever written, compiled between the eighth century B.C.E. and the fourth century C.E. Its stories lay the foundation for Hinduism and provide a way to interpret the modern world. For example, in one story an unborn baby listens to his father, a famous warrior, explain valuable military strategies to his mother. When the child is born, he already has this knowledge. This suggests the cultural idea that a fetus is a thinking, feeling entity instead of a collection of cells. In another story, the Kaurava Brothers, all 100 of them, are born after their mother has been pregnant for two years. They are born as a single fleshy mass that a holy man divides into 100 pots, each of which is then nurtured individually. Some point to this story as evidence of Hinduism's tolerance of IVF and even cloning.[32]

In Asia, smallpox has been a scourge since ancient times. The Chinese physician Ko Hung was the first to discuss the disease in his book *Chou hou pei chi fang* (Handy therapies for emergencies), written about 304 C.E. A much later Chinese book, *Tou chen ting hun* (Definitive discussion of smallpox), written in 1703, describes an inoculation procedure that had apparently been practiced in India for centuries and made its way to China in the 11th century. The practice, called *tikah*, was widespread in India before it was banned by the British government in 1803.[33] *Tikah* involved inserting pus from a mature pox under the skin of a person's upper arm or forehead with a needle. The puncture mark was then covered with a paste of boiled rice.

India's modern biotech industry can be traced back to 1980, with the drafting of the country's sixth five-year plan, which specifically addressed genetics. The plan called for the creation of the National Biotechnology Board, which issued its "Long Term Plan in Biotechnology for India" in 1983. In the private sector, Kiran Mazumdar-Shaw, daughter of a brewmaster, formed India's first biotech company, Biocon India, after returning from Australia where she became a brewmaster herself. Her company's early projects included isolation and extraction of a papaya enzyme that tenderizes meat and a fish collagen that improves beer.[34]

HINDU BELIEFS COMPATIBLE WITH BIOTECHNOLOGY

The cow is sacred in India and to Hindus in particular. The animal is respected as a beast of burden in this highly agricultural culture and for its ability to provide nourishing milk for humans. Reincarnation is an important tenet of Hinduism, and cows are believed to possess souls. Hindu deities have been known to take the form of cows, and killing a cow would be a terrible deed if the animal was actually a deity. These beliefs have been central to India becoming the world leader in milk production, although most dairy farms are small, family-run operations.

Ethical debates surrounding embryonic stem cell research are few in India. In fact, the government has seized on Western ambivalence on the issue to lure researchers to new, state-of-the-art research laboratories. Yet this embrace of cutting-edge biotechnology stands in stark contrast to the vast numbers of Indian citizens who lack basic health care, clean water, and adequate food. Some fear that these conditions could lead to the country's indigent population becoming willing guinea pigs for companies seeking to test new techniques and drugs. A reporter for *India Today* noted that these circumstances created a veritable "gold mine" of "the world's largest population of naïve sick patients, on whom no medicine has ever been tried. India's distinct communities and large families are ideal subjects for genetic and clinical research."[35]

Poverty has already led to a boom in illegal organ harvesting for transplants. With plenty of poverty-stricken Indian citizens willing to undergo surgery to remove a kidney in exchange for money to feed their families, and plenty of Western patients willing to pay upward of $100,000 for a kidney that will prevent a long (and possibly fatal) wait for a legitimate donor, India has become a leader in for-profit organ donation. Although the practice was explicitly deemed illegal by the Transplantation of Human Organs Act of 1994, a buoyant black market has evolved. In 2004, a *National Geographic* reporter traveled to "kidney village," an unidentified town in India where many residents have sold one of their kidneys for the going rate of $800.[36] Those needing a kidney engaged in "transplant tourism," traveling to India to undergo the transplant surgery. Those who had undergone surgery to remove a kidney appeared to be healthy, although the money they received was not enough to permanently alter their economic status. Additionally, the long-term health of these subjects is at risk, especially without proper follow-up medical attention. "When you encounter folks who are so poverty-stricken, it's a gruesome option for them, but it is an option. It certainly raises a lot of ethical questions," said reporter Lisa Ling.[37] The World Health Organization (WHO) discourages illicit transplant tourism and monitors how organs are trafficked.

INDIA'S GREEN REVOLUTION

By the middle of the 20th century, it was apparent that India would soon face significant problems securing sufficient food and water for its surging population. Many forecast an era of mass starvation as resources were taxed beyond their limits. Fortunately, a crisis was avoided, largely because of what came to be called the Green Revolution. The term refers to the leap of technology that resulted in seeds that yielded more tons of grain than was previously possible. Additionally, more food was made available on the world market to make up the difference. India's ability to feed itself was met.

Fast-forward to the 21st century. India's population continues to climb, and its agriculture is impeded by a decrepit irrigation system. More than 70 percent of the population is directly involved in subsistence agriculture. Too many people have too little access to water, and the specter of starvation once again looms. The Indian Agricultural Research Institute, founded in 1905 and central to the success of the Green Revolution, is developing ways to implement sustainable agricultural practices. GM crops modified to include vitamins, minerals, and protein, research scientists say, could provide millions of poor people with valuable nutrients—an easy, inexpensive way to address a hefty problem. Other products could include biofuels grown from native sugarcane (all the better to fuel the record number of automobiles on the road), and a vaccine for leprosy. As Prime Minister Manmohan Singh said in his speech to the Global Agro Industries Forum in 2008:

> *We need a Second Green Revolution. We need new technologies, new organizational structures, new institutional responses and, above all, a new compact between farmers, technologists, scientists, administrators, businessmen, bankers and consumers. The global community and global agencies must fashion a collective response that leads to a quantum leap in agricultural productivity and output so that the spectre of food shortages is banished from the horizon once again.*[38]

DEPARTMENT OF BIOTECHNOLOGY

The National Biotechnology Board was replaced in 1986 by the Department of Biotechnology under the Ministry of Science and Technology, which oversees many national programs related to genetics, research, medicine, energy, technology transfer, safety, and ethics. The department funds much of the research in India's universities and laboratories, most of which is concentrated in medicine rather than agriculture. Department officials interact with scientists, laboratories, and universities. It has established a peer-reviewing mechanism to insure quality research, assists individuals with the patent process, and facilitates technology transfer from academia to industry. Its projects have included silkworm genome analysis, research into human genetic disorders, and development of recombinant vaccines.

In 2007, the Department of Biotechnology and the Indian Council of Medical Research drafted guidelines for stem cell research and created organizations of oversight for research using human embryonic stem cells. The rules allow stem cells from human embryos in research, along with tissue and other material derived from pregnancy terminations.[39] Ethical considerations such as informed consent and research protocol are addressed in detail.

Still, biotechnology has been slow to take hold in the country. One reason for this is the divide between academics and business: Where partnerships between the two are common in other countries, in India those in the academic realm often do not pursue the patent registration their discoveries require for transferring technology to the business sector.[40] The divide also means that research may take place independently of what the marketplace needs. Steps to alleviate this discrepancy were addressed by the government in the mid-1990s, particularly through patent law reform and a push toward joint research ventures between universities and pharmaceutical companies that sponsor open-ended research.

India's potential, as seen with its explosive information technology industry, is huge. Ernst & Young's Global Biotechnology Report 2004 forecast that India's biotechnology companies would grow tenfold by 2010, creating a million new jobs.[41] One example of such a company is Reliance Life Sciences, India's largest private biotech company, which is working toward new treatments for diabetes, Parkinson's disease, and Alzheimer's disease.

Current Situation

BT COTTON

India is the world's third-largest producer of cotton, behind China and the United States. Although only 5 percent of cropland is devoted to cotton, 50 percent of the pesticides used in the country go to combating the cotton bollworm.[42] Bt cotton, introduced in a joint venture between U.S.-based Monsanto and Indian agribusiness Mahyco, came to India in 1995. The idea was to cross Monsanto's Bt seed with local varieties to create a hybrid that would be well suited to India's climate and eliminate the bollworm. Commercial production of this Bt cottonseed began legally in 2002, but prior to that the seed had been planted in Gujarat due to a lack of regulatory oversight. The unsanctioned Bt cottonseed may have cross-pollinated with non–GM cotton and compromised the effectiveness of Monsanto and Mahyco's final product.[43] Some worry that imprecise cultivation will hasten the bollworm's resistance to the GM seed because it takes only one mutation in the bollworm for it to become resistant to the Bt toxin. That would put farmers back where they started.

While controversy over GM crops in India is not as fervent as in Europe, scientists are proceeding with caution. In 2003, the Indian government rejected U.S.-grown soy and corn shipments because U.S. officials could not say for sure whether they had been genetically modified. The incident highlighted a major difference between the United States and the rest of the world: Lax labeling laws in the United States stand in opposition to many countries' desire to know the origin of imported food. Despite this, India has remained pragmatic. By 2008, it had many other GM crops in field trials, including rice, eggplant, okra, potatoes, tomatoes, and groundnuts.

GENETIC SCREENING FOR THALASSEMIA

Thalassemia is the world's most common genetic blood disorder; it is prevalent in Mediterranean people, Arabs, and Asians. It is becoming increasingly common in India (estimates range from 3.5 percent to 15 percent of the population, or 100 million people[44]), and officials are enacting programs to offer genetic screening and counseling to make people aware of it. The disease exists in several forms, and the most severe can cause death by heart failure in infancy if not treated with frequent blood transfusions. In previous generations, the disease was incurable, but new treatments using stem cell replacement, including a bone marrow transplant, cord blood transfusion using the umbilical cord blood from a newborn sibling, and gene therapy that involves inserting a normal beta-globin gene into the patient's stem cells look promising. The disease prohibits the stem cells in a person's bone marrow from producing enough red blood cells and also causes problems with the spleen and gallstones. Some 60 to 80 million people worldwide are estimated to carry the beta thalassemia trait, meaning they possess the recessive gene.

India has established programs to test people to determine if they are thalassemia carriers. A simple blood test is all that is required, and the government encourages all adults to have the test before marriage or at the very latest when they decide to have a child. If both partners are found to be thalassemia carriers and still decide to have a child, a prenatal test can be performed to see if the child will have thalassemia, and abortion for thalassemia-positive fetuses is allowed. Children born with thalassemia are given free blood transfusions.

BIOTECH BACKLASH: VANDANA SHIVA

Vandana Shiva is a Western-educated Indian physicist and activist who opposes agricultural genetic engineering. She maintains a high international profile through public appearances, publication of books, and demonstrations against GM food, her overall belief in the sanctity of localized farming, equal rights for women, and independence from corporate meddling in rural life. She believes that the biggest danger from GM food is loss of biodiversity. If a few corporations succeed in disseminating a few varieties of seed over large swaths of the Earth, traditional crops may disappear, along with the livelihoods of independent farmers. GM seed, Shiva maintains, may not be as suited to the ecosystems as the varieties they replace.

Near-subsistence-level farmers who are forced to use the more expensive seed risk falling into debt in order to buy it and would also need to purchase the herbicides and irrigation systems it requires. Once they sow the seed, they may find that it depletes the nutrients in the soil.

To protect farmers from these dangers, Shiva founded the organization Navdanya (Nine Seeds), a New Delhi–based nonprofit that assists farmers in cultivating hardy, non–GM seed using time-proven, inexpensive organic practices. These farmers experience yields similar to their GM–growing neighbors but with fewer associated costs and resulting in more sustainable practices for the ecosystem.[45] Navdanya also banks seed as an insurance policy against man-made or natural catastrophe.

In her book *Stolen Harvest: The Hijacking of the Global Food Supply,* Shiva focuses on how poor Indian farmers are adversely affected by corporate dominance of agriculture. In essence, she contends that dependence on GM seed turns clients into tenant farmers dispossessed of their land, which then may become susceptible to the type of ecological disaster that caused the dust bowl during the Great Depression in the United States. Farmers who become indebted to banks in order to buy seed, fertilizer, and pesticides risk becoming unable to provide for their families in an already underdeveloped economy. One result is a surge in suicide among poor Indian farmers—with some 200,000 more suicides in 2008 than in 1998, according to some estimates. "Corporate seeds aren't about increasing productivity—they are about increasing debt. When moneylenders come to repossess the land the farmer cannot bear it, and consumes pesticides to end his life," Shiva told journalist Rowenna Davis.[46] The system is encouraged by governments, which succumb to lobbying pressure from corporations. Soon farmers discover their traditional crops have been outlawed and new systems put in place that require GM seed and registration processes that are foreign to their way of life.

Counterstrategies

WOMBS FOR RENT

India has capitalized on its surplus of well-trained doctors by becoming a prime destination for medical tourists. As health care costs soar in the United States, many people find it cost effective to travel overseas to undergo procedures such as heart surgery, back surgery, and hip and knee replacements. Many U.S. health insurers support this trend, and as medical tourism has gained in popularity, India has added new procedures to the mix, including IVF and surrogate motherhood. Advocates call this "reproductive outsourcing" a win-win situation; expenses for the couples are lower in India than in the United States, and payment to Indian surrogates amounts to more than most of the women could hope to make in over a decade in a conventional job. Of course commercial surrogacy—paying a woman to carry a child that will be handed over at birth—is illegal in the West (although couples are allowed to pay for a surrogate's health expenses), but it was made legal in India in 2002.[47]

In one instance, a gay couple from Tel Aviv traveled to Rotunda—the Center for Human Reproduction, in Mumbai, where a doctor performed IVF between one of the men's sperm and an egg from a Mumbai housewife.[48] The embryo was then implanted in a different woman. The couple paid $30,000 in expenses, of which the surrogate received about $7,500. This is a substantial amount in India. It may be enough to send a child to a good school or even buy a house. Additionally, a woman may elect to stay at a hostel affiliated with the medical center and receive free meals and medical care. In this case, the Israeli couple returned to India to pick up their child upon its birth.

One doctor said that Indian surrogates are popular even with Western couples who can afford higher costs in their home countries because Indian women can almost be guaranteed to be drug-free, nonsmoking teetotalers. The procedure is regulated by the Indian Council of Medical Research, which prohibits the surrogate mother from retaining any rights to the baby and does not even allow her name to appear on the birth certificate. As the practice has become more popular, however, many have called for tighter regulations. The Ministry of Women and Child Development as of 2008 was closely monitoring the situation, but legislation did not appear to be forthcoming.[49]

SYSTEM OF RICE INTENSIFICATION

The system of rice intensification, or SRI, is seen by some scientists as the best hope for reversing India's agricultural misfortune. SRI does not involve the use of GM seed but rather garners higher yields by using less water, planting younger seedlings, planting fewer seedlings farther apart, using organic fertilizer, and weeding manually. SRI is low-tech but appears to be producing good results. Research of the technique was undertaken by V. K. Ravichandran, a professor at Tamil Nadu Agricultural University, and is supported by the World Bank. "SRI produces higher yields (40–80 percent) with less seed (85 percent) and water use (32 percent saving)."[50]

GERMANY

Since the formation of the EU in 1993, Germany's attitudes toward biotechnology and genetic engineering, while still somewhat unique, have fallen more in line with those of the other EU countries. Reunification in 1989, following decades of separation between European West Germany and Soviet East Germany, created a country of dichotomies. Despite the economic turmoil caused by the process, reunification was aided by a world-class educational system, a high regard for the sciences, and a common history and culture that the iron curtain could not erase.

History

ZYMOTECHNOLOGY AND REINHEITSGEBOT

Zymotechnology is the science of fermentation, and it is often considered the forerunner of modern biotechnology. By the 19th century, Germany was a world leader in zymotechnology thanks to its highly developed brewing industry; by then beer had long been one of the country's most lucrative export commodities and a significant contributor to the government's tax coffers. The small town of Weihenstephan in Bavaria is home to the world's oldest brewery still in operation—the Bavarian State Brewery Weihenstephan—which was founded at the Benedictine abbey there in 1040. Today, the brewery operates in conjunction with the life sciences campus of the Munich University of Technology.

In 1516, Germany established Reinheitsgebot (German beer purity law), rules that standardized the process and ingredients of beermaking. The Reinheitsgebot required that beer be comprised of only three ingredients: water, barley, and hops. That Germans standardized this process at the time of the Reformation speaks to their ability to understand the science of fermentation enough to regulate it. Throughout the centuries, the laws were modified periodically until they were repealed in 1987, although many breweries still adhere to them. The laws helped establish Germany's reputation for finely crafted beer and distinguished between ales (top-fermenting beer) and lagers (bottom-fermenting beer). Even today, the 1993 Vorläufiges Deutsches Biergesetz (provisional German beer law of 1993) regulates ingredients and brewing processes.

RACIAL HYGIENE

In the 1930s, German politicians were inspired to enact their own policies on racial hygiene after witnessing the rise of eugenics in the United States. Hans Schemm, the Nazi Minister of Education in Bavaria, stated in 1935 that "National Socialism is applied biology," a claim that was later often repeated by members of the Nazi Party. The party promulgated the idea of the Übermensch, translated as "superman" or "overman," adopted from Friedrich Nietzsche's 1883 book *Also sprach Zarathustra* (*Thus Spoke Zarathustra*). The Übermensch is an individual who transcends worldly morality and exhibits the will to power, one who has reached the apex of human potential.

At the same time, racial discrimination against Jewish people was codified in the Nuremberg Laws of 1935, also known as the Law for the Protection of German Blood and German Honor, which stripped Jews of their German citizenship and forbade Germans from marrying Jews. As in the United States, eugenics was divided into positive and negative components. Positive eugenics manifested itself in the concept of an Aryan race consisting of full-blooded, healthy, heterosexual Germans. People were encouraged to have large families, and their children were encouraged to excel in athletics and academics and to join the Hitler Youth in order to prepare to be valiant soldiers. Others, including Jews, Roma (gypsies), homosexuals, alcoholics, and the physically and mentally impaired were considered "life unworthy of life," and many of them were forcibly sterilized in accordance with the 1933 Law for the Prevention of Hereditarily Diseased Offspring. The law created 200 eugenics courts and required physicians to report all patients who met the criteria for sterilization. Over the next 12 years, more than 400,000 individuals were sterilized; many others were euthanized.[51] The law was superseded in 1949 with the enactment of the Grundgesetz, or basic law, which refuted the idea of the master race and put into place a code of human rights.

Current Situation

AGRICULTURE

Agriculture accounts for only a small portion of Germany's GDP. Most farms are small and family owned. Beer is still brewed in Bavaria; vineyards dot the landscape in the southern and western portions of the country; the hills and mountains are home to dairy farms and cattle ranches; on the northern plains cereal crops and sugar beets are grown. Because Germany is part of the EU, its agricultural policy is formed in Brussels. Nevertheless, in 1994 the German parliament passed the Genetic Engineering Act, which protects farmers from contamination from GM crops they have not planted; it is among the strictest agricultural laws in Europe. If a farmer's crops are found to be contaminated with GM seed not planted by the farmer, then the farmer who did plant the seed, if discovered, is financially responsible for the damages. Additionally, the European Commission has determined that any crop found to contain more than 0.9 percent GM material must be labeled as genetically modified. Environmentalists, especially members of the Green Party, are active in Germany, and their involvement in the political process has led to a high level of awareness among citizens of the issues involved in GM food. Biotech companies campaigned against the Genetic Engineering Act, stating that it would make Germany less competitive in the world market.

Given Germans' mostly wary attitude toward GM food, it is no surprise that organic food is popular. In 2001, the government established the "Bio-Siegel," a seal granted to certified organic products. By 2006, the market for organically certified foods topped 4.5 billion euros, and stores had difficulty keeping up with demand.[52] As of 2008, more than 35,000 German products displayed the Bio-Siegel,[53] and "bio supermarkets" that specialize in organic foods and products of all types were becoming increasingly common, as were restaurants that serre organic food.

STEM CELLS

Although religion does not exert a major influence on daily life in Germany, embryonic stem cell research is prohibited under the Embryo Protection Law. Passed in 1991, the law protects all human embryos from destruction and regulates IVF practices. No more than three embryos can be created during an IVF cycle, and all three are required to be implanted into a woman's womb at the same time. No embryo can be frozen or discarded, and this stipulation alone precludes embryonic stem cell research. In 2002, parliament permitted stem cells created before 2002 to be imported from other countries, allowing scientists a way around the law. Although such stem cells are derived from destroyed embryos, their destruction does not occur on German soil. The ethical compromise satisfied both sides.[54] Despite this, the Green Party leader Voker Beck has called embryonic stem cell research "veiled cannibalism."[55]

UNFAVORABLE BIOTECH CLIMATE

Christiane Nusslein-Volhard is the head of the Max Planck Institute for Developmental Biology in Tübingen, Germany, and she received the 1995 Nobel Prize in physiology or medicine for her research on the genetic control of embryonic development (for which she used the ever-popular model organism, *Drosophila melanogaster*). She believes that Germany's laws against GM food and stem cell research are rooted in a popular mistrust of science that dates back to the Nazi era. This public mistrust is hindering scientific advancement in the country, even though "there is not much rationality behind these decisions."[56] Nusslein-Volhard believes that more knowledge of science among laypeople is necessary to dispel fears about genetic engineering. "People think if you have deciphered the genome of humans that you can change everything. But you cannot change everything, because you do not know what the genes mean, and you have no methods for changing them, and you can't do experiments with humans like you can with animals. And therefore it's totally unrealistic to have fears about this," she told *Smithsonian* interviewer Amy Crawford.[57]

GERMAN EDUCATION, PHARMACEUTICAL COMPANIES

German universities have long prized academic freedom and scientific experimentation. Thanks to the Enlightenment philosopher and educational theorist Wilhelm von Humboldt, by the early 19th century German universities had adopted practices that came to be known as the Humbold-tian model of higher education. In accordance with Enlightenment ideals, the model encouraged academic freedom, seminars, hands-on laboratories, and self-directed discovery. It proved successful and was imported to the United States in the mid-19th century with the founding of some of the country's most venerable institutions, including Johns Hopkins University and the University of Chicago. The Humboldtian model was crucial in mak-ing Germany a world leader in science and pharmaceuticals in the 19th century. Within a generation, the country's pharmaceutical industry was the envy of the developed world, and it remains so today.

Bayer, Boehringer Ingelheim, and Merck are just three of the multibillion dollar corporations that can trace their roots back to the 19th century (or ear-lier) and continue to research and develop drugs in 21st-century Germany. Boehringer Ingelheim, for instance, began operations in 1885 by making tartaric acid (an antioxidant) from wine yeast. It now concentrates on bio-pharmaceuticals. Bayer was formed in 1863 and manufactured aspirin—the first commercial company to do so—which is the world's most successful pharmaceutical. The company also discovered heroin, methadone, mustard gas, Ciprofloxacin (the antibiotic effective against anthrax), and Levitra. It is also responsible for developing polyurethane and polycarbonate, both plastics with many manufacturing uses. In 2002, the company spun off Bayer CropScience, which focuses on agricultural science, GM seed, and pesticides. Bayer CropScience developed both Liberty Link Corn and Liberty Link Rice, two of the first widely used GM seed varieties.

Merck was founded even earlier than Bayer and Boehringer Ingelheim, when Friedrich Jacob Merck bought the Angel Pharmacy in the town of Darmstadt in 1668. The pharmacy was passed down through generations of the family, and in 1827 Heinrich Emanuel Merck expanded the apothecary to include chemical and research operations. From these humble beginnings, the pharmaceutical giant was born, eventually pioneering the manufacture of morphine, cocaine, and MDMA (ecstasy). In the 20th century, like so many of its competitors, Merck began developing chemicals and pesticides that initially helped with its country's war effort and then were transferred to the agricultural sector.

Counterstrategies

THE GERMAN NATIONAL ETHICS COUNCIL

The German National Ethics Council was formed in 2001 to initiate a dialogue between scientists and the state regarding ethical issues in the life sciences. The council has 25 members, who are appointed by the federal chancellor to four-year terms. Members hold regular meetings at which they promote an interdisciplinary view of medicine, theology, and philosophy. One of the first cases they weighed in on was in 2003, when a teacher was denied a position because she refused to be genetically tested for Huntington's disease. The disease ran in her family, and the state felt it had a right to know if she was likely to develop it. The teacher sued, claiming that her right to privacy included the right to keep her genetic information to herself, and the court eventually ruled in her favor.[58] The case brought to light the need for a national policy on genetic testing.

The German National Ethics Council seeks to protect German citizens from discrimination in the workplace due to the results of genetic tests, although it acknowledges a difference between workers in the public sector and those in the private sector. The difference stems from the fact that in Germany civil service workers are hired for life. If a civil employee is likely to develop a disease, the state wants to know about it because it will be responsible for possible related medical expenses. In a 2005 statement, the Ethics Council conveyed that no individual should be forced to obtain information about his or her predisposition for a certain disease against his or her will. But it also acknowledged that employers have a right to know if a prospective employee will become a liability in the near future.[59]

Since its formation, the Ethics Council has issued a number of statements covering a variety of biotechnical issues, including biobanks (repositories of biological specimens for research, typically blood and tissue), cloning, end-of-life care, patenting of biotech inventions, prenatal diagnosis, and stem cell research. With regard to prenatal testing, for example, if a fetus is found to have a genetic defect, abortion is allowed within a certain time frame. In contrast, preimplantation genetic diagnosis (PGD) is forbidden in Germany under the Embryo Protection Law. Where PDG is used, embryos created through IVF are tested for genetic abnormalities and implanted in a woman's womb only if they are disease free. In 2007, the Ethics Council and the German Research Foundation both advocated that the Embryo Protection Law and the Stem Cell Act of 2002 (which allows the use of stem cells imported from other countries and created before 2002) be revised to allow more stem cell research. Many of those who want the law updated are concerned that Germany is falling behind the research curve, especially compared to Asian nations.

NO CLONES

By the early 21st century, it became clear that restrictions on cloning would need to be international. Otherwise, the global economy would accommodate scientists in one way or another—much as it had stem cell research, with scientists lured to countries willing to give them generous incentives to undertake possibly lucrative research. Thus in 2001, the issue of cloning was taken up by the United Nations (UN).

Germany and France sought an international agreement in the UN to prevent reproductive cloning of human beings but wanted a discussion on therapeutic, or research, cloning. Therapeutic cloning is that in which scientists clone certain cells or parts of a person in the hope of replicating tissue that has been destroyed by disease; reproductive cloning seeks to create a copy of an entire person. The United States and the Vatican objected to France and Germany's desire to separate the issues of reproductive and therapeutic cloning. In 2005, the UN Declaration on Human Cloning was approved by a vote of the General Assembly, with Germany and the United States voting for it. The nonbinding resolution called for member states to ban all forms of cloning, both reproductive and therapeutic, because it is "incompatible with human dignity and the protection of human life." Countries with significant therapeutic stem cell research programs, such as Great Britain, voted against the resolution. Many Islamic states abstained.[60]

In late 2008, the International Bioethics Committee met at the United Nations Educational, Scientific and Cultural Organization (UNESCO) headquarters in Paris to revisit the issue. Progress in science and ethics was discussed at the request of member states that wanted to reassess the distinction between reproductive and therapeutic cloning.[61]

SOUTH AFRICA

In December 1967, a charismatic South African cardiologist named Christiaan Barnard performed the first human-to-human heart transplant in a nine-hour operation at Groote Schuur Hospital in Cape Town. The patient lived for 18 days before dying of pneumonia brought on by the immunosuppressive drugs the procedure required. The operation was considered a success that could only be improved upon. A month later Barnard performed his next heart transplant on a patient who survived for 19 months, and the following year he gave Dorothy Fisher a new heart that lasted for more than 12 years.

That a medical milestone as significant as the world's first human heart transplant took place in South Africa is telling. The country has long been a tale of two worlds. While it leads all African countries in education and technology, it was also shunned by the international community for decades because of apartheid, its legal code of racial segregation. Apartheid was abolished in 1994, but the highly dedicated contingent of scientists who seek to make the country a relevant force in biotechnology and genetic engineering is still hampered by a "brain drain." Many scientists emigrate to leading institutions abroad in order to conduct research that will transcend what would be possible at home.

NOBEL PRIZE WINNERS

Despite South Africa's brain drain, the country produced four Nobel Prize winners in physiology or medicine between 1951 and 2002. All received their primary and much of their secondary education in South Africa, and most were from lower-class immigrant families. Max Theiler won the Nobel in 1951 for his work on developing a vaccine for yellow fever. Allan Cormack won the award in 1979 for his development of X-ray computed tomography (CT). Sir Aaron Klug received the Nobel Prize in 1982 and worked with Rosalind Franklin on the tobacco mosaic virus and X-ray crystallography that led to Watson and Crick's discovering the DNA double-helix structure. Finally, Sydney Brenner won the Nobel Prize in 2002 for his discoveries concerning "genetic regulation of organ development and programmed cell death." These scientists' success indicates that South Africa has had world-class research facilities for some time.

Despite the political progress of the past 20 years, South Africa is still beset by many social problems. While these problems do not preclude the emergence of a biotech industry, they limit resources that scientists take for granted in other countries. South Africa has one of the world's highest rates of HIV infection, for example, a fact that colors all aspects of health care and diverts attention from issues not related to HIV/AIDS. That said, the combination of resources and crisis in South Africa uniquely situates it to deal with one of the most pressing humanitarian crises of the 21st century.

As Barnard's success in developing the human heart transplant demonstrates, in some ways South Africa is a daring outpost of technology, a crossroads of the first world and the third, where knowledge and risk can pay off handsomely. Raymond Hoffenberg, a colleague of Barnard's working at Groote Schuur at the time of that first transplant, explained:

The first point to make is that the standard of medicine in Cape Town in the 1960s was advanced and sophisticated. There were well-equipped research laboratories and an ethos in which research and initiative were encouraged. . . . What inhibited U.S. surgeons [from performing the first human heart transplant] were ethical and legal considerations rather than technical skill. Opinion in South Africa was more permissive, the removal of the heart did not arouse such strong feelings of abhorrence, there was less likelihood of criticism that this would, in fact, 'kill' the donor. . . . It has been postulated that the reason why the operation could so easily take place in South Africa was the climate of relative disregard for human life.

While this might have been true in certain contexts, it did not exist to any material degree in the medical world and certainly not at Groote Schuur Hospital, where all races received treatment of the highest standard.[62]

History
PANOPLY OF CULTURES

The area that is now South Africa has been inhabited continuously since the dawn of humanity. It is home to dozens of indigenous peoples with distinct cultures and languages. Europeans first sailed past the Cape of Good Hope—the southern point of the continent that separates the Atlantic Ocean from the Indian Ocean—in 1487 in their bid to establish trade routes with the east. In 1820, after several centuries of colonization by the Dutch and others, it became a British colony. The British abolished slavery of native Africans in 1833, but institutionalized racism remained.

In 2007, South Africa's population stood at 47.9 million, of which nearly 80 percent was black and 9.2 percent was white; the remaining percentage was split between Asians, "coloreds," and Indians. Four growing urban areas contribute to the economy: Cape Town, Port Elizabeth, Durban, and Pretoria/Johannesburg, but outside these cities poverty is endemic and economic development sparse. South Africa is ranked 20th in the world in terms of GDP, and it is the eighth-largest wine producer in the world. Agriculture is a large market sector; the country is a net exporter of several crops, including sugar, grapes, and nectarines.

Despite the abolishment of apartheid, income inequality is still very much divided along racial lines. White households have incomes four times higher than black households. Yet black South Africans are almost uniformly better off financially than black Africans in other countries.

EUGENICS AND APARTHEID

Like many other governments in the late 19th century, the South African government promoted eugenics as a way to justify racist attitudes and policies. South African intellectuals such as H. B. Fantham, a Cambridge-educated professor of zoology and comparative anatomy at the University of Witwatersrand, promoted eugenics as a way of maintaining "racial purity." The South African Association for the Advancement of Science (SAAAS) was the leading organization for promoting eugenics, and Fantham was head of its eugenics and genetics standing committee.[63] Committee members believed that black Africans were not as mentally fit as white Africans and that mixing between the races should therefore be prohibited. As Fantham stated: "When once chromosomes of Bantu origin get mingled in white families, they cannot

112

be bred out, as is so often popularly supposed, but will exhibit themselves in unfortunate ways and at unfortunate times throughout the ages."[64]

Impoverished whites were considered to be "feebleminded" as a way to rationalize their economic failure. But even these whites were considered more mentally advanced than black Africans. Eugenicists believed that impoverished white children possessed the genetics necessary to improve their living conditions as long as they received the right opportunities; the same was not true for blacks. Fantham's eugenics committee affiliated itself with the Eugenics Education Society in London and took it upon itself to address "the Native question," which boiled down to segregating blacks and depriving them of the political and educational rights given to whites.

In 1930, the interdepartmental committee on mental deficiency, also led by Fantham, advocated voluntary sterilization as a way to "eliminate the undesirable," but forced sterilization was never enacted. "Heredity is all powerful," wrote the Rhodes University professor of zoology J. E. Duerden in 1921; "The blood of the labourer produces the labourer, and outstanding individuals do not arise from ancestrally poor stock."[65] Other academics agreed, including G. Eloff, head of the genetics department at the University of the Orange Free State, who melded the Christian-nationalist religious viewpoint with the idea of positive eugenics in hopes of improving the "Boer race."[66]

The SAAAS eugenics committee was active until 1933, when all records of it suddenly cease. This date coincides with the demise of eugenics programs in other countries in response to the rise of Germany's Nazi Party.

BIOWEAPONS IN APARTHEID SOUTH AFRICA: PROJECT COAST

In 1979, the largest anthrax outbreak in history struck tribal lands in Rhodesia (now Zimbabwe), a self-declared independent state within South Africa, where a civil war raged between black Africans and the ruling white minority. The anthrax outbreak decimated cattle herds belonging to black farmers, sickened 10,000 people, and killed 182. While anthrax outbreaks occur naturally from time to time among animal populations, Rhodesia had never before been affected. Could the incident have been an act of biowarfare? Some think so.[67]

For years, white South African government officials were sympathetic to white Rhodesian government officials and hostile toward the Rhodesian guerrilla insurgents who assembled within South Africa's borders. Efforts to contain the guerrillas and other political dissenters led to the development of South Africa's covert chemical and biological weapons program, Project Coast. Project Coast was headed by Wouter Basson, the personal physician to Prime Minister P. W. Botha, and it officially ran from 1981 through 1995. Its existence was in direct violation of the UN Biological and Toxin Weapons Convention, signed by South Africa in 1972, hence its covert status. Through Project Coast, South Africa developed tear gas, mustard gas, anthrax, cholera, and sterility vaccines to use for assassination of political opponents and crowd control purposes.[68] The project was explained by Laurie Nathan and Patricia Lewis:

> *Project Coast was to develop a range of chemical and biological agents designed to control, poison, and kill people within and outside South Africa. Large quantities of riot gas were produced, as were methaqualone and MDMA. Other chemical and biological agents were produced in small quantities and were used in the covert murder and attempted murders of individuals who were seen as a threat to the apartheid government. This included members of the police and the armed forces and at least once, an organism was used with the intention of deliberately infecting a whole community.[69]*

Project Coast was disbanded at the same time apartheid was abolished. Basson was arrested for dealing the drug MDMA (Ecstasy) in 1997, tried, and eventually acquitted. But the drug investigation unearthed many secret documents about Project Coast, some of which hinted at its involvement in the 1979 anthrax outbreak. During a 1996 Truth and Reconciliation Commission hearing, scientists testified that they had released anthrax spores into the environment through envelopes, cigarettes, and chocolate.[70] Since then, many scientists worldwide have called for greater efforts to ensure compliance with the 1975 Biological and Toxin Weapons Convention, especially in the era of bioinformatics, the Internet, and public genetic databases because "trends suggest that before long the world may simultaneously face biological weapons threats from naturally occurring pathogens and genetically modified organisms. Governments need to develop new approaches to monitor and manage this still poorly understood class of threats."[71]

Current Situation

THE QUAGGA: HALF ZEBRA, HALF HORSE

The quagga was an animal whose striped head and torso looked like that of a zebra and whose solid-colored hindquarters looked like that of a horse. It once roamed the plains of southern Africa in large numbers but was driven to extinction by poachers. The last known quagga died in an Amsterdam zoo in 1883. Today, about 20 taxidermied quagga are held in natural history museums around the world. The South African taxidermist Reinhold Rau has devoted his professional career to studying the quagga and attempting to resurrect them through selective breeding.

Beginning in the 1970s, Rau obtained DNA samples from taxidermied quagga and compared them to the DNA of existing plains zebra. Analysis led Rau and others to believe that the quagga was not a unique species but rather a variant of the plains zebra—a cousin, essentially. This means that the quagga DNA still exists within the DNA of the plains zebra.

Rau is the first to admit that the actual, specific quagga genome is lost to history. But he wondered, would it be possible to recreate an animal that looks like the quagga by selectively breeding successive generations of carefully chosen plains zebras? This process is called breeding back and is the idea behind Rau's Quagga Project, launched in 1987.[72] Rau and his colleagues are quick to point out that an animal that looks like a quagga is not actually a quagga. A selectively bred zebra may have physical traits of its cousin, but it would still be, genetically speaking, a plains zebra. In essence, Rau seeks to bring back an animal long extinct without resorting to cloning or genetic manipulation of any kind, but rather by using old-fashioned breeding techniques of the kind that Mendel used on his pea plants.

For the Quagga Project, Rau set up a special breeding camp in Namibia's Etosha National Park and selected nine plains zebras from a population of 2,500 that most resembled the extinct species. It took only 13 years—instead of the several decades he expected—to breed two zebras that, like the quagga, lack black and white stripes on their hindquarters.[73] Since then, those zebra have produced offspring that resemble their extinct cousins even more closely. Although the project was funded privately, many of the foals born in the breeding camp have been released into South African national parks. The quaggalike zebras represent a low-tech way of expanding the biodiversity of a region in order to regain what was lost.

DAWN OF COMMERCIAL BIOTECH

The Council for Scientific and Industrial Research (CSIR) was formed by parliament in 1945 and is located on a campuslike setting in Pretoria. As of 2009, it employed 3,000 researchers in all areas of scientific inquiry, making it by far the largest research and development organization in South Africa; it receives 10 percent of all research funding. Many of its initiatives concern the biosciences and nanotechnology, and as of 2010 it is making progress in developing generic antiretroviral drugs for the treatment of AIDS.

In 1978, the South African Committee for Genetic Experimentation was formed to regulate recombinant DNA technology and to advise the National Department of Agriculture. The committee was superseded in 1997 by the Genetically Modified Organisms Act. In 1989, Delta Pine and Land, a U.S. seed company, was the first commercial concern to introduce GMOs to the country, with trials for its Bt cotton. Eight years later, South Africa was the first country on the continent to produce GM crops, and it now produces Bt cotton, maize (corn), and Roundup Ready soybeans, courtesy of seed provided by international companies such as Monsanto, Syngenta, and Pioneer-Hi-Bred. Pannar Seed was founded in 1958 in Greytown, in the KwaZulu-Natal midlands, to sell corn seed, and it still carries out research in the country and in its satellite locations in the United States and Argentina. Its work includes genetic mapping and marker-assisted breeding, and its products include GM maize, fruits, and vegetables.

Home-grown biotechnology firms are few and far between, and they are hampered by the lack of scientists willing to work for them. For example, Ph.D. holders in South Africa earn only about 40 percent of what their counterparts abroad make.[74] But roughly four dozen small firms exist, such as Infruitec, which has developed a herbicide-resistant strawberry. To give the industry a boost, government-sponsored organizations and academic faculties have come together to promote the country's economic goals. This has resulted in the Forestry and Agricultural Biotechnology Institute at the University of Pretoria, which works on cereal genomics, fungal genetics, and citrus, bananas, and mangos; the Department of Molecular and Cell Biology at the University of Cape Town, which has developed a transgenic maize variety that is resistant to the African maize streak virus and is researching tobacco-produced vaccines for humans; and the South African Sugar Experimental Station, which has developed an herbicide-tolerant variety of sugarcane.

The Genetically Modified Organisms Act 15 of 1997 regulates the cultivation of GM crops in South Africa and requires farmers to have a permit to grow, sell, and export GM food. It also requires all GM seeds to undergo a risk assessment process.[75] In August 2002, the World Summit on Sustainable Development was held in Johannesburg in recognition of the pivotal role South Africa could play in the future of GM farming. The event became a battleground that pitted pro–GM forces against anti–GM forces. The former included the Ubongwa Farmers Union, which sees GM crops as crucial to their "right not to starve"[76] and which protested the appearance of Indian activist Vandana Shiva, who was present to protest GMOs. As Susanne Freidberg and Leah Horowitz said:

> *The debate over GM in South Africa, as elsewhere, is not just extraordinarily polarized, but also complicated by increasingly blurry divisions between the traditional categories of political-economic actors. At the summit, it was not always apparent who stood for what, or who represented whom, or what authority they possessed. This blurring was often strategic—as when corporations promoted their cause via NGOs [non-*

government organizations]—but it also reflected much broader changes in the roles and relationships between the public, private, and voluntary actors engaged in agro-food politics and policymaking.[77]

In 2003, South Africa signed the Cartagena Protocol on Biosafety and issued regulations on GMOs, which stated that GM food must be labeled "if its composition, nutritional value, or mode of storage or cooking is significantly different from conventional food." This was in line with the precautionary approach outlined in the Rio Declaration, from which the Cartagena Protocol evolved; it also meant that South Africa's only GM crops, corn and soy, did not need to be labeled.[78]

The country's legacy of apartheid partly accounts for the grassroots support of GM foods. Because anti–GMO groups are overwhelmingly white, they tend to be viewed skeptically by the country's black majority, "who have long associated environmentalism with the racist conservation policies of the apartheid era,"[79] despite the fact that the anti–GMO groups seek to limit the influence of large multinational corporations in order to protect subsistence farmers.

Farmers themselves are aware of the need to keep GM crops separate from non–GM ones because South Africa exports corn to countries that have banned GM varieties, including Japan, Zimbabwe, Malawi, and Kenya.[80] Within South Africa, there has been little resistance to GM foods among the general population, and roughly 4,000 smallholder farmers began cultivating GM cotton in 2002, especially in the Makhatini Flats area of KwaZulu-Natal. Nationwide, in 2006, some 1.4 million hectares of GM crops were planted. Evidence exists to support the claim that Bt cotton is advantageous to small farmers because although the seed costs more than traditional varieties, it requires fewer pesticides and results in larger yields.[81] Others disagree. The initial Bt cotton "results from the Makhatini cannot serve as a model for Africa," wrote Noah Zerbe, because the system of agricultural subsidies for it were eliminated within the first three years of the program, resulting in small landowners becoming debt-ridden. Thus, poor farmers did not receive the help they needed to make the program a success. This ultimately benefited large landowners, who were now able to outperform small landowners in every way.[82]

The largest GM crop in South Africa is white maize, followed by yellow maize, then soybeans and cotton. However, while GM cotton is grown on 22,000 hectares, it represents 92 percent of the country's cotton crop and is grown primarily by small-scale farmers. Moreover, much cotton is a double-trait GM crop, meaning it is genetically engineered to resist bollworms *and* herbicides. GM maize, beginning in 2007, also incorporated two GM traits—resistance to stem borers and herbicides.

Counterstrategies
COMING TO TERMS WITH THE AIDS EPIDEMIC

South Africa's gravest concern in terms of the health and welfare of its people is AIDS. South Africa has more HIV-positive people than any other country in the world: at least 5.7 million according to a 2007 estimate.[83] That amounts to almost 22 percent of the adult population. A majority of these people are black; less than 1 percent of white South Africans are HIV positive. The only countries with higher rates of infection are those that surround South Africa: Swaziland, Botswana, Lesotho, and Zimbabwe. One thousand people a day die from AIDS in South Africa; about 1.2 million children have been orphaned, overwhelming the limited public health and social services systems. Health care is scarce for most South Africans, including those with HIV. Of the millions who are HIV positive, only a fraction receive antiretroviral drugs that can extend their lives. Current research within South Africa focuses not on antiretroviral drugs, which are expensive and can be complicated to use, but on developing a vaccine to prevent the sexual transmission of HIV. This would allow women to prevent becoming infected with HIV from their boyfriends or husbands, which is the primary means by which the disease has spread in Africa. Spearheading the research into the vaccine—still in clinical trials as of 2009—was the Centre for the AIDS Programme of Research in South Africa (CAPRISA). Another organization conducting similar research is the South African AIDS Vaccine Initiative (SAAVI) based in Cape Town, which has made significant progress in testing a vaccine for the HIV-1 C subtype, the dominant strain of the virus in Africa and Asia.

For years, as HIV/AIDS spiraled out of control, President Thabo Mbeki insisted the country's surging death rate was due to poverty, not AIDS. "The world's biggest killer and the greatest cause of ill health and suffering across the globe, including South Africa, is extreme poverty," Mbeki said at the 2000 International AIDS Conference. "As I listened and heard the whole story about our own country, it seemed to me that we could not blame everything on a single virus."[84] In 2003 he was widely quoted as remarking that "personally, I don't know anyone who has died of AIDS."[85] During this same time the country's health minister, Manto Tshabalala-Msimang, promoted a diet of beetroot, garlic, olive oil, and lemon as a cure for the disease, and the government refused offers of international aid grants to obtain antiretroviral drugs. These drugs, long the standard treatment for AIDS patients worldwide, did not become available through public health channels in South Africa until 2003, and even then access was severely limited.

AIDS denialism revealed that it was not medical biotechnology in South Africa that lagged behind the curve but rather the misguided beliefs of a few key individuals who prevented biotechnology from achieving important gains. The scientific community was so appalled by Mbeki's denialism that they responded with the Durban Declaration, signed by 5,000 physicians (including 11 Nobel Prize winners) at the 2000 International AIDS Conference in Durban, South Africa, affirming that HIV causes AIDS. The declaration is "one of the saddest documents in modern scientific history," according to Michael Specter, for the fact that it was necessary to state something that had been an established medical fact for decades.[86] Both Mbeki and Tshabalala-Msimang dismissed the document and urged the country's doctors not to sign it.

Because of this and other political reasons, Mbeki was forced to resign as president on September 24, 2008. His successor, Kgalema Motlanthe, promptly removed Tshabalala-Msimang and appointed Barbara Hogan as health minister. Although Hogan was not a physician, she had long been an AIDS activist, and upon her appointment she declared that "the era of denialism is over completely in South Africa." A month later, Harvard researchers published a study in which they estimated the death toll of Mbeki's misguided policy of withholding antiretroviral drug therapy from AIDS patients for the years 2000 through 2005 at 330,000.[87]

PREVENTING MORE PANDEMICS

AIDS in South Africa is divided along racial lines; it primarily affects heterosexual blacks. That has caused some black Africans to view expensive antiretroviral drug regimens, the provenance of a white medical elite, as just another form of apartheid. The Durban Declaration was followed by a similar declaration from the South African Congress, which was crucial; without the acknowledgment of HIV, the government could put no effective treatment system in place. Only in 2002 did the government succumb to international pressure and begin providing antiretroviral drugs to AIDS patients. By 2006, 140,000 AIDS patients received such treatment, still far short of the estimated 500,000 who needed it.[88]

The AIDS pandemic in southern Africa has resulted in a corresponding tuberculosis pandemic; TB is the leading cause of death for those suffering from AIDS. TB is becoming increasingly difficult to treat because many strains are becoming resistant to known antibiotics. Malaria presents a similar situation. Malaria disproportionately kills women and children, at higher rates in southern Africa than anywhere else in the world. Drugs used to treat malaria are losing their effectiveness; the disease is quick to adapt to new drugs, requiring a coordinated treatment. Efforts to combat these public health crises are international. WHO and Western NGOs have initiated numerous programs designed to ease disease, and thus poverty, in southern Africa.

NATIONAL BIOTECHNOLOGY STRATEGY FOR SOUTH AFRICA, 2001

Recognizing that technology is the key to improving the country's economy, South African officials developed the National Biotechnology Strategy in 2001. The strategy earmarked US$75 million for biotech initiatives and created the National Innovation Centre for Plant Biotechnology and three biotechnology regional innovation centres, which work to improve the health of crops, livestock, and humans. The South African government hosted the Human Genome and Africa Conference in March 2003, and Cape Town is the home for the UN-initiated International Center for Genetic Engineering and Biotechnology.

The South African National Bioinformatics Institute (SANBI) was founded in 1996 at the University of the Western Cape with help from the National Research Foundation, U.S. pharmaceutical company Glaxo, and the U.S. Department of Energy (DOE). The institute conducts research and education, led the country's genome initiative in 2000, and helped develop the National Biotechnology Strategy of 2001. SANBI's first scientific breakthrough was discovering a genetic cause for retinitis pigmentosa, a type of progressive tunnel vision, in 1999.

CONCLUSION

Though countries may differ in their approaches to biotechnology, what emerges from the preceding analysis is the fact that the issues are increasingly debated on an international stage. Scientists will try to circumvent restrictions they feel hamper important research, and citizens will continue to debate issues that affect their beliefs. Meanwhile, many countries understand the need for cooperation between research institutions and businesses in order to facilitate technology transfer, thus ensuring that citizens benefit from the latest biotechnical breakthroughs.

A country's economic standing, history of agriculture and science education, and the moral views of its citizens combine to create a unique atmosphere in which the biotechnology debate takes place. All of the countries examined here have conducted their debates in light of their national history; all have contributed to the international dialogue. The conversation is driven as much by hope and knowledge as it is by fear. As scientists come closer to being able to manipulate the elements that make us human, others work to make sure our humanity is not obliterated.

---

[1] Thomas Grose. "The New Icelandic Saga." *Time International* (9/29/97). Available online. URL: http://www.time.com/time/magazine/1997/int/970929/business.the_new_icela.html. Accessed September 9, 2009.

2 Renate Gertz. "An Analysis of the Icelandic Supreme Court Judgment on the Health Sector Database Act." *SCRIPTed: A Journal of Law, Technology, and Society* (June 2004). Available online. URL: http://www.law.ed.ac.uk/ahrc/script-ed/issue2/iceland.pdf. Accessed May 15, 2009.

3 Jim Motavalli. "Agrarian Nation: Can Iceland Become the First All-Organic Country?" *E* 5.6 (December 1994): 25.

4 Iceland's Ministry of Foreign Affairs Web site. Available online. URL: http://www.iceland. is/economy-and-industry/agriculture//nr/29. Accessed May 15, 2009.

5 CIA Factbook. "Singapore." Available online. URL: https://www.cia.gov/library/publications/ the-world-factbook/geos/sn.html. Accessed May 15, 2009.

6 Chan Sue Ling. "Singapore's Biotechnology Push: Stem Cells and Urine Power Hint at Innovations Under Way." *International Herald Tribune* (9/19/05).

7 Wayne Arnold. "Singapore Acts as Haven for Stem Cell Research." *New York Times* (8/17/06). Available online. URL: http://www.nytimes.com/2006/08/17/business/worldbusiness/ 17stem.html. Accessed September 9, 2009.

8 "Singapore Biotech Drive Loses Star UK Scientists." Reuters (9/21/07). Available online. URL: http://www.reuters.com/article/scienceNews/idUSSIN11079420070921. Accessed May 15, 2009.

9 Robert Booth. "Charles Warns GM Farming Will End in Ecological Disaster." *Guardian* (8/13/08). Available online. URL: http://www.guardian.co.uk/environment/2008/aug/13/ prince.charles.gm.farming. Accessed May 15, 2009.

10 "GM Foods and Denial of Rights and Choices: Interview with Arpad Pusztai." *Frontline* 17.22 (10/28/00). Available online. URL: http://www.hinduonnet.com/fline/fl1722/17220860. htm. Accessed May 15, 2009.

11 Karen Charman. "Spinning Science into Gold." *Sierra* 86.4 (July 2001): 40. Available online. URL: http://www.sierraclub.org/sierra/200107/charman.asp. Accessed September 9, 2009.

12 John Gatehouse. "Pusztai's Misconceptions." *Chemistry and Industry* (5/22/00).

13 Andrew Pollack. "Japan's Biotechnology Effort." *New York Times* (8/28/84). Available online. URL: http://www.nytimes.com/1984/08/28/business/japan-s-biotechnology-effort. html. Accessed September 9, 2009.

14 Pollack. "Japan's Biotechnology Effort."

15 David Stipp. "Biological Warfare: How the U.S. Triumphed and Japan Beat Itself." *Fortune* (4/1/96).

16 Jennifer Robertson. "Japan's First Cyborg? Miss Nippon, Eugenics and Wartime Technologies of Beauty, Body, and Blood." *Body and Society* 7.1 (2001): 1–34.

17 K. Hirosima. "Essay on the history of population policy in modern Japan. 2. Population policy on quality and quantity in National Eugenic Law." *Jinko Mondai Kenkyu* [Journal of population problems] 160 (October 1981): 61–77. Abstract available online. URL: http:// www.ncbi.nlm.nih.gov/pubmed/12155095?dopt=Abstract. Accessed May 15, 2009.

18 Michio Miyasaka. "An Ethical Analysis of Leprosy Control Policy in Japan." Paper read at the eighth Tsukuba International Bioethics Roundtable. Tsukuba, Japan, February 15–18, 2003. Available online. URL: http://www.clg.niigata-u.ac.jp/~miyasaka/hansen/leprosypolicy.html. Accessed May 15, 2009.

[19] Christopher Hudson. "Doctors of Depravity." *Daily Mail* (3/2/07). Available online. URL: http://www.dailymail.co.uk/news/article-439776/Doctors-Depravity.html. Accessed May 15, 2009.

[20] "Samples Kept by Researchers." *Japan Times* (3/29/01). Available online. URL: http://search.japantimes.com.jp/cgi-bin/nn20010329a3.html. Accessed September 9, 2009.

[21] Darryl R. J. Macer. "Bioethics in Japan and East Asia." *Turkish Journal of Medical Ethics* 9 (2001): 70–77.

[22] Macer. "Bioethics in Japan and East Asia."

[23] Takao Takahashi. *Taking Life and Death Seriously: Bioethics from Japan.* Amsterdam: Elsevier, 2005.

[24] The information in the preceding paragraph on a fetus's right to life and IVF in Japan and the information in this paragraph come from Macer. "Bioethics in Japan and East Asia."

[25] Andrew Pollack. "In Lean Times, Biotech Grains Are Less Taboo." *New York Times* (4/21/08). Available online. URL: http://www.nytimes.com/2008/04/21/business/21crop.html. Accessed September 9, 2009.

[26] Michael Fitzpatrick. "Japan's GM Dilemma." Just-Food.com.

[27] Fitzpatrick. "Japan's GM Dilemma."

[28] Martin Fackler. "Risk Taking Is in His Genes." *New York Times* (12/11/07). Available online. URL: http://www.nytimes.com/2007/12/11/science/11prof.html. Accessed September 9, 2009.

[29] "Japan's Biotech Market." Available online. URL: http://www.pref.chiba.lg.jp/syozoku/f_rich/bio/aboutbiotech.html. Accessed May 15, 2009.

[30] Junichi Taira. "Blueprint for Biotech." *Look Japan* 49.568 (July 2003): 26.

[31] Tomita Masahiro, Katsuhiko Shimizu, and Katsutoshi Yoshizato. "Transgenic Silkworms That Weave Recombinant Human Collagen into Coccoons." In *Transgenic Silkworms*, ed. Katsutoshi Yoshizato. Austin, Texas: Landes Bioscience, 2009. Available online. URL: http://www.nytimes.com/2005/08/21/weekinreview/21mishra.html. Accessed September 9, 2009.

[32] Pankaj Mishra. "How India Reconciles Hindu Values and Biotech." *New York Times* (8/21/05).

[33] D. P. Agrawal and Lalit Tiwari. "Did You Know?" *Indian Science.* Available online. URL: http://www.indianscience.org/dyk/t_dy_Q14.shtml. Accessed May 15, 2009.

[34] Amy Waldman. "How India's Mother of Invention Built an Industry." *New York Times* (8/16/03). Available online. URL: http://www.nytimes.com/2003/08/16/world/the-saturday-profile-how-india-s-mother-of-invention-built-an-industry.html. Accessed September 9, 2009.

[35] Pankaj Mishra. "How India Reconciles Hindu Values and Biotech." *New York Times* (8/21/05).

[36] Brian Handwerk. "Organ Shortage Fuels Illicit Trade in Human Parts." *National Geographic Ultimate Explorer* (1/16/04). Available online. URL: http://news.nationalgeographic.com/news/pf/30179519.html. Accessed May 15, 2009.

[37] Handwerk. "Organ Shortage Fuels Illicit Trade in Human Parts."

[38] Manmohan Singh. Text of speech given at the 2008 Global Agro Industries Forum in New Delhi (4/10/08). Available online. URL: http://www.icar.org.in/news/PM-ADDRESSES-GLOBAL-AGRO-INDUSTREIS-FORUM.htm. Accessed May 15, 2009.

[39] Guidelines for Stem Cell Research and Therapy. Department of Biotechnology and Indian Council of Medical Research, 2007. Available online. URL: http://www.icmr.nic.in/stem_cell/stem_cell_guidelines.pdf. Accessed May 15, 2009.

[40] Ganapati Mudhur. "New Rules Push Researchers Closer to Biotech Industry." *Science* 269.5222 (7/21/95): 297.

[41] Pankaj Mishra. "How India Reconciles Hindu Values and Biotech." *New York Times* (8/21/05).

[42] "GM in India: The Battle over Bt Cotton." SciDevNet (12/20/06). Available online. URL: http://www.scidev.net/en/features/gm-in-india-the-battle-over-bt-cotton.html. Accessed May 15, 2009.

[43] "GM in India: The Battle over Bt Cotton."

[44] S. S. Ambeka et al. "Pattern of Hemoglobinopathies in Western Maharashtra." *Indian Pediatrics* 38 (2001): 530–534. Available online. URL: http://www.indianpediatrics.net/may2001/may-530-534.htm. Accessed May 15, 2009.

[45] Meenakshi Ganguly. "Heroes for the Green Century: Vandana Shiva, Seeds of Self-Reliance." Time.com. Available online. URL: http://www.time.com/time/2002/greencentury/heroes/index_shiva.html. Accessed May 15, 2009.

[46] Rowenna Davis. "Interview with Vandana Shiva: Environmentalist Extraordinaire. (Making Waves)." *New Internationalist* 410 (April 2008): 29.

[47] "Wombs for Rent: Commercial Surrogacy Growing in India." Foxnews.com. Available online. URL: http://www.foxnews.com/story/0,2933,319106,00.html. Accessed May 15, 2009.

[48] Amelia Gentleman. "India Nurtures Business of Surrogate Motherhood." *New York Times* (3/10/08). Available online. URL: http://www.nytimes.com/2008/03/10/world/asia/10surrogate.html. Accessed September 9, 2009.

[49] Gentleman. "India Nurtures Business of Surrogate Motherhood."

[50] The World Bank. "India: A New Way of Cultivating Rice." Available online. URL: http://go.worldbank.org/0WRVDPNJM0. Accessed May 15, 2009.

[51] Ian Kershaw. *Profile in Power: Hitler.* London: Longman, 1991.

[52] "Food Crunch as German Appetite for Organic Products Grows." *Deutsche Welle* (5/1/07). Available online. URL: http://www.dw-world.de/dw/article/0,2144,2301319,00.html. Accessed May 15, 2009.

[53] "Spotlight 2008: Germany's Organic Foods." Germanfoods.org. Available online. URL: http://www.germanfoods.org/consumer/facts/organicfoods.cfm. Accessed May 15, 2009.

[54] "World Briefing Europe: Germany: Imports of Stem Cells Approved." *New York Times* (1/21/02). Available online. URL: http://www.nytimes.com/2002/01/31/world/world-briefing-europe-germany-imports-of-stem-cells-approved.html. Accessed September 9, 2009.

[55] John Tierney. "Are Scientists Playing God? It Depends on Your Religion." *New York Times* (11/20/07). Available online. URL: http://www.nytimes.com/2007/11/20/science/20tier.html. Accessed September 9, 2009.

[56] "Stem Cells, Ethics and the Nazi Past." Claudia Dreifus interview with Christiane Nusslein-Volhard. *New York Times* video. (7/4/06). Available online. URL: http://video.nytimes.com/video/2006/06/28/science/1194817096743/stem-cells-ethics-and-the-nazi-past.html. Accessed September 9, 2009.

[57] Amy Crawford. "Christiane Nusslein-Volhard: A Nobel Laureate Holds Forth on Flies, Genes and Women in Science." *Smithsonian* (June 2006). Available online. URL: http://www.smithsonianmag.com/science-nature/10022146.html?page=1. Accessed May 15, 2009.

[58] Jane Burgermeister. "Teacher Was Refused Job Because Relatives Have Huntington's Disease." *British Medical Journal* 327.7419 (10/11/03): 827.

[59] German National Ethics Council Web site. URL: http://www.ethikrat.org/_english/main_topics/advance_directive.html. Accessed May 15, 2009.

[60] "General Assembly Adopts United Nations Declaration on Human Cloning by Vote of 84=34=37." Press release (3/8/05). Available online. URL: http://www.un.org/News/Press/docs/2005/ga10333.doc.htm. Accessed May 15, 2009.

[61] "UNESCO's International Bioethics Committee Resumes Debate on Human Cloning." Available online. URL: http://portal.unesco.org/shs/en/ev.php-URL_ID=12473&URL_DO-DO_TOPIC&URL_SECTION=201.html. Accessed May 15, 2009.

[62] Raymond Hoffenberg. "Christiaan Barnard: His First Transplants and Their Impact on Concepts of Death." *British Medical Journal* 323.7327 (December 22, 2001): 1,478–1,480. Available online. URL: http://www.pubmedcentral.nih.gov/articlerender.fcgi?artid=1121917. Accessed May 15, 2009.

[63] Johann Louw. "Social Context and Psychological Testing in South Africa, 1918–1939." *Theory and Psychology* 7.2 (1997): 235–256.

[64] H. B. Fantham. "Some Thoughts on Biology and Race." *South African Journal of Science* 24 (1927): 10.

[65] J. E. Duerden. "Social Anthropology in South Africa: Problems of Race and Nationality." *South African Journal of Science* 18 (1921): 29.

[66] Louw. "Social Context and Psychological Testing in South Africa, 1918–1939," 246.

[67] Meryl Nass. "Anthrax Epizootic in Zimbabwe, 1978–1980: Due to Deliberate Spread?" Physicians for Social Responsibility (1992). Available online. URL: http://www.anthraxvaccine.org/zimbabwe.html. Accessed May 15, 2009.

[68] Chandré Gould and Peter Folb. *Project Coast: Apartheid's Chemical and Biological Warfare Programme.* Foreword by Desmond Tutu. Centre for Conflict Resolution, United Nations Publications, 2002.

[69] Gould and Folb. *Project Coast: Apartheid's Chemical and Biological Warfare Programme.*

[70] David Martin. "Human Anthrax Scares Were a Major Reality in Southern Africa." *South African News Features* (11/16/01).

[71] Helen Purkitt and Virgen Wells. "Evolving Bioweapon Threats Require New Countermeasures." *Chronicle of Higher Education* 53.7 (10/6/06).

[72] The Quagga Project Web site. Available online. URL: http://www.quaggaproject.org. Accessed May 15, 2009.

[73] Edward R. Winstead. "In South Africa, the Quagga Project Breeds Success." Genome News Network (10/20/00). Available online. URL: http://www.genomenewsnetwork.org/articles/10_00/Quagga_project.shtml. Accessed May 15, 2009.

74 Marnus Grouse. "South Africa: Revealing the Potential and Obstacles, the Private Sector Model and Reaching the Traditional Sector." In *The Gene Revolution: GM Crops and Unequal Development,* ed. Sakiko Fukuda-Parr. London: Earthscan, 2007, pp. 175–195.

75 Department of Agriculture. "Genetically Modified Organisms Act, 1997 (Act 15 of 1997)." Available online. URL: http://www.nda.agric.za/docs/geneticresources/GMO_R576.pdf. Accessed May 15, 2009.

76 Susanne Friedberg and Leah Horowitz. "Converging Networks and Clashing Stories: South Africa's Agricultural Biotechnology Debate." *Africa Today* 51.1 (fall 2004): 2.

77 Friedberg and Horowitz. "Converging Networks and Clashing Stories: South Africa's Agricultural Biotechnology Debate."

78 Department of Health, Republic of South Africa. "Regulations Governing the Labeling of Foodstuffs Obtained through Certain Techniques of Genetic Modification" (1/16/04). Available online. URL: http://www.doh.gov.za/docs/regulations/2004/reg0025.html. Accessed May 15, 2009.

79 Friedberg and Horowitz. "Converging Networks and Clashing Stories: South Africa's Agricultural Biotechnology Debate."

80 Marnus Grouse. "South Africa: Revealing the Potential and Obstacles, the Private Sector Model and Reaching the Traditional Sector."

81 Yousouf Ismael, Richard Bennett, and Stephen Morse. "Benefits from Bt Cotton Use by Smallholder Farmers in South Africa." *AgBioForum* 5.1 (2002): 1–5. Available online. URL: http://www.agbioforum.missouri.edu/v5n1/v5n1a01-morse.pdf. Accessed May 15, 2009.

82 Noah Zerbe. "Assessing the Impacts of Biotech Agriculture in South Africa: The Case of Bt Cotton in Makhatini Flats." Paper presented at the annual meeting of the International Studies Association, Le Centre Sheraton Hotel, Montreal, Quebec, Canada (3/17/04).

83 UNAIDS. *2008 Report on the Global AIDS Epidemic.* Available online. URL: http://www.unaids.org/en/KnowledgeCentre/HIVData/GlobalReport/2008/2008_Global_report.asp. Accessed May 15, 2009.

84 "Controversy Dogs AIDS Forum." BBC News (7/10/00). Available online. URL: http://news.bbc.co.uk/2/hi/africa/826742.stm. Accessed May 15, 2009.

85 Barry Bearak. "South Africa's President to Quit Under Pressure." *New York Times* (9/20/08). Available online. URL: http://www.nytimes.com/2008/09/21/world/africa/21safrica.html. Accessed September 9, 2009.

86 Michael Specter. "The Denialists." *New Yorker* (3/12/07).

87 Pride Chigwedere et al. "Estimating the Lost Benefits of Antiretroviral Drug Use in South Africa." *Acquired Immune Deficiency Syndrome* 49.4 (12/1/08): 410. Available online. URL: http://aids.harvard.edu/Lost_Benefits.pdf. Accessed May 15, 2009.

88 Terry Leonard. "Eighty Scientists Condemn South Africa's AIDS Policies." *Washington Post* (9/7/06). Available online. URL: http://www.washingtonpost.com/wp-dyn/content/article/2006/09/06/AR2006090601964.html. Accessed May 15, 2009.

PART II

Primary Sources

4

United States Documents

The following primary sources are arranged in chronological order and reflect the evolution of the life sciences in the United States over the past century. The country's earliest policies were designed to help farmers because the United States was primarily a rural, agrarian nation well into the 20th century. Much research focused on crop science and hybridization. As immigration and urbanization began to change the equation, eugenics arose as a "scientific" way to cure the vexing social problems of poverty, overcrowding, and crime. Eugenics was embraced by many of the country's brightest minds, and their decisions spawned a legacy that casts a long shadow over bioethics in the 21st century. As the promises of the Human Genome Project become a reality and scientists identify the genes responsible for specific conditions, those on the front lines of bioethics will continue to question the morality of rearranging the building blocks of life. Documents that have been excerpted are identified as such; all others are reproduced in full.

William J. Beal: "Some Reasons for Plant Migration" from *Seed Dispersal* (1898) (excerpt)

William J. Beal (1833–1924) was a professor of botany at Michigan State University and studied indigenous plants in their natural habitats, creating a substantial body of work on how plants interact with their environment. In Seed Dispersal, *Beal identifies all the ways seeds propagate their species. The book's final chapter, "Some Reasons for Plant Migration," notes that plants are purely selfish in nature—their main goal is to reproduce. Toward that end, they employ a number of tactics to ensure their seeds are distributed throughout the environment. Scientists who understand these tactics will realize how difficult it is to contain a plant species in the wild. As genetically modified crops become more common, understanding the issue of transgenic contamination is essential. Beal's book serves as a primary lesson on what can happen when seeds escape the confines of the laboratory. It also explains how plants "rebel" against monoculture by succumbing to disease and consistently having their seeds dispersed far and wide in order to escape pests and take advantage of new territory.*

Plants are not charitable beings: Man uses to his advantage a large number of plants, but there appears to be no evidence that the schemes for their dispersion were designed for anything except to benefit the plants themselves. The elegant foliage and beautiful flowers, the great diversity of attractive seeds and fruits, all point to plants as strictly selfish beings, if I may so use the term; and not to plants as works of charity, to be devoured by animals without any compensation. By fertilizing flowers, by distributing plants, and by other helpful acts, animals pay for at least a portion of the damage they do.

By an almost infinite number of devices, we have seen that seeds and fruits flee from the parental spot on the wings of the wind, float on currents of ocean, lake, and river. They are shot by bursting pods and capsules in every direction. With hooks, barbs, and glands they cling to the covering of animals. Allured by brilliant colors, birds and other animals seek and devour the fruits of many plants, the seeds of which are preserved from harm by a solid armor; these seeds are then sown broadcast over the land, ready to start new colonies. Nuts are often carried by squirrels for long distances, and there securely buried, a few in a place. By a slow process, which, however, covers a considerable space, in a few years many plants send forth roots, rootstalks, stolons, and runners, and thus increase their possessions or find new homes.

Plants migrate to improve their condition: The various devices by which plants are shifted from place to place are not merely to extend and multiply the species, and reach a fertile soil, but to enable them to flee from the great number of their own kind, and from their enemies among animals and parasitic plants. The adventurers among plants often meet with the best success, not because the seeds are larger, or stronger, or better, but because they find, for a time, more congenial surroundings. We must not overlook the fact, so well established, that one of the greatest points to be gained by plant migration is to enable different stocks of a species to be cross fertilized, and thereby improved in vigor and productiveness.

Fruit grown in a new country is often fair: Every horticulturist knows that apples grown in a new country, that is suited to them, are healthy and fair; but, sooner or later, the scab, and codling moth, and bitter rot, and bark louse arrive, each to begin its particular mode of attack. Peach trees in new places, remote from others, are often easily grown and free from dangers; but soon will arrive the yellows, borers, leaf curl, rot, and other enemies. For a few years plums may be grown, in certain new localities, without danger from curculio, or rot, or shot-hole fungus. It has long been known that the nicest way to grow a few cabbages, radishes, squashes, cucumbers, or potatoes is to plant a few here and there in good soil, at considerable distances from where any have heretofore been grown. For a time enemies are not likely to find them. I have often noticed that, while pear-blight decimated or swept large portions of a pear orchard, a few isolated trees, scattered about the neighborhood, usually remain healthy. The virgin soil of the Dakotas produced, at a trifling cost, healthy, clean wheat, but it was not long before the Russian thistle, false flax, and other pests followed, to contest their rights to the soil.

As animals starve out, in certain seasons when food is scarce, or more likely migrate to regions which can afford food, so plants desert worn-out land and seek fresh fields. As animals retreat to secluded and isolated spots to escape their enemies, so, likewise, many plants accomplish the same thing by sending out scouts in all directions to find the best places; these scouts, it is needless to say, are seeds, and when they have found a good place, they occupy it, without waiting for further instructions.

Much remains to be discovered: "In this, as in other branches of science, we have made a beginning. We have learned just enough to perceive how little we know. Our great masters in natural history have immortalized themselves by their discoveries, but they have not exhausted the field; and if seeds and fruits cannot vie with flowers in the brilliance and color with which they decorate our gardens and our fields, still they surely rival them—it would be impossible to excel them—in the almost infinite variety of the problems they present to us, the ingenuity, the interest, and the charm of the beautiful contrivances which they offer for our study and our admiration." [*Flowers, Fruits, and Leaves,* by Sir John Lubbock]

Frequent rotations seem to be the rule for many plants, when left to themselves in a state of nature. Confining to a permanent spot invites parasites and other enemies, and a depleted soil, while health and vigor are secured by frequent migrations. The more we study in detail the methods of plant dispersion, the more we shall come to agree with a statement made by Darwin concerning the devices for securing cross-fertilization of flowers, that they "transcend, in an incomparable degree, the contrivances and adaptations which the most fertile imagination of the most imaginative man could suggest with unlimited time at his disposal." [*Fertilization of Orchids* by Charles Darwin]

Let no reader think that the topics here taken up are treated exhaustively, for if he will go over any part of this work and verify any observation or experiment, he will be sure to find something new, and very likely something different from what is here stated.

Source: William J. Beal. "Some Reasons for Plant Migration." In *Seed Dispersal.* Boston: Ginn & Co., 1898. Available online. URL: http://www.gutenberg.org/etext/26158. Accessed May 28, 2009.

Upton Sinclair: *The Jungle* (1906) (excerpt)

The following excerpt from Upton Sinclair's novel The Jungle *describes the grim life of slaughterhouse workers in Chicago at the turn of the 20th century. Sinclair wrote the story as an act of socialist agitation, hoping to inspire exploited workers into organizing unions to improve their pay and working conditions. Shortly after the publication of the book, President Theodore Roosevelt sent labor commissioner Charles P. Neill and social worker James Bronson Reynolds to visit the Chicago meatpacking houses to determine if conditions were as bad as Sinclair claimed. Neill and Reynolds stated, emphatically, that they were worse. The Neill-Reynolds Report prompted Roosevelt to sign the Federal Meat Inspection Act of 1906 in hopes of putting an end to the squalor that Sinclair so vividly described. Note that the processing of a "downer" cow is comparable to the situation that led to the outbreak of mad cow disease in Great Britain in the 1990s and to several similar scares in the United States in the past decade.*

One day a man slipped and hurt his leg; and that afternoon, when the last of the cattle had been disposed of, and the men were leaving, Jurgis was ordered to remain and do some special work which this injured man had usually done. It was late, almost dark, and the government inspectors had all gone, and there were only a dozen or two of men on the floor. That day they had killed about four thousand cattle, and these cattle had come in freight trains from far states, and some of them had got hurt. There were some with broken legs, and some with gored sides; there were some that had died, from what cause no one could say; and they were all to be disposed of, here in darkness and silence. "Downers," the men called them; and the packing house had a special elevator upon which they were raised to the killing beds, where the gang proceeded to handle them, with an air of businesslike nonchalance which said plainer than any words that it was a matter of everyday routine. It took a couple of hours to get them out of the way, and in the end Jurgis saw them go into the chilling rooms with the rest of the meat, being carefully scattered here and there so that they could not be identified. When he came home that night he was in a very somber mood, having begun to see at last how those might be right who had laughed at him for his faith in America. . . .

There was no heat upon the killing beds; the men might exactly as well have worked out of doors all winter. For that matter, there was very little heat anywhere in the building, except in the cooking rooms and such places—and it was the men who worked in these who ran the most risk of all, because whenever they had to pass to another room they had to go through ice-cold corridors, and sometimes with nothing on above the waist except a sleeveless undershirt. On the killing beds you were apt to be covered with blood, and it would freeze solid; if you leaned against a pillar, you would freeze to that, and if you put your hand upon the blade of your knife, you would run a chance of leaving your skin on it. The men would tie up their feet in newspapers and old sacks, and these would be soaked in blood and frozen, and then soaked again, and so on, until by nighttime a man would be walking on great lumps the size of the feet of an elephant. Now and then, when the bosses were not looking, you would see them plunging their feet and ankles into the steaming hot carcass of the steer, or darting across the room to the hot-water jets. The cruelest thing of all was that nearly all of them—all of those who used knives—were unable to wear gloves, and their arms would be white with frost and their hands would grow numb, and then of course there would be accidents. Also the air would be full of steam, from the hot water and the hot blood, so that you could not see five feet before you; and then, with men rushing about at the speed they kept up on the killing beds, and all with butcher knives, like razors, in their hands—well, it was to be counted as a wonder that there were not more men slaughtered than cattle. . . .

There were the men in the pickle rooms, for instance, where old Antanas had gotten his death; scarce a one of these that had not some spot of horror on his person. Let a man so much as scrape his finger pushing a truck in the pickle rooms, and he might have a sore that would put him out of the world; all the joints in his fingers might be eaten by the acid, one by one. Of the butchers and floorsmen, the beef-boners and trimmers, and all those who used knives, you could scarcely find a person who had the use of his thumb; time and time again the base of it had been slashed, till it was a mere lump of flesh against which the man pressed the knife to hold it. The hands of these men would be criss-crossed with cuts, until you could no longer pretend to count them or to trace them. They would have no nails,—they had worn them off pulling hides; their knuckles were swollen so that their fingers spread out like a fan. There were men who worked in the cooking rooms, in the midst of steam and sickening odors, by artificial light; in these rooms the germs of tuberculosis might live for two years, but the supply was renewed every hour. There were the beef-luggers, who carried two-hundred-pound quarters into the refrigerator-cars; a fearful kind of work, that began at four o'clock in the morning, and that wore out the most powerful men in a few years. There were those who worked in the chilling rooms, and whose special disease was rheumatism; the time limit that a man could work in the chilling rooms was said to be five years. There were the wool-pluckers, whose hands went to pieces even sooner than the hands of the pickle men; for the pelts of the sheep had to be painted with acid to loosen the wool, and then the pluckers had to pull out this wool with their bare hands, till the acid had eaten their fingers off. There were those who made the tins for the canned meat; and their hands, too, were a maze of cuts, and each cut represented a chance for blood poisoning. Some worked at the stamping machines, and it was very seldom that one could work long there at the pace that was set, and not give out and forget himself and have a part of his hand chopped off. There were the "hoisters," as they were called, whose task it was to press the lever which lifted the dead cattle off the floor. They ran along upon a rafter, peering down through the damp and the steam; and as old Durham's architects had not built the killing room for the convenience of the hoisters, at every few feet they would have to stoop under a beam, say four feet above the one they ran on; which got them into the habit of stooping, so that in a few years they would be walking like chimpanzees. Worst of any, however, were the fertilizer men, and those who served in the cooking rooms. These people could not be shown to the visitor,—for the odor of a fertilizer man would scare any ordinary visitor at a hundred yards, and as for the other men, who worked in tank rooms full of steam, and in some of which there were open vats near the level of the floor, their peculiar trouble was that they fell into the vats; and when they were fished out, there was never enough of them left to be worth exhibiting,—sometimes they would be overlooked for days, till all but the bones of them had gone out to the world as Durham's Pure Beef Lard!

Source: Upton Sinclair. *The Jungle.* New York: Doubleday, 1906. Available online. URL: http://www.gutenberg.org/etext/140. Accessed May 28, 2009.

Harry Hamilton Laughlin: *Eugenical Sterilization in the United States* (1922) (excerpt)

Harry Hamilton Laughlin was the founder of the American Eugenics Society and the assistant director of the Eugenics Record Office at Cold Spring Harbor Laboratory. The laboratory still exists, but its focus has shifted from eugenics to genetics, and it counts among its researchers seven Nobel laureates. Laughlin's book Eugenical Sterilization in the United States *included the following model law, which he encouraged states to adapt for their own use. Indeed, 18 states passed a version of it, in addition to the 14 states that had already enacted similar laws. Carrie Buck was prosecuted under Connecticut's version of the law, which was upheld by the U.S. Supreme Court in the landmark decision* Buck v. Bell. *Laughlin himself suffered from epilepsy, a condition that met the criteria for sterilization. Though he was not sterilized, he never had any children. While Laughlin takes great pains to define who is "degenerate" or the "potential parent of a socially inadequate offspring" and thus subject to forced sterilization, none of his criteria are measurable in any quantitative way.*

Full Text for a Model State Law.

An act to prevent the procreation of persons socially inadequate from defective inheritance, by authorizing and providing for the eugenical sterilization of certain potential parents carrying degenerate hereditary qualities.

Be It Enacted by the People of the State of . . . . that:

Section 1. Short Title. This Act shall be known as the "Eugenical Sterilization Law."

Section 2. Definitions. For the purpose of this Act, the terms (a) socially inadequate person, (b) socially inadequate classes, (c) heredity, (d) potential parent, (e) to procreate, (f) potential parent of socially inadequate offspring, (g) cacogenic person, (h) custodial institution, (i) inmate, and (j) eugenical sterilization, are hereby defined as follows:

(a) A socially inadequate person is one who by his or her own effort, regardless of etiology or prognosis, fails chronically in comparison with normal persons, to maintain himself or herself as a useful member of the organized social life of the state; provided that the term socially inadequate shall not be applied to any person whose individual or social ineffectiveness is due to the normally expected exigencies of youth, old age, curable injuries, or temporary physical or mental illness, in case such ineffectiveness is adequately taken care of by the particular family in which it occurs.

(b) The socially inadequate classes, regardless of etiology or prognosis, are the following: (1) Feeble-minded; (2) Insane (including the psychopathic); (3) Criminalistic (including the delinquent and wayward); (4) Epileptic; (5)

Inebriate (including drug-habitués); (6) Diseased (including the tuberculo-sis, the syphilitic, the leprous, and others with chronic, infectious and legally segregable diseases); (7) Blind (including those with seriously impaired vision); (8) Deaf (including those with seriously impaired hearing); (9) Deformed (including the crippled); and (10) Dependent (including orphans, ne'er-do-wells, the homeless, tramps and paupers).

(c) Heredity in the human species is the transmission, through sperma-tozoön and ovum, of physical, physiological and psychological qualities, from parents to offspring; by extension it shall be interpreted in this Act to include also the transmission post-conceptionally and antenatally of physio-logical weakness, poisons or infections from parent or parents to offspring.

(d) A potential parent is a person who now, or in the future course of devel-opment, may reasonably be expected to be able to procreate offspring.

(e) To procreate means to beget or to conceive offspring, and applies equally to males and females.

(f) A potential parent of socially inadequate offspring is a person who, regardless of his or her own physical, physiological or psychological person-ality, and of the nature of the germ-plasm of such person's co-parent, is a potential parent at least one-fourth of whose possible offspring, because of the certain inheritance from said parent of one or more inferior or degener-ate physical, physiological or psychological qualities would, on the average, according to the demonstrated laws of heredity, most probably function as socially inadequate persons; or at least one-half of whose possible offspring would receive from said parent, and would carry in the germ-plasm but would not necessarily show in the personality, the genes or genes-complex for one or more inferior or degenerate physical, physiological or psychologi-cal qualities, the appearance of which quality or qualities in the personality would cause the possessor thereof to function as a socially inadequate per-son, under the normal environment of the state.

(g) The term cacogenic person, as herein used, is a purely legal expression, and shall be applied only to persons declared, under the legal procedure pro-vided by this Act, to be potential parents of socially inadequate offspring.

(h) A custodial institution is a habitation which, regardless of whether its authority or support be public or private, provides (1) food and lodging, and (2) restraint, treatment, training, care or residence for one or more socially inadequate inmates; provided that the term custodial institution shall not apply to a private household in which the socially inadequate member or members are close blood-kin or marriage relations to, or legally adopted by, an immediate member of the care-taking family.

(i) An inmate is a socially inadequate person who is a prisoner, patient, pupil, or member of, or who is otherwise held, treated, trained, cared for, or resident within a custodial institution, regardless of whether the relation of such person to such institution be voluntary or involuntary, or that of pay or charity.

(j) Eugenical Sterilization is a surgical operation upon or the medical treatment of the reproductive organs of the human male or female, in consequence of which the power to procreate offspring is surely and permanently nullified; provided, that as used in this Act the term eugenical sterilization shall imply skillful, safe and humane medical and surgical treatment of the least radical nature necessary to achieve permanent sexual sterility and the highest possible therapeutic benefits depending upon the exigencies of each particular case.

Section 3. Office of State Eugenicist. There is hereby established for the State of . . . . the office of State Eugenicist, the function of which shall be to protect the state against the procreation of persons socially inadequate from degenerate or defective physical, physiological or psychological inheritance.

Section 4. Qualifications of State Eugenicist. The State Eugenicist shall be a trained student of human heredity, and shall be skilled in the modern practice of securing and analyzing human pedigrees; and he shall be required to devote his entire time and attention to the duties of his office as herein contemplated.

Section 5. Term of Office, Appointment, and Responsibility. The State Eugenicist shall be appointed by the Governor, with the consent of the Senate, shall be responsible directly to the Governor, and shall hold office until removed by death, resignation, or until his successor shall have been duly appointed.

Section 6. Seal. The Governor of the State shall cause a seal to be fashioned and made for the Office of the State Eugenicist, which seal shall be duly entrusted to the State Eugenicist and shall constitute the evidence of authority under this Act.

Section 7. Duties of State Eugenicist. It shall be the duty of the State Eugenicist:

(a) To conduct field-surveys seeking first-hand data concerning the hereditary constitution of all personals in the State who are socially inadequate personally or who, although normal personally, carry degenerate or defective hereditary qualities of a socially inadequate nature, and to cooperate with, to hear the complaints of, and to seek information from individuals and public and private social-welfare, charitable and scientific organizations possessing special acquaintance with and knowledge of such persons, to the end that the State shall possess equally accurate data in reference to the personal and family histories of all persons existing in the State, who are potential parents of socially inadequate offspring, regardless of whether such potential parents by members of the population at large or inmates of custodial institutions, regardless also of the personality, sex, age, marital condition, race or possessions of such persons.

(b) To examine further into the natural physical, physiological and psychological traits, the environment, the personal histories, and the family pedigrees of all persons existing in the State, whether in the population at large or as inmates of custodial institutions, who reasonably appear to be potential parents of socially inadequate offspring, with the view to determining more definitely whether in each particular case the individual is a cacogenic person within the meaning of this Act.

(c) To maintain a roster of all public and private custodial institutions in the state, and to require from the responsible head of each such institution, a record by full names and addresses, social and medical diagnosis and other pertinent data in reference to all accessions and losses of inmates as such occur from time to time; the said State Eugenicist may require a copy of any record which the particular institution may possess in reference to the case, family or institutional histories of any inmate which the State Eugenicist may name.

(d) To follow up, so far as possible, the case-histories of persons eugenically sterilized under this Act, with special reference to their social, economic, marital and health records, and to investigate the specific effects of eugenical sterilization. . . .

Section 8. Cooperation by Custodial Institutions. For the purpose of securing the facts essential to the determination required by this Act, the responsible head of any public or private custodial institution within the State shall, on demand, render promptly to the State Eugenicist all reports herein contemplated, and shall extend to said Officer and his duly appointed agents ready access to all records and inmates of the particular institution.

Section 9. Power to Administer Oaths and to Make Arrests. The State Eugenicist and his assistants appointed in writing by him for the purpose, shall have power to administer oaths, to subpoena and to examine witnesses under oath, and to make arrests. . . .

C. The Federal Government and Eugenical Sterilization. . . .

Up to the present time, the Federal Government has not enacted any legislation bearing either directly or indirectly upon eugenical sterilization. The matter of segregating, sterilizing, or otherwise rendering non-reproductive the degenerate human strains in America is, in accordance with the spirit of our institutions, fundamentally a matter for each state to decide for itself. There is, however, a specialized field in which the Federal Government must cooperate with the several states, if the human breeding stock in our population is to be purged of its defective parenthood.

The relation between the inheritable qualities of our immigrants and the destiny of the American nation is very close. Granting that the fecundity of native and immigrant stock will run evenly, then it is clear that from generation to generation the natural qualities of our present human parenthood will more and more assume the character of the natural qualities of immigrant parents. Thus, if the American nation desires to upbuild or even to maintain its standard of natural qualities, it must forbid the addition through immigration to our breeding stock of persons of a lower natural hereditary constitution than that which constitutes the desired standard.

If our standard of physical, mental and moral qualities for parenthood strike more heavily against one race than another, then we should be willing to enforce laws which take on the appearance of racial discrimination but which indeed would not be such, because in every race, even the very lowest, there are some individuals who through natural merit could conform to our standards of admission.

The immigration policy of the eugenicist, who has at heart the preservation, upbuilding and specialization of our better family stocks, is to base the criterion for admission of would-be immigrants primarily upon the possession of sterling natural qualities, regardless of race, language, or present social or economic condition.

It is suggested that a Federal Eugenicist, attached to the Public Health Service, or to the Children's Bureau, aided by an ample corps of assistants, would constitute an effective administrative agency for sterilization under federal authority. Some of the assistants of the office of Federal Eugenicist should be delegated to cooperate with the Immigration Service of the Department of Labor, and the Bureaus of Criminal Identification, and of Prisons, of the Department of Justice, and possibly with the Bureau of Education of the Department of the Interior. If the projected plan for examining the admissibility of immigrants in their native homes before the purchase of transportation, or even upon the steamships before landing, were adopted, it would be possible to pass satisfactorily upon the eugenical qualifications of the particular immigrant. . . .

Source: Harry Hamilton Laughlin. "Eugenical Sterilization in the United States." Chicago: Psychopathic Laboratory of the Municipal Court of Chicago, 1922. Available online. URL: http://www.people.fas.harvard.edu/~wellerst/laughlin/Laughlin_Model_Law.pdf. Accessed May 28, 2009.

Oliver Wendell Holmes, Jr.: *Buck v. Bell* (1927)

Oliver Wendell Holmes, Jr., was appointed to the U.S. Supreme Court by President Theodore Roosevelt in 1902. He served for 30 years, writing some of the court's most well-known opinions, and he was lauded for his evenhandedness and his protections of the Bill of Rights. Buck v. Bell, in hindsight, is one of his most notorious opinions because he diverged from his usual principles of justice and fairness in upholding the philosophy of eugenics by advocating sterilization of a woman thought to be feebleminded. In his determination that "three generations of imbeciles are enough," Holmes gave voice to the widely prevailing opinion of the day—that medical doctors had the right to surgically prevent women from having children but that women themselves were not allowed to prevent pregnancy by means of birth control. Sterilization in cases such as Buck's was accomplished by removal of the fallopian tubes, a procedure called a salpingectomy. In upholding Buck v. Bell, Holmes took for granted that Buck's admission to the State Colony for Epileptics and Feeble Minded was legitimate, when in reality her foster family placed her there to cover up the fact that she had become pregnant after having been raped by a relative. Buck's feeblemindedness, as well as that of her mother and daughter, was considered an established fact despite a lack of verifiable proof. Holmes's assertion that "heredity plays an important part in the transmission of insanity [and] imbecility" was not questioned despite the fact that scientists at the time had little understanding of genetics.

This is a writ of error to review a judgment of the Supreme Court of Appeals of the State of Virginia, affirming a judgment of the Circuit Court of Amherst County, by which the defendant in error, the superintendent of the State Colony for Epileptics and Feeble Minded, was ordered to perform the operation of salpingectomy upon Carrie Buck, the plaintiff in error, for the purpose of making her sterile. 143 Va. 310, 130 S. E. 516. The case comes here upon the contention that the statute authorizing the judgment is void under the Fourteenth Amendment as denying to the plaintiff in error due process of law and the equal protection of the laws.

Carrie Buck is a feeble-minded white woman who was committed to the State Colony above mentioned in due form. She is the daughter of a feeble-minded mother in the same institution, and the mother of an illegitimate feeble-minded child. She was eighteen years old at the time of the trial of her case in the Circuit Court in the latter part of 1924. An Act of Virginia approved March 20, 1924 (Laws 1924, c. 394) recites that the health of the patient and the welfare of society may be promoted in certain cases by the sterilization of mental defectives, under careful safeguard, etc.; that the sterilization may be effected in males by vasectomy and in females by salpingectomy, without serious pain or substantial danger to life; that the Commonwealth is supporting in various institutions many defective persons who if now discharged would become [274 U.S. 200, 206] a menace but if incapable of procreating might be discharged with safety and become self-supporting with benefit to themselves and to society; and that experience has shown that heredity plays an important part in the transmission of insanity, imbecility, etc. The statute then enacts that whenever the superintendent of certain institutions including the above named State Colony shall be of opinion that it is for the best interest of the patients and of society that an inmate under his care should be sexually sterilized, he may have the operation performed upon any patient afflicted with hereditary forms of insanity, imbecility, etc., on complying with the very careful provisions by which the act protects the patients from possible abuse.

The superintendent first presents a petition to the special board of directors of his hospital or colony, stating the facts and the grounds for his opinion, verified by affidavit. Notice of the petition and of the time and place of the hearing in the institution is to be served upon the inmate, and also upon his guardian, and if there is no guardian the superintendent is to apply to the Circuit Court of the County to appoint one. If the inmate is a minor notice also is to be given to his parents, if any, with a copy of the petition. The board is to see to it that the inmate may attend the hearings if desired by him or his guardian. The evidence is all to be reduced to writing, and after the board has made its order for or against the operation, the superintendent, or the inmate, or his guardian, may appeal to the Circuit Court of the County. The Circuit Court may consider the record of the board and the evidence before it and such other admissible evidence as may be offered, and may affirm, revise, or reverse the order of the board and enter such order as it deems just. Finally any party may apply to the Supreme Court of Appeals, which, if it grants the appeal, is to hear the case upon the record of the trial [274 U.S. 200, 207] in the Circuit Court and may enter such order as it thinks the Circuit Court should have entered. There can be no doubt that so far as procedure is concerned the rights of the patient are most carefully considered, and as every step in this case was taken in scrupulous compliance with the statute and after months of observation, there is no doubt that in that respect the plaintiff in error has had due process at law.

The attack is not upon the procedure but upon the substantive law. It seems to be contended that in no circumstances could such an order be justified. It certainly is contended that the order cannot be justified upon the existing grounds. The judgment finds the facts that have been recited and that Carrie Buck "is the probable potential parent of socially inadequate offspring, likewise afflicted, that she may be sexually sterilized without detriment to her general health and that her welfare and that of society will be promoted by her sterilization," and thereupon makes the order. In view of the general declarations of the Legislature and the specific findings of the Court obviously we cannot say as matter of law that the grounds do not exist, and if they exist they justify the result. We have seen more than once that the public welfare may call upon the best citizens for their lives. It would be strange if it could not call upon those who already sap the strength of the State for these lesser sacrifices, often not felt to be such by those concerned, in order to prevent our being swamped with incompetence. It is better for all the world, if instead of waiting to execute degenerate offspring for crime, or to let them starve for their imbecility, society can prevent those who are manifestly unfit from continuing their kind. The principle that sustains compulsory vaccination is broad enough to cover cutting the Fallopian tubes. *Jacobson v. Massachusetts*, 197 U.S. 11, 25 S. Ct. 358, 3 Ann. Cas. 765. Three generations of imbeciles are enough. [274 U.S. 200, 208] But, it is said, however it might be if this reasoning were applied generally, it fails when it is confined to the small number who are in the institutions named and is not applied to the multitudes outside. It is the usual last resort of constitutional arguments to point out shortcomings of this sort. But the answer is that the law does all that is needed when it does all that it can, indicates a policy, applies it to all within the lines, and seeks to bring within the lines all similarly situated so far and so fast as its means allow. Of course so far as the operations enable those who otherwise must be kept confined to be returned to the world, and thus open the asylum to others, the equality aimed at will be more nearly reached.

Judgment affirmed.

Source: Oliver Wendell Holmes, Jr. *Buck v. Bell,* 274 U.S. 200 (1927). Available online. URL: http://laws.findlaw. com/us/274/200.html. Accessed May 28, 2009.

Summary Statement of the Asilomar Conference on Recombinant DNA Molecules (1975)

The Asilomar Conference was organized by the Stanford biochemist Paul Berg for the purpose of developing guidelines for biotechnology experiments with unforeseen consequences, especially those in which DNA from two or more organisms would be combined. In an early application of the precautionary principle (later developed in the Montreal Protocol, the Rio Declaration, and the Cartagena Protocol, among numerous other international treaties), the writers of the Summary Statement determined that experiments must be designed and conducted as ethically and safely as possible. The authors advise that DNA experiments must be performed in contained areas that conform to standards for minimal-risk, low-risk, moderate-risk, or high-risk procedures. They denote what substances and procedures qualify for minimal-, low-, moderate-, and high-risk circumstances and caution that new knowledge will inform revisions to the procedures as scientific inquiry continues. The statement has provided the framework for research design that has stood the test of time for more than 30 years.

I. Introduction and General Conclusions

This meeting was organized to review scientific progress in research on recombinant DNA molecules and to discuss appropriate ways to deal with the potential biohazards of this work. Impressive scientific achievements have already been made in this field and these techniques have a remarkable potential for furthering our understanding of fundamental biochemical processes in pro- and eukaryotic cells. The use of recombinant DNA methodology promises to revolutionize the practice of molecular biology. While there has as yet been no practical application of the new techniques, there is every reason to believe that they will have significant practical utility in the future.

Of particular concern to the participants at the meeting was the issue of whether the pause in certain aspects of research in this area, called for by the Committee on Recombinant DNA Molecules of the National Academy of Sciences, U.S.A. in the letter published in July, 1974, should end; and, if so, how the scientific work could be undertaken with minimal risks to workers in laboratories, to the public at large and to the animal and plant species sharing our ecosystems.

The new techniques, which permit combination of genetic information from very different organisms, place us in an area of biology with many unknowns. Even in the present, more limited conduct of research in this field, the evaluation of potential biohazards has proved to be extremely difficult. It is this ignorance that has compelled us to conclude that it would be wise to exercise considerable caution in performing this research. Nevertheless, the participants at the Conference agreed that most of the work on construction of recombinant DNA molecules should proceed provided that appropriate safeguards, principally biological and physical barriers adequate to contain the newly created organisms, are employed. Moreover, the standards of protection should be greater at the beginning and modified as improvements in the methodology occur and assessments of the risks change. Furthermore, it was agreed that there are certain experiments in which the potential risks are of such a serious nature that they ought not to be done with presently available containment facilities. In the longer term serious problems may arise in the large scale application of this methodology in industry, medicine and agriculture. But it was also recognized that future research and experience may show that many of the potential biohazards are less serious and/or less probable than we now suspect.

II. Principles Guiding the Recommendations and Conclusions
Though our assessments of the risks involved with each of the various lines of research on recombinant DNA molecules may differ, few, if any, believe that this methodology is free from any risk. Reasonable principles for dealing with these potential risks are: 1) that containment be made an essential consideration in the experimental design and, 2) that the effectiveness of the containment should match, as closely as possible, the estimated risk. Consequently, whatever scale of risks is agreed upon, there should be a commensurate scale of containment. Estimating the risks will be difficult and intuitive at first but this will improve as we acquire additional knowledge; at each stage we shall have to match the potential risk with an appropriate level of containment. Experiments requiring large scale operations would seem to be riskier than equivalent experiments done on a small scale, and, therefore, require more stringent containment procedures. The use of cloning vehicles or vectors (plasmids, phages) and bacterial hosts with a restricted capacity to multiply outside of the laboratory would reduce the potential biohazard

of a particular experiment. Thus, the ways in which potential biohazards and different levels of containment are matched may vary from time to time particularly as the containment technology is improved. The means for assessing and balancing risks with appropriate levels of containment will need to be reexamined from time to time. Hopefully, through both formal and informal channels of information within and between the nations of the world, the way in which potential biohazards and levels of containment are matched would be consistent.

Containment of potentially biohazardous agents can be achieved in several ways. The most significant contribution to limiting the spread of the recombinant DNAs, is the use of biological barriers. These barriers are of two types: 1) fastidious bacterial hosts unable to survive in natural environments, and 2) non-transmissible and equally fastidious vectors (plasmids, bacteriophages or other viruses) able to grow only in specified hosts. Physical containment, exemplified by the use of suitable hoods, or where applicable, limited access or negative pressure laboratories, provides an additional factor of safety. Particularly important is strict adherence to good microbiological practices which, to a large measure can limit the escape of organisms from the experimental situation, and thereby increase the safety of the operation. Consequently, education and training of all personnel involved in the experiments is essential to the effectiveness of all containment measures. In practice these different means of containment will complement one another and documented substantial improvements in the ability to restrict the growth of bacterial hosts and vectors could permit modifications of the complementary physical containment requirements.

Stringent physical containment and rigorous laboratory procedures can reduce but not eliminate the possibility of spreading potentially hazardous agents. Therefore, investigators relying upon "disarmed" hosts and vectors for additional safety must rigorously test the effectiveness of these agents before accepting their validity as biological barriers.

III. Specific Recommendations for Matching
Types of Containment with Types of Experiments

No classification of experiments as to risk and no set of containment procedures can anticipate all situations. Given our present uncertainties about the hazards, the parameters propose here are broadly conceived and meant to provide provisional guidelines for investigators and agencies concerned with research on recombinant DNAs. However, each investigator bears a responsibility for determining whether, in his particular case, special circumstances warrant a higher level of containment than is suggested here.

BIOTECHNOLOGY AND GENETIC ENGINEERING

A. Types of Containment

1. Minimal Risk: This type of containment is intended for experiments in which the biohazards may be accurately assessed and are expected to be minimal. Such containment can be achieved by following the operating procedures recommended for clinical microbiological laboratories. Essential features of such facilities are no drinking, eating or smoking in the laboratory, wearing laboratory coats in the work area, the use of cotton-plugged pipettes or preferably mechanical pipetting devices and prompt disinfection of contaminated materials.

2. Low Risk: This level of containment is appropriate for experiments which generate novel biotypes but where the available information indicates that the recombinant DNA cannot alter appreciably the ecological behavior of the recipient species, increase significantly its pathogenicity, or prevent effective treatment of any resulting infections. The key features of this containment (in addition to the minimal procedures mentioned above) are a prohibition on mouth pipetting, access limited to laboratory personnel, and the use of biological safety cabinets for procedures likely to produce aerosols (e.g., blending and sonication). Though existing vectors may be used in conjunction with low risk procedures, safer vectors and hosts should be adopted as they become available.

3. Moderate Risk: Such containment facilities are intended for experiments in which there is a probability of generating an agent with a significant potential for pathogenicity or ecological disruption. The principle features of this level of containment, in addition to those of the two preceding classes, are that transfer operations should be carried out in biological safety cabinets (e.g., laminar flow hoods), gloves should be worn during the handling of infectious materials, vacuum lines must be protected by filters and negative pressure should be maintained in the limited access laboratories. Moreover, experiments posing a moderate risk must be done only with vectors and hosts that have an appreciably impaired capacity to multiply outside of the laboratory.

4. High Risk: This level of containment is intended for experiments in which the potential for ecological disruption or pathogenicity of the modified organism could be severe and thereby pose a serious biohazard to laboratory personnel or the public. The main features of this type of facility, which was designed to contain highly infectious microbiological agents, are its isolation from other areas by air locks, a negative pressure environment, a requirement for clothing changes and showers for entering personnel and laboratories fitted with treatment systems to inactivate or remove biological agents that may be contaminants in exhaust air, liquid and solid wastes. All persons occupying these areas should wear protective laboratory clothing and shower at each exit from the containment facility. The handling of agents should be confined to biological safety cabinets in which the exhaust air is incinerated or passed through Hepa filters. High risk containment includes, beside the physical and procedural features described above, the use of rigorously tested vectors and hosts whose growth can be confined to the laboratory.

B. Types of Experiments

Accurate estimates of the risks associated with different types of experiments are difficult to obtain because of our ignorance of the probability that the anticipated dangers will manifest themselves. Nonetheless, experiments involving the construction and propagation of recombinant DNA molecules using DNAs from 1) prokaryotes, bacteriophages and other plasmids, 2) animal viruses, and 3) eukaryotes have been characterized as minimal, low, moderate and high risks to guide investigators in their choice of the appropriate containment. These designations should be viewed as interim assignments which will need to be revised upward or downward in the light of future experience.

The recombinant DNA molecules themselves, as distinct from cells carrying them, may be infectious to bacteria or higher organisms. DNA preparations from these experiments, particularly in large quantities, should be chemically inactivated before disposal.

1. Prokaryotes, bacteriophages and bacterial plasmids: Where the construction of recombinant DNA molecules and their propagation involves prokaryotic agents that are known to exchange genetic information naturally, the experiments can be performed in minimal risk containment facilities. Where such experiments pose a potential hazard, more stringent containment may be warranted.

Experiments involving the creation and propagation of recombinant DNA molecules from DNAs of species that ordinarily do not exchange genetic information, generate novel biotypes. Because such experiments may pose biohazards greater than those associated with the original organisms, they should be performed, at least, in low risk containment facilities. If the experiments involve either pathogenic organisms, or genetic determinants that may increase the pathogenicity of the recipient species, or if the transferred DNA can confer upon the recipient organisms new metabolic activities not native to these species and thereby modify its relationship with the environment, then moderate or high risk containment should be used.

Experiments extending the range of resistance of established human pathogens to therapeutically useful antibiotics or disinfectants should be undertaken only under moderate or high risk containments depending upon the virulence of the organism involved.

2. Animal Viruses: Experiments involving linkage of viral genomes or genome segments to prokaryotic vectors and their propagation in pro-karyotic cells should be performed only with vector-host systems having demonstrably restricted growth capabilities outside the laboratory and with moderate risk containment facilities. Rigorously purified and characterized segments of non-oncogenic viral genomes or of the demonstrably non-transforming regions of oncogenic viral DNAs can be attached to presently existing vectors and propagated in moderate risk containment facilities; as safer vector-host systems become available such experiments may be performed in low risk facilities.

Experiments designed to introduce or propagate DNA from non-viral or other low risk agents in animal cells should use only low risk animal DNAs as vectors (e.g., viral, mitochondrial) and manipulations should be confined to moderate risk containment facilities.

3. Eukaryotes: The risks associated with joining random fragments of eukaryote DNA to prokaryotic DNA vectors and the propagation of these recombinant DNAs in prokaryotic hosts are the most difficult to assess.

A priori, the DNA from warm-blooded vertebrates is more likely to contain cryptic viral genomes potentially pathogenic for many than is the DNA from other eukaryotes. Consequently, attempts to clone segments of DNA from such animal and particularly primate genomes should be performed only with vector-host systems having demonstrably restricted growth capabilities outside the laboratory and in a moderate risk contain-ment facility. Until cloned segments of warm-blooded vertebrate DNA are completely characterized, they should continue to be maintained in the most restricted vector-host system in moderate risk containment laborato-ries; when such cloned segments are characterized, they may be propagated as suggested above for purified segments of virus genomes.

Unless the organism makes a product known to be dangerous (e.g., toxin, virus), recombinant DNAs from cold-blooded vertebrates and all other lower eukaryotes can be constructed and propagated with the safest vector-host system available in low risk containment facilities.

Purified DNA from any source that performs known functions and can be judged to be non-toxic, may be cloned with currently available vectors in low risk containment facilities. (Toxic here includes potentially oncogenic products or substances that might perturb normal metabolism if produced in an animal or plant by a resident microorganism.)

4. Experiments to Be Deferred: There are feasible experiments which present such serious dangers that their performance should not be undertaken at this time with the currently available vector-host systems and the presently available containment capability. These include the cloning of recombinant DNAs derived from highly pathogenic organisms (i.e., Class III, IV, V etiologic agents as classified by the United States Department of Health, Education and Welfare), DNA containing toxin genes and large scale experiments (more than 10 liters of culture) using recombinant DNAs that are able to make products potentially harmful to man, animals or plants.

IV. Implementation

In many countries steps are already being taken by national bodies to formulate codes of practice for the conduct of experiments with known or potential biohazard. Until these are established, we urge individual scientists to use the proposals in this document as a guide. In addition, there are some recommendations which could be immediately and directly implemented by the scientific community.

A. Development of Safer Vectors and Hosts

An important and encouraging accomplishment of the meeting was the realization that special bacteria and vectors can be constructed genetically, which have a restricted capacity to multiply outside the laboratory, and that the use of these organisms could enhance the safety of recombinant DNA experiments by many orders of magnitude. Experiments along these lines are presently in progress and in the near future, variants of λ bacteriophage, non-transmissible plasmids and special strains of *E. coli* will become available. All of these vectors could reduce the potential biohazards by very large factors and improve the methodology as well. Other vector-host systems, particularly modified strains of *Bacillus subtilis* and their relevant bacteriophages and plasmids, may also be useful for particular purposes. Quite possibly safe and suitable vectors may be found for eukaryotic hosts such as yeast and readily cultured plant and animal cells. There is likely to be a continuous development in this area and the participants at the meeting agreed that improved vector-host systems which reduce the biohazards of recombinant DNA research will be made freely available to all interested investigators.

B. Laboratory Procedures

It is the clear responsibility of the principal investigator to inform the staff of the laboratory of the potential hazards of such experiments, before they are initiated. Free and open discussion is necessary so that each individual participating in the experiment fully understands the nature of the experiment and any risk that might be involved. All workers must be properly trained in the containment procedures that are designed to control the hazard, including emergency actions in the event of a hazard. It is also recommended that appropriate health surveillance of all personnel, including serological monitoring, be conducted periodically.

C. Education and Reassessment

Research in this area will develop very quickly and the methods will be applied to many different biological problems. At any given time it is impossible to foresee the entire range of all potential experiments and make judgments on them. Therefore, it is essential to undertake a continuing reassessment of the problems in the light of new scientific knowledge. This could be achieved by a series of annual workshops and meetings, some of which should be at the international level. There should also be courses to train individuals in the relevant methods since it is likely that the work will be taken up by laboratories which may not have had extensive experience in this area. High priority should also be given to research that could improve and evaluate the containment effectiveness of new and existing vector-host systems.

V. New Knowledge

This document represents our first assessment of the potential biohazards in the light of current knowledge. However, little is known about the survival of laboratory strains of bacteria and bacteriophages in different ecological niches in the outside world. Even less is known about whether recombinant DNA molecules will enhance or depress the survival of their vectors and hosts in nature. These questions are fundamental to the testing of any new organism that may be constructed. Research in this area needs to be undertaken and should be given high priority. In general, however, molecular biologists who may construct DNA recombinant molecules do not undertake these experiments and it will be necessary to facilitate collaborative research between them and groups skilled in the study of bacterial infection or ecological microbiology. Work should also be undertaken which would enable us to monitor the escape or dissemination of cloning vehicles and their hosts.

Nothing is known about the potential infectivity in higher organisms of phages or bacteria containing segments of eukaryotic DNA and very little about the infectivity of the DNA molecules themselves. Genetic transformation of bacteria does occur in animals suggesting that recombinant DNA molecules can retain their biological potency in this environment. There are many questions in this area, the answers to which are essential for our assessment of the biohazards of experiments with recombinant DNA molecules. It will be necessary to ensure that this work will be planned and carried out; and it will be particularly important to have this information before large scale applications of the use of recombinant DNA molecules is attempted.

Source: Paul Berg, David Baltimore, Sydney Brenner, Richard O. Roblin III, and Maxine F. Singer. "Summary Statement of the Asilomar Conference on Recombinant DNA Molecules." *Proceedings of the National Academy of Sciences* 72.6 (June 1975): 1,981–1,984.

The Belmont Report (1979) (excerpt)

The Belmont Report, also known as "Ethical Principles and Guidelines for the Protection of Human Subjects of Research," was written to address the violation of human rights that occurred during "The Tuskegee Study of Untreated Syphilis in the Negro Male," in which 400 subjects were intentionally misled and denied medical treatment for their disease from 1932 to 1974. When the details of the study came to light, medical professionals and members of the public alike were appalled, and the National Research Act, which created the National Commission for the Protection of Human Subjects of Biomedical and Behavioral Research, was signed into law. Its principles of respect for test subjects and informed consent are the cornerstone of modern-day research in biotechnology and other fields.

Basic Ethical Principles

The expression "basic ethical principles" refers to those general judgments that serve as a basic justification for the many particular ethical prescriptions and evaluations of human actions. Three basic principles, among those generally accepted in our cultural tradition, are particularly relevant to the ethics of research involving human subjects: the principles of respect of persons, beneficence and justice.

1. Respect for Persons. Respect for persons incorporates at least two ethical convictions: first, that individuals should be treated as autonomous agents, and second, that persons with diminished autonomy are entitled to protection. The principle of respect for persons thus divides into two separate moral requirements: the requirement to acknowledge autonomy and the requirement to protect those with diminished autonomy. . . .

[N]ot every human being is capable of self-determination. The capacity for self-determination matures during an individual's life, and some individuals lose this capacity wholly or in part because of illness, mental disability, or circumstances that severely restrict liberty. Respect for the immature and the incapacitated may require protecting them as they mature or while they are incapacitated. . . .

In most cases of research involving human subjects, respect for persons demands that subjects enter into the research voluntarily and with adequate information. In some situations, however, application of the principle is not obvious. The involvement of prisoners as subjects of research provides an instructive example. On the one hand, it would seem that the principle of respect for persons requires that prisoners not be deprived of the opportunity to volunteer for research. On the other hand, under prison conditions they may be subtly coerced or unduly influenced to engage in research activities for which they would not otherwise volunteer. Respect for persons would then dictate that prisoners be protected. Whether to allow prisoners to "volunteer" or to "protect" them presents a dilemma. Respecting persons, in most hard cases, is often a matter of balancing competing claims urged by the principle of respect itself.

2. Beneficence. Persons are treated in an ethical manner not only by respecting their decisions and protecting them from harm, but also by making efforts to secure their well-being. Such treatment falls under the principle of beneficence. The term "beneficence" is often understood to cover acts of kindness or charity that go beyond strict obligation. In this document, beneficence is understood in a stronger sense, as an obligation. Two general rules have been formulated as complementary expressions of beneficent actions in this sense: (1) do no harm and (2) maximize possible benefits and minimize possible harms. . . .

3. Justice. Who ought to receive the benefits of research and bear its burdens? This is a question of justice, in the sense of "fairness in distribution" or "what is deserved." An injustice occurs when some benefit to which a person is entitled is denied without good reason or when some burden is imposed unduly. Another way of conceiving the principle of justice is that equals ought to be treated equally. However, this statement requires explication. Who is equal and who is unequal? What considerations justify departure from equal distribution? Almost all commentators allow that distinctions based on experience, age, deprivation, competence, merit and position do sometimes constitute criteria justifying differential treatment for certain purposes. It is necessary, then, to explain in what respects people should be treated equally. There are several widely accepted formulations of just ways to distribute burdens and benefits. Each formulation mentions some relevant property on the basis of which burdens and benefits should be distributed. These formulations are (1) to each person an equal share, (2) to each person according to individual need, (3) to each person according to individual effort, (4) to each person according to societal contribution, and (5) to each person according to merit. . . .

[D]uring the 19th and early 20th centuries the burdens of serving as research subjects fell largely upon poor ward patients, while the benefits of improved medical care flowed primarily to private patients. Subsequently, the exploitation of unwilling prisoners as research subjects in Nazi concentration camps was condemned as a particularly flagrant injustice. In this country, in the 1940's, the Tuskegee syphilis study used disadvantaged, rural black men to study the untreated course of a disease that is by no means confined to that population. These subjects were deprived of demonstrably effective treatment in order not to interrupt the project, long after such treatment became generally available. . . .

C. Applications

Applications of the general principles to the conduct of research leads to consideration of the following requirements: informed consent, risk/benefit assessment, and the selection of subjects of research.

1. Informed Consent. Respect for persons requires that subjects, to the degree that they are capable, be given the opportunity to choose what shall or shall not happen to them. This opportunity is provided when adequate standards for informed consent are satisfied. . . . [T]here is widespread agreement that the consent process can be analyzed as containing three elements: information, comprehension and voluntariness.
Information. Most codes of research establish specific items for disclosure intended to assure that subjects are given sufficient information. These items generally include: the research procedure, their purposes, risks and anticipated benefits, alternative procedures (where therapy is involved),

and a statement offering the subject the opportunity to ask questions and to withdraw at any time from the research. Additional items have been proposed, including how subjects are selected, the person responsible for the research, etc. . . .

A special problem of consent arises where informing subjects of some pertinent aspect of the research is likely to impair the validity of the research. In many cases, it is sufficient to indicate to subjects that they are being invited to participate in research of which some features will not be revealed until the research is concluded. In all cases of research involving incomplete disclosure, such research is justified only if it is clear that (1) incomplete disclosure is truly necessary to accomplish the goals of the research, (2) there are no undisclosed risks to subjects that are more than minimal, and (3) there is an adequate plan for debriefing subjects, when appropriate, and for dissemination of research results to them. Information about risks should never be withheld for the purpose of eliciting the cooperation of subjects, and truthful answers should always be given to direct questions about the research. . . .

Comprehension. The manner and context in which information is conveyed is as important as the information itself. For example, presenting information in a disorganized and rapid fashion, allowing too little time for consideration or curtailing opportunities for questioning, all may adversely affect a subject's ability to make an informed choice.

Because the subject's ability to understand is a function of intelligence, rationality, maturity and language, it is necessary to adapt the presentation of the information to the subject's capacities. Investigators are responsible for ascertaining that the subject has comprehended the information. While there is always an obligation to ascertain that the information about risk to subjects is complete and adequately comprehended, when the risks are more serious, that obligation increases. On occasion, it may be suitable to give some oral or written tests of comprehension.

Special provision may need to be made when comprehension is severely limited—for example, by conditions of immaturity or mental disability. Each class of subjects that one might consider as incompetent (e.g., infants and young children, mentally disabled patients, the terminally ill and the comatose) should be considered on its own terms. Even for these persons, however, respect requires giving them the opportunity to choose to the extent they are able, whether or not to participate in research. . . .

Voluntariness. An agreement to participate in research constitutes a valid consent only if voluntarily given. This element of informed consent requires conditions free of coercion and undue influence. Coercion occurs when an overt threat of harm is intentionally presented by one person to another in order to obtain compliance. Undue influence, by contrast, occurs through an offer of an excessive, unwarranted, inappropriate or improper reward or other overture in order to obtain compliance. Also, inducements that would ordinarily be acceptable may become undue influences if the subject is especially vulnerable. . . .

2. Assessment of Risks and Benefits. The assessment of risks and benefits requires a careful arrayal of relevant data, including, in some cases, alternative ways of obtaining the benefits sought in the research. Thus, the assessment presents both an opportunity and a responsibility to gather systematic and comprehensive information about proposed research. For the investigator, it is a means to examine whether the proposed research is properly designed. For a review committee, it is a method for determining whether the risks that will be presented to subjects are justified. For prospective subjects, the assessment will assist the determination whether or not to participate. . . .

[A]ssessment of the justifiability of research should reflect at least the following considerations: (i) Brutal or inhumane treatment of human subjects is never morally justified. (ii) Risks should be reduced to those necessary to achieve the research objective. It should be determined whether it is in fact necessary to use human subjects at all. Risk can perhaps never be entirely eliminated, but it can often be reduced by careful attention to alternative procedures. (iii) When research involves significant risk of serious impairment, review committees should be extraordinarily insistent on the justification of the risk (looking usually to the likelihood of benefit to the subject—or, in some rare cases, to the manifest voluntariness of the participation). (iv) When vulnerable populations are involved in research, the appropriateness of involving them should itself be demonstrated. A number of variables go into such judgments, including the nature and degree of risk, the condition of the particular population involved, and the nature and level of the anticipated benefits. (v) Relevant risks and benefits must be thoroughly arrayed in documents and procedures used in the informed consent process.

3. Selection of Subjects. Just as the principle of respect for persons finds expression in the requirements for consent, and the principle of beneficence in risk/benefit assessment, the principle of justice gives rise to moral requirements that there be fair procedures and outcomes in the selection of research subjects.

155

Justice is relevant to the selection of subjects of research at two levels: the social and the individual. Individual justice in the selection of subjects would require that researchers exhibit fairness: thus, they should not offer potentially beneficial research only to some patients who are in their favor or select only "undesirable" persons for risky research. Social justice requires that distinction be drawn between classes of subjects that ought, and ought not, to participate in any particular kind of research, based on the ability of members of that class to bear burdens and on the appropriateness of placing further burdens on already burdened persons. . . . One special instance of injustice results from the involvement of vulnerable subjects. Certain groups, such as racial minorities, the economically disadvantaged, the very sick, and the institutionalized may continually be sought as research subjects, owing to their ready availability in settings where research is conducted. Given their dependent status and their frequently compromised capacity for free consent, they should be protected against the danger of being involved in research solely for administrative convenience, or because they are easy to manipulate as a result of their illness or socioeconomic condition.

Source: The Belmont Report. National Commission for the Protection of Human Subjects of Biomedical and Behavioral Research (4/18/79). Available online. URL: http://ohsr.od.nih.gov/guidelines/belmont.html. Accessed September 9, 2009.

Diamond v. Chakrabarty, 447 U.S. 303 (1980) (excerpt)

In Diamond v. Chakrabarty, *the U.S. Supreme Court decided that manufactured organisms can be patented. Ananda Chakrabarty was a genetic engineer for General Electric who developed a bacteria capable of breaking down crude oil. The bacteria was intended for use in cleaning up oil spills, and General Electric wanted to protect its proprietary technology, as developed by Chakrabarty, through a patent. The patent was initially rejected because it was thought that living organisms were not patentable. The Court of Customs and Patent Appeals reversed this decision, and the U.S. Supreme Court agreed in this opinion written by Chief Justice Warren Burger.*

. . . In 1972, respondent Chakrabarty, a microbiologist, filed a patent application, assigned to the General Electric Co. The application asserted 36 claims related to Chakrabarty's invention of "a bacterium from the genus Pseudomonas containing therein at least two stable energy-generating plasmids, each of said plasmids providing a separate hydrocarbon degradative pathway." This human-made, genetically engineered bacterium is capable of breaking down multiple components of crude oil. Because of this property, which is possessed by no naturally occurring bacteria, Chakrabarty's invention is believed to have significant value for the treatment of oil spills.

Chakrabarty's patent claims were of three types: first, process claims for the method of producing the bacteria; [447 U.S. 303, 306] second, claims for an inoculum comprised of a carrier material floating on water, such as straw, and the new bacteria; and third, claims to the bacteria themselves. The patent examiner allowed the claims falling into the first two categories, but rejected claims for the bacteria. His decision rested on two grounds: (1) that micro-organisms are "products of nature," and (2) that as living things they are not patentable subject matter under 35 U.S.C. 101.

Chakrabarty appealed the rejection of these claims to the Patent Office Board of Appeals, and the Board affirmed the examiner on the second ground. Relying on the legislative history of the 1930 Plant Patent Act, in which Congress extended patent protection to certain asexually reproduced plants, the Board concluded that 101 was not intended to cover living things such as these laboratory created micro-organisms. . . .

The Constitution grants Congress broad power to legislate to "promote the Progress of Science and useful Arts, by securing for limited Times to Authors and Inventors the exclusive Right to their respective Writings and Discoveries." Art. I, 8, cl. 8. The patent laws promote this progress by offering inventors exclusive rights for a limited period as an incentive for their inventiveness and research efforts. *Kewanee Oil Co. v. Bicron Corp.,* 416 U.S. 470, 480–481 (1974); *Universal Oil Co. v. Globe Co.,* 322 U.S. 471, 484 (1944). The authority of Congress is exercised in the hope that "[t]he productive effort thereby fostered will have a positive effect on society through the introduction of new products and processes of manufacture into the economy, and the emanations by way of increased employment and better lives for our citizens." *Kewanee,* supra, at 480.

The question before us in this case is a narrow one of statutory interpretation requiring us to construe 35 U.S.C. 101, which provides:

"Whoever invents or discovers any new and useful process, machine, manufacture, or composition of matter, or any new and useful improvement thereof, may obtain a patent therefore, subject to the conditions and requirements of this title."

Specifically, we must determine whether respondent's micro-organism constitutes a "manufacture" or "composition of matter" within the meaning of the statute. . . .

[T]his Court has read the term "manufacture" in 101 in accordance with its dictionary definition to mean "the production of articles for use from raw or prepared materials by giving to these materials new forms, qualities, properties, or combinations, whether by hand-labor or by machinery." *American Fruit Growers, Inc. v. Brogdex Co.*, 283 U.S. 1, 11 (1931). Similarly, "composition of matter" has been construed consistent with its common usage to include "all compositions of two or more substances and . . . all composite articles, whether they be the results of chemical union, or of mechanical mixture, or whether they be gases, fluids, powders or solids." *Shell Development Co. v. Watson*, 149 F. Supp. 279, 280 (DC 1957) (citing 1 A. Deller, Walker on Patents 14, p. 55 (1st ed. 1937)). In choosing such expansive terms as "manufacture" and "composition of matter," modified by the comprehensive "any," Congress plainly contemplated that the patent laws would be given wide scope.

The relevant legislative history also supports a broad construction. The Patent Act of 1793, authored by Thomas Jefferson, defined statutory subject matter as "any new and useful art, machine, manufacture, or composition of matter, or any new or useful improvement [thereof]." Act of Feb. 21, 1793, 1, 1 Stat. 319. The Act embodied Jefferson's philosophy that "ingenuity should receive a liberal encouragement." [447 U.S. 303, 309] 5 Writings of Thomas Jefferson 75–76 (Washington ed. 1871). See *Graham v. John Deere Co.*, 383 U.S. 1, 7–10 (1966). Subsequent patent statutes in 1836, 1870, and 1874 employed this same broad language. In 1952, when the patent laws were recodified, Congress replaced the word "art" with "process," but otherwise left Jefferson's language intact. The Committee Reports accompanying the 1952 Act inform us that Congress intended statutory subject matter to "include anything under the sun that is made by man." S. Rep. No. 1979, 82d Cong., 2d Sess., 5 (1952); H. R. Rep. No. 1923, 82d Cong., 2d Sess., 6 (1952).

This is not to suggest that 101 has no limits or that it embraces every discovery. The laws of nature, physical phenomena, and abstract ideas have been held not patentable. . . . Einstein could not patent his celebrated law that $E=mc^2$.; nor could Newton have patented the law of gravity. . . .

Judged in this light, respondent's micro-organism plainly qualifies as patentable subject matter. His claim is not to a hitherto unknown natural phenomenon, but to a nonnaturally occurring manufacture or composition of matter—a product of human ingenuity "having a distinctive name, character [and] [447 U.S. 303, 310] use." . . .

[T]he patentee has produced a new bacterium with markedly different characteristics from any found in nature and one having the potential for significant utility. His discovery is not nature's handiwork, but his own; accordingly it is patentable subject matter under 101. . . .

Two contrary arguments are advanced, neither of which we find persuasive.

(A) The petitioner's first argument rests on the enactment of the 1930 Plant Patent Act, which afforded patent protection to certain asexually reproduced plants, and the 1970 Plant [447 U.S. 303, 311] Variety Protection Act, which authorized protection for certain sexually reproduced plants but excluded bacteria from its protection. In the petitioner's view, the passage of these Acts evidences congressional understanding that the terms "manufacture" or "composition of matter" do not include living things; if they did, the petitioner argues, neither Act would have been necessary.

We reject this argument. Prior to 1930, two factors were thought to remove plants from patent protection. The first was the belief that plants, even those artificially bred, were products of nature for purposes of the patent law. This position appears to have derived from the decision of the Patent Office in Ex parte Latimer, 1889 Dec. Com. Pat. 123, in which a patent claim for fiber found in the needle of the Pinus australis was rejected. . . . The second obstacle to patent protection for plants was the fact that plants were thought not amenable to the "written description" requirement of the patent law. See 35 U.S.C. 112. Because new plants may differ from old only in color or perfume, differentiation by written description was often impossible. See Hearings on H. R. 11372 before the House Committee on Patents, 71st Cong., 2d Sess., 7 (1930) (memorandum of Patent Commissioner Robertson).

In enacting the Plant Patent Act, Congress addressed both of these concerns. It explained at length its belief that the work of the plant breeder "in aid of nature" was patentable invention. S. Rep. No. 315, 71st Cong., 2d Sess., 6–8 (1930); H. R. Rep. No. 1129, 71st Cong., 2d Sess., 7–9 (1930). And it relaxed the written description requirement in favor of "a description . . . as complete as is reasonably possible." 35 U.S.C. 162. No Committee or Member of Congress, however, expressed the broader view, now urged by the petitioner, that the terms "manufacture" or "composition of matter" exclude living things. The sole support for that position in the legislative history of the 1930 Act is found in the conclusory statement of Secretary of Agriculture Hyde, in a letter to the Chairmen of the House and Senate Committees considering the 1930 Act, that "the patent laws . . . at the present time are understood to cover only inventions or discoveries in the field of inanimate nature." See S. Rep. No. 315, supra, at Appendix A; H. R. Rep. No. 1129, supra, at Appendix A. . . .

Congress thus recognized that the relevant distinction was not between living and inanimate things, but between products of nature, whether living or not, and human-made inventions. Here, respondent's micro-organism is the result of human ingenuity and research. Hence, the passage of the Plant Patent Act affords the Government no support.

Nor does the passage of the 1970 Plant Variety Protection Act support the Government's position. As the Government acknowledges, sexually reproduced plants were not included under the 1930 Act because new varieties could not be reproduced true-to-type through seedlings. Brief for Petitioner 27, n. 31. By 1970, however, it was generally recognized that true-to-type reproduction was possible and that plant patent protection was therefore appropriate. The 1970 Act extended that protection. There is nothing in its language or history to suggest that it was enacted because 101 did not include living things. . . .

(B) The petitioner's second argument is that micro-organisms cannot qualify as patentable subject matter until Congress expressly authorizes such protection. His position rests on the fact that genetic technology was unforeseen when Congress enacted 101. From this it is argued that resolution of the patentability of inventions such as respondent's should be left to Congress. The legislative process, the petitioner argues, is best equipped to weigh the competing economic, social, and scientific considerations involved, and to determine whether living organisms produced by genetic engineering should receive patent protection. . . .

It is, of course, correct that Congress, not the courts, must define the limits of patentability; but it is equally true that once Congress has spoken it is "the province and duty of the judicial department to say what the law is." *Marbury v. Madison*, 1 Cranch 137, 177 (1803). Congress has performed its constitutional role in defining patentable subject matter in 101; we perform ours in construing the language Congress has employed. In so doing, our obligation is to take statutes as we find them, guided, if ambiguity appears, by the legislative history and statutory purpose. Here, we perceive no ambiguity. The subject-matter provisions of the patent law have been cast in broad terms to fulfill the constitutional and statutory goal of promoting "the Progress of Science and the useful Arts" with all that means for the social and economic benefits envisioned by Jefferson. Broad general language is not necessarily ambiguous when congressional objectives require broad terms. . . .

To buttress his argument, the petitioner, with the support of amicus, points to grave risks that may be generated by research endeavors such as respondent's. The briefs present a gruesome parade of horribles. Scientists, among them Nobel laureates, are quoted suggesting that genetic research may pose a serious threat to the human race, or, at the very least, that the dangers are far too substantial to permit such research to proceed apace at this time. We are told that genetic research and related technological developments may spread pollution and disease, that it may result in a loss of genetic diversity, and that its practice may tend to depreciate the value of human life. These arguments are forcefully, even passionately, presented; they remind us that, at times, human ingenuity seems unable to control fully the forces it creates—that, with Hamlet, it is sometimes better "to bear those ills we have than fly to others that we know not of."

It is argued that this Court should weigh these potential hazards in considering whether respondent's invention is [447 U.S. 303, 317] patentable subject matter under 101. We disagree. The grant or denial of patents on micro-organisms is not likely to put an end to genetic research or to its attendant risks. The large amount of research that has already occurred when no researcher had sure knowledge that patent protection would be available suggests that legislative or judicial fiat as to patentability will not deter the scientific mind from probing into the unknown any more than Canute could command the tides. Whether respondent's claims are patentable may determine whether research efforts are accelerated by the hope of reward or slowed by want of incentives, but that is all.

What is more important is that we are without competence to entertain these arguments—either to brush them aside as fantasies generated by fear of the unknown, or to act on them. The choice we are urged to make is a matter of high policy for resolution within the legislative process after the kind of investigation, examination, and study that legislative bodies can provide and courts cannot. That process involves the balancing of competing values and interests, which in our democratic system is the business of elected representatives. Whatever their validity, the contentions now pressed on us should be addressed to the political branches of the Government, the Congress and the Executive, and not to the courts. [447 U.S. 303, 318]

We have emphasized in the recent past that "[o]ur individual appraisal of the wisdom or unwisdom of a particular [legislative] course . . . is to be put aside in the process of interpreting a statute." *TVA v. Hill,* 437 U.S. 194. Our task, rather, is the narrow one of determining what Congress meant by the words it used in the statute; once that is done our powers are exhausted. Congress is free to amend 101 so as to exclude from patent protection organisms produced by genetic engineering. Cf. 42 U.S.C. 2181 (a), exempting from patent protection inventions "useful solely in the utilization of special nuclear material or atomic energy in an atomic weapon." Or it may choose to craft a statute specifically designed for such living things. But, until Congress takes such action, this Court must construe the language of 101 as it is. The language of that section fairly embraces respondent's invention.

Accordingly, the judgment of the Court of Customs and Patent Appeals is Affirmed.

Source: Warren Burger. *Diamond v. Chakrabarty,* 447 U.S. 303 (1980). Available online. URL: http://laws.findlaw. com/us/447/303.html. Accessed May 13, 2009.

President's Council on Bioethics: "Better Children" (2003) (excerpt)

The President's Council on Bioethics was established by George W. Bush on November 28, 2001, in order to "advise the President on bioethical issues that may emerge as a consequence of advances in biomedical science and technology." The council has published a number of documents on current topics, including stem cell research, human cloning, and reproductive technology. In this excerpt from Beyond Therapy: Biotechnology and the Pursuit of Happiness, *the council discusses the ethical issues that may confront prospective parents who have the choice of using biotechnology to alter their child's genetic destiny.*

What father or mother does not dream of a good life for his or her child? What parents would not wish to enhance the life of their children, to make them better people, to help them live better lives? Such wishes and intentions guide much of what all parents do for and to their children. To help our children on their way and to make them strong in body and in mind, we feed and clothe them, see that they get rest, fresh air, and exercise, and take great pains regarding their education. Beyond ordinary schooling, we give them swimming and piano lessons, enroll them in Scouts or Little League, and help them acquire a variety of skills—artistic, intellectual, and social. In addition, we try to develop their character, educate their tastes and sensibilities, and nurture their spiritual growth. In all of these efforts we are guided, whether consciously or not, by some notion or other of what it *means* to improve our children, of what it means to make them *better.* . . .

In most of our efforts to assist our children's development, we proceed through speech and symbolic deed, using praise and blame, reward and punishment, encouragement and admonition, as well as habituation, training, and ritualized activities. Yet nature sets limits on what can be accomplished by education and training alone. No matter how much we try to help, the tone-deaf will need more training to learn to carry a tune, the short will be less likely to excel at basketball, the irascible will have trouble restraining their tempers, and the insufficiently smart will remain handicapped for competitive college admissions. If the inborn "equipment" is faulty, or even only normally limited and hence inadequate for realizing some human purposes, it is inviting to think about improving the native powers or the efficacy of their expression and use. For whether we like it or not, certain desired improvements in our children will be possible, if at all, only by improving their native equipment.

Even before the coming of the present age of biotechnology, we have used technological adjuncts to improve upon nature's gifts. We give our children supplementary vitamins, fluoridated toothpaste, and, where necessary, corrective lenses or hearing aids. We even use biological means of improving their limited human capacity to resist disease: we immunize our children against polio, diphtheria, and measles, among other infectious diseases, by injecting them with attenuated viruses and bacteria in the form of vaccines. But the scope of these now-routine kinds of biomedical improvement has until now been limited to restoring or protecting our children's health in a quite straightforward sense. . . .

Improving Native Powers: Genetic Knowledge and Technology

A. An Overview

The possibility of using genetic knowledge and genetic engineering to improve the human race and its individual members has been discussed for many years, especially in the heady decades immediately following Watson and Crick's discovery, in 1953, of the structure of DNA. New life was breathed into old eugenic dreams, which had been temporarily discredited by the Nazi pursuits of a "superior race." As late as the early 1970s, serious scientists talked optimistically about humankind's new opportunity to take the reins of its own evolution, thanks to the predicted confluence of genetic engineering and reproductive technologies. But as scientists have learned just how difficult it is to engineer precise genetic change—even to treat individuals with genetic diseases caused by a simple one-gene mutation—explicit talk about improving the species has largely faded. Instead recent years have seen, in its place, much talk about coming prospects for "designer babies," children born with improved genetic endowments, the result either of careful screening and selecting of embryos carrying desirable genes, or of directed genetic change ("genetic engineering") in gametes or embryos. . . .

BIOTECHNOLOGY AND GENETIC ENGINEERING

Genetic Engineering of Desired Traits ("Fixing Up")

With directed genetic change aimed at producing certain desired improvements, we enter the futuristic realm of "designer babies." Proponents have made this prospect look straightforward, and, on a theory of strict genetic determinism, it is. One would first need to identify all (or enough) of the specific variants of genes whose presence (or absence) correlates with certain desired traits: higher intelligence, better memory, perfect pitch, calmer temperament, sunnier disposition, greater ambitiousness, etc. Once identified, the requisite genes could be isolated, replicated or synthesized, and then inserted into the early embryo (or perhaps into the egg or sperm) in ways that would eventually contribute to the desired phenotypic traits. In the limit, there is talk of babies "made to order," embodying a slew of desirable qualities acquired with such genetic engineering. But in our considered judgment, these dreams of fully designed babies, based on directed genetic change, are for the foreseeable future pure fantasies. There are huge obstacles, both to accurate knowing and to effective doing. One of these obstacles—the reality that these traits are heavily influenced by environment—will not be overcome by better technology.

Most of the traits for which parents might wish to engineer improvements in their children—appearance, intelligence, memory—are most certainly polygenic, that is, traits (or phenotypes) that depend on specific genes or their variants at several, perhaps many, distinct loci. In such cases the relationships and interactions among these genes (and between one's genes and the environment) are certain to be enormously complex. Isolating all the relevant genetic variants, and knowing how to work with them to produce the desired result, will therefore prove immensely difficult. To be sure, not every trait for which parents might wish to select need turn out to be highly polygenic: for example, height, skin color, eye color, or even the genetic contributions to sexual orientation or basic temperament might be heavily influenced by a very few genes. . . . [O]ne mutation in a single gene has been shown to result in enormous increases in the lifespan of flies, worms, and mice, and the same gene has been identified in humans. Yet even here there would be no guarantee that the predisposing genes, even if correctly and safely introduced into the zygote or early embryo, would necessarily express themselves as desired, to yield the sought-for improvement.

Even more of an obstacle to successful genetic engineering is the practical difficulty of inserting genes into embryos (or gametes) in ways that would produce the desired result and *only* the desired result. Getting the genes into the right place in the cell, able to function yet without disturbing regular cellular functions, is an enormously challenging task. Insertion of genes into the host genome can cause abnormalities, either by activating harmful genes or by inactivating useful ones. Recently, for example, children undergoing experimental gene therapy for immune system deficiencies have developed leukemia after retroviral gene transfer into bone marrow stem cells, very likely the result of activation of a cancer-producing gene by the virus used to transfer the therapeutic genes into the cell. And should introduced genes become inserted into inappropriate locations, normal host genes could be inactivated. Moreover, because many genes are pleiotropic—that is, they influence many traits, not just one—even a properly inserted gene introduced to enhance a particular trait would often have multiple effects, not all of them for the better.

Running such risks might be justified in *gene therapy* efforts for already existing individuals, where the genes hold out the only hope of cure for an otherwise deadly disease. But these safety risks will pose formidable obstacles to all interventions in gametes or embryos, especially *nontherapeutic* interventions aimed at producing children who would allegedly be, in one respect or another, "better than well." It is difficult to see how such an intervention could ever be considered ethical, especially since the negative effects might extend to future generations.

As a possible way around the hazards of gene insertion, some researchers have proposed the assembly and injection of artificial chromosomes: the new "better" genes could be packaged in small, manufactured chromosomal elements that, on introduction into cells, would not integrate into any of the normal forty-six human chromosomes. Such artificial chromosomes could, in theory, be introduced into ova or zygotes without fear of causing new mutations. But methods would have to be found to guarantee the synchronized replication and normal segregation of such artificial chromosomes. Otherwise, the package of improved genes, once introduced into the embryo, would not be conserved in all cells after normal mitotic division. Even more dauntingly, any gene introduced on such a chromosome would now be present in three copies (one from mother, one from father, and one on the extra chromosome) instead of the usual two, throwing off the normal balance of gene copies among all the genes. The consequences of such "triploidy" can be deleterious (for example, Down syndrome). All in all, safety and efficacy standards would seem to preclude doing such experiments with human subjects, at least in the United States, for the foreseeable future. It is true that research along these lines might be undertaken in other countries (for example, China), by scientists unconstrained by these considerations, with eventual success in effecting directed genetic change in human embryos. But, at least for the time being, we believe that we may set this prospect safely to the side.

BIOTECHNOLOGY AND GENETIC ENGINEERING

Selecting Embryos for Desired Traits ("Choosing In")

Unlike the prospect for precise genetic engineering through directed genetic change, the possibility of genetic enhancement of children through embryo selection cannot be easily dismissed. This approach, less radical or complete in its power to control, would not introduce new genes but would merely select positively among those that occur naturally. It depends absolutely on IVF, as augmented by the screening of the early embryos for the presence (or absence) of the desired genetic markers, followed by the selective transfer of those embryos that pass muster. This would amount to an "improvement-seeking" extension of the recently developed practice of preimplantation genetic diagnosis (PGD), now in growing use as a way to detect the presence or absence of genetic or chromosomal abnormalities *before* the start of a pregnancy.

As currently practiced, PGD works as follows: Couples at risk for having a child with a chromosomal or genetic disease undertake IVF to permit embryo screening before transfer, obviating the need for later prenatal diagnosis and possible abortion. A dozen or more eggs are fertilized and the embryos are grown to the four-cell or the eight-to-ten-cell stage. One or two of the embryonic cells (blastomeres) are removed for chromosomal analysis and genetic testing. Using a technique called polymerase chain reaction to amplify the tiny amount of DNA in the blastomere, researchers are able to detect the presence of genes responsible for one or more genetic disorders. Only the embryos free of the genetic or chromosomal determinants for the disorders under scrutiny are made eligible for transfer to the woman to initiate a pregnancy. The use of IVF and PGD to move from disease avoidance to baby improvement is conceptually simple, at least in terms of the techniques of screening, and would require no change in the procedure. Indeed, PGD has already been used to serve two goals unrelated to the health of the child-to-be: to pre-select the sex of a child, and to produce a child who could serve as a compatible bone-marrow or umbilical-cord-blood donor for a desperately ill sibling. (In the former case, chromosomal analysis of the blastomere identifies the embryo's sex; in the latter case, genetic analysis identifies which embryos are immunocompatible with the needy recipient.) It is certainly likely that blastomere testing can be adapted to look for specific genetic variants at *any* locus of the human genome. And even without knowing the precise function of specific genes, statistical correlation of the presence of certain genetic variants with certain phenotypic traits (say, with an increase in IQ points or with perfect pitch) could lead to testing for these genetic variants, with selection following on this basis. As Dr. Francis Collins, director of the National Human Genome Research Institute, noted in his presentation to the Council, the time may soon arrive in which PGD

166

is practiced for the purpose of selecting embryos with desired genotypes, even in the absence of elevated risk of particular genetic disorders. Dr. Yury Verlinsky, director of the Reproductive Genetics Institute in Chicago, has recently predicted that soon "there will be no IVF without PGD." Over the years, more and more traits will presumably become identifiable with the aid of PGD, including desirable genetic markers for intelligence, musicality, and so on, as well as undesirable markers for obesity, nearsightedness, color-blindness, etc.

Yet, as Dr. Collins also pointed out to the Council, there are numerous practical difficulties with this scenario. For one thing, neither of the parents may carry the genetic variant they are most interested in selecting for. Also, selecting for highly polygenic traits would require screening a large number of embryos in order to find one that had the desirable complement. With only a dozen or so embryos to choose from, it will not be possible to optimize for the many necessary variants.

The practice of PGD and selective transfer is still quite new, and fewer than 10,000 children have been born with its aid. How likely or widespread such a practice might become is difficult to predict. As we have already indicated, a number of practical issues would need to be addressed before PGD could be extended to permit selection of desirable traits beyond the absence of genetic disorders. First are questions of possible harm caused by removing blastomeres for testing (up to a sixth or even a quarter of the embryo's cells are taken). Although current evidence (from limited practice) suggests that the procedure inflicts neither any immediately visible harm on the early embryos, nor any obvious harm on the child that results, more attention to long-term risks to the child born following PGD is needed before many people would consider using it for "improvement" purposes only. Because many of the desirable human phenotypic traits are very likely polygenic, the contribution of any single gene identifiable by blastomere testing is likely to be small, and the likelihood of finding all the "desired" genetic variants in a single embryo is exponentially smaller still. Testing for multiple genetic variants using the DNA from a single blastomere is likely to be limited—for a time—by the quantities of DNA available, the sensitivity of the genetic tests, and the ability to perform multiple tests on the same sample. But it seems only a matter of time before techniques are perfected that will permit simultaneous screening of IVF embryos for multiple genetic variants. And should some of the "desirable" genes come grouped in clusters, selection for at least some desired traits might well be possible.

Finally, even if PGD could be used successfully to select an embryo with a number of desirable genetic variants, there is simply no guarantee that the child born after this procedure would grow up with the desired traits. The interplay of nature and nurture (genes and environment) in human development is too complex and too little understood to make such results predictable. Given that IVF combined with PGD is an inconvenient and expensive alternative to normal procreation, and given that success is doubtful at best, the purely elective use of this procedure seems unlikely to become widespread in the foreseeable future. As Professor Steven Pinker put it, in his presentation to the Council:

> The choice that parents would face in a hypothetical future in which even genetic enhancement were possible would not be the one that's popularly portrayed, namely, "Would you opt for a procedure that would give you a happier, more talented child?" When you put it like that, well, who would say no to that question? More realistically, the question that parents would face would be something like this: "Would you opt for a traumatic and expensive procedure that might give you a very slightly happier and more talented child, might give you a less happy, less talented child, might give you a deformed child, and probably would do nothing?"[1]

Nevertheless, we think it would be imprudent to ignore completely this approach to "better children." More and more people are turning to assisted reproduction technologies (ART): in parts of western Europe, roughly five percent of all births involve ART; in the United States, it is roughly one percent and climbing, as the average maternal age of childbirth keeps rising and family size keeps declining. More and more people are using IVF not merely to overcome infertility but to screen and select embryos free of certain genetic defects. Women who plan to delay childbearing are being encouraged to consider early removal and cryopreservation of their own youthful ovarian tissue, to be reintroduced into their bodies at sites easily accessible for egg harvesting when they decide to have children. Other novel methods of obtaining supplies of eggs for IVF—possibly including deriving them in bulk from stem cells—would make the procedure less burdensome, and would, in theory, permit the creation of a large enough population of embryos to make screening for polygenic traits feasible.

The anticipated vast extension of genetic screening will make many more couples aware of the risks they run in natural reproduction, and they may choose to turn to IVF to reduce them—especially if obtaining eggs became easy. Once more and more couples start screening embryos for disease-related concerns, and once scientists have identified those genes that correlate with various admirable traits, the anticipated expansion of improved and more precise screening techniques might enable users of

IVF to screen for "desirable genes" as well. People already using PGD to screen for disease markers might seek information also about other traits, as they have with sex or histocompatibility. And if, once screening becomes automated, its cost comes down, or if society decides to reimburse for PGD (regarding it as less expensive than the care of genetically diseased children), the use of this approach toward "better children" might well become the practice of at least a significant minority. Under these circumstances, should genuine and significant improvements be achieved for a few highly desired attributes (say, in maximum lifespan), one can easily imagine that there would be an increased demand for the practice, inconvenient or not. In the meantime, we would do well to consider the ethical implications not only of such future prospects but also of our current practices that make use of genetic knowledge.

Source: President's Council on Bioethics. "Better Children" from *Beyond Therapy: Biotechnology and the Pursuit of Happiness.* Washington, D.C.: 2003. Available online. URL: http://bioethics.gov/reports/beyondtherapy/index.html. Accessed May 28, 2009.

[1] Pinker, S. "Human Nature and Its Future." Presentation at the March 2003 meeting of the President's Council on Bioethics. Washington, D.C. Transcript available on the Council's Web site at www.bioethics.gov.

5

International Documents

This chapter begins with a selection of documents that laid the foundation for modern bioethics, beginning with an excerpt from Mary Shelley's cautionary tale, *Frankenstein, or the Modern Prometheus* (1818), which is as relevant today as it was when it was written. An excerpt from Pasteur's *Germ Theory* (1878) presents the dawn of a new era in medicine, in which it becomes evident that microscopic living organisms play a crucial role in every aspect of biology, for good and for bad. Galton and Chesterton present opposite views of eugenics, and the Nuremberg Code (1949) stands as the final arbiter in the bitter eugenics debate that erupted after Nazi atrocities during World War II came to light.

Also included are documents relating to the biotechnology policies of Japan, India, Germany, and South Africa. Selections were chosen for illustrating how government officials seek to make their countries leaders in biotechnology in order to ensure economic growth as well as to improve the health and well-being of their citizens.

The documents are arranged in chronological order within each section. Documents that have been excerpted are identified as such; all others are reproduced in full.

Mary Wollstonecraft Shelley: *Frankenstein, or the Modern Prometheus* (1818) (excerpt)

Shelley wrote this novel for many reasons, one of which was to caution against humans' desire to create life in their own image. Dr. Victor Frankenstein, driven by questionable motives and an unstoppable ego, creates a living being (called "the Creature") out of body parts stolen from cadavers. The Creature resembles a monster; he is large, strong, ugly, and Dr. Frankenstein is repulsed at the sight of him. Over the years, Frankenstein has come to mean something of unnatural origin, and Shelley's novel has become a warning to researchers and scientists who tinker with the building blocks of life. The story suggests that motives and ethics are strong factors in determining whether or not an experiment should be conducted. What if the result of the experiment cannot be contained? The novel's influence is evident in the term Frankenfood *as applied to genetically modified food and in ethical concerns about recombinant DNA and cloning. In this excerpt, Victor Frankenstein, a medical student, writes of being inspired by his professor, M. Waldman.*

Partly from curiosity and partly from idleness, I went into the lecturing room, which M. Waldman entered shortly after. This professor was very unlike his colleague. He appeared about fifty years of age, but with an aspect expressive of the greatest benevolence; a few grey hairs covered his temples, but those at the back of his head were nearly black. His person was short but remarkably erect and his voice the sweetest I had ever heard. He began his lecture by a recapitulation of the history of chemistry and the various improvements made by different men of learning, pronouncing with fervour the names of the most distinguished discoverers. He then took a cursory view of the present state of the science and explained many of its elementary terms. After having made a few preparatory experiments, he concluded with a panegyric upon modern chemistry, the terms of which I shall never forget: "The ancient teachers of this science," said he, "promised impossibilities and performed nothing. The modern masters promise very little; they know that metals cannot be transmuted and that the elixir of life is a chimera but these philosophers, whose hands seem only made to dabble in dirt, and their eyes to pore over the microscope or crucible, have indeed performed miracles. They penetrate into the recesses of nature and show how she works in her hiding-places. They ascend into the heavens; they have discovered how the blood circulates, and the nature of the air we breathe. They have acquired new and almost unlimited powers; they can command the thunders of heaven, mimic the earthquake, and even mock the invisible world with its own shadows."

Such were the professor's words—rather let me say such the words of the fate—enounced to destroy me. As he went on I felt as if my soul were grappling with a palpable enemy; one by one the various keys were touched which formed the mechanism of my being; chord after chord was sounded, and soon my mind was filled with one thought, one conception, one purpose. So much has been done, exclaimed the soul of Frankenstein—more, far more, will I achieve; treading in the steps already marked, I will pioneer a new way, explore unknown powers, and unfold to the world the deepest mysteries of creation.

BIOTECHNOLOGY AND GENETIC ENGINEERING

I closed not my eyes that night. My internal being was in a state of insurrection and turmoil; I felt that order would thence arise, but I had no power to produce it. By degrees, after the morning's dawn, sleep came. I awoke, and my yesternight's thoughts were as a dream. There only remained a resolution to return to my ancient studies and to devote myself to a science for which I believed myself to possess a natural talent. On the same day I paid M. Waldman a visit. His manners in private were even more mild and attractive than in public, for there was a certain dignity in his mien during his lecture which in his own house was replaced by the greatest affability and kindness. I gave him pretty nearly the same account of my former pursuits as I had given to his fellow professor. He heard with attention the little narration concerning my studies and smiled at the names of Cornelius Agrippa and Paracelsus, but without the contempt that M. Krempe had exhibited. He said that "These were men to whose indefatigable zeal modern philosophers were indebted for most of the foundations of their knowledge. They had left to us, as an easier task, to give new names and arrange in connected classifications the facts which they in a great degree had been the instruments of bringing to light. The labours of men of genius, however erroneously directed, scarcely ever fail in ultimately turning to the solid advantage of mankind." I listened to his statement, which was delivered without any presumption or affectation, and then added that his lecture had removed my prejudices against modern chemists; I expressed myself in measured terms, with the modesty and deference due from a youth to his instructor, without letting escape (inexperience in life would have made me ashamed) any of the enthusiasm which stimulated my intended labours. I requested his advice concerning the books I ought to procure.

"I am happy," said M. Waldman, "to have gained a disciple; and if your application equals your ability, I have no doubt of your success. Chemistry is that branch of natural philosophy in which the greatest improvements have been and may be made; it is on that account that I have made it my peculiar study; but at the same time, I have not neglected the other branches of science. A man would make but a very sorry chemist if he attended to that department of human knowledge alone. If your wish is to become really a man of science and not merely a petty experimentalist, I should advise you to apply to every branch of natural philosophy, including mathematics." He then took me into his laboratory and explained to me the uses of his various machines, instructing me as to what I ought to procure and promising me the use of his own when I should have advanced far enough in the science not to derange their mechanism. He also gave me the list of books which I had requested, and I took my leave. Thus ended a day memorable to me; it decided my future destiny. . . .

One of the phenomena which had peculiarly attracted my attention was the structure of the human frame, and, indeed, any animal endued with life. Whence, I often asked myself, did the principle of life proceed? It was a bold question, and one which has ever been considered as a mystery; yet with how many things are we upon the brink of becoming acquainted, if cowardice or carelessness did not restrain our inquiries. I revolved these circumstances in my mind and determined thenceforth to apply myself more particularly to those branches of natural philosophy which relate to physiology. Unless I had been animated by an almost supernatural enthusiasm, my application to this study would have been irksome and almost intolerable. To examine the causes of life, we must first have recourse to death. I became acquainted with the science of anatomy, but this was not sufficient; I must also observe the natural decay and corruption of the human body. In my education my father had taken the greatest precautions that my mind should be impressed with no supernatural horrors. I do not ever remember to have trembled at a tale of superstition or to have feared the apparition of a spirit. Darkness had no effect upon my fancy, and a churchyard was to me merely the receptacle of bodies deprived of life, which, from being the seat of beauty and strength, had become food for the worm. Now I was led to examine the cause and progress of this decay and forced to spend days and nights in vaults and charnel-houses. My attention was fixed upon every object the most insupportable to the delicacy of the human feelings. I saw how the fine form of man was degraded and wasted; I beheld the corruption of death succeed to the blooming cheek of life; I saw how the worm inherited the wonders of the eye and brain. I paused, examining and analysing all the minutiae of causation, as exemplified in the change from life to death, and death to life, until from the midst of this darkness a sudden light broke in upon me—a light so brilliant and wondrous, yet so simple, that while I became dizzy with the immensity of the prospect which it illustrated, I was surprised that among so many men of genius who had directed their inquiries towards the same science, that I alone should be reserved to discover so astonishing a secret. . . .

When I found so astonishing a power placed within my hands, I hesitated a long time concerning the manner in which I should employ it. Although I possessed the capacity of bestowing animation, yet to prepare a frame for the reception of it, with all its intricacies of fibres, muscles, and veins, still remained a work of inconceivable difficulty and labour. I doubted at first whether I should attempt the creation of a being like myself, or one of simpler organization; but my imagination was too much exalted by my first success to permit me to doubt of my ability to give life to an animal as complex and wonderful as man. The materials at present within my command hardly appeared adequate to so arduous an undertaking, but I doubted not that I should ultimately succeed. I prepared myself for a multitude of reverses; my operations might be incessantly baffled, and at last my work be imperfect, yet when I considered the improvement which every day takes place in science and mechanics, I was encouraged to hope my present attempts would at least lay the foundations of future success. Nor could I consider the magnitude and complexity of my plan as any argument of its impracticability. It was with these feelings that I began the creation of a human being. As the minuteness of the parts formed a great hindrance to my speed, I resolved, contrary to my first intention, to make the being of a gigantic stature, that is to say, about eight feet in height, and proportionably large. After having formed this determination and having spent some months in successfully collecting and arranging my materials, I began.

No one can conceive the variety of feelings which bore me onwards, like a hurricane, in the first enthusiasm of success. Life and death appeared to me ideal bounds, which I should first break through, and pour a torrent of light into our dark world. A new species would bless me as its creator and source; many happy and excellent natures would owe their being to me. No father could claim the gratitude of his child so completely as I should deserve theirs. Pursuing these reflections, I thought that if I could bestow animation upon lifeless matter, I might in process of time (although I now found it impossible) renew life where death had apparently devoted the body to corruption.

These thoughts supported my spirits, while I pursued my undertaking with unremitting ardour. My cheek had grown pale with study, and my person had become emaciated with confinement. Sometimes, on the very brink of certainty, I failed; yet still I clung to the hope which the next day or the next hour might realize. One secret which I alone possessed was the hope to which I had dedicated myself; and the moon gazed on my midnight labours, while, with unrelaxed and breathless eagerness, I pursued nature to her hiding-places. Who shall conceive the horrors of my secret toil as I dabbled among the unhallowed damps of the grave or tortured the living animal to animate the lifeless clay? My limbs now tremble, and my eyes swim with the remembrance; but then a resistless and almost frantic impulse urged me forward; I seemed to have lost all soul or sensation but for this one pursuit. It was indeed but a passing trance, that only made me feel with renewed acuteness so soon as, the unnatural stimulus ceasing to operate, I had returned to my old habits. I collected bones from charnel-houses and disturbed, with profane fingers, the tremendous secrets of the human frame. In a solitary chamber, or rather cell, at the top of the house, and separated from all the other apartments by a gallery and staircase, I kept my workshop of filthy creation; my eyeballs were starting from their sockets in attending to the details of my employment. The dissecting room and the slaughter-house furnished many of my materials; and often did my human nature turn with loathing from my occupation, whilst, still urged on by an eagerness which perpetually increased, I brought my work near to a conclusion. . . .

It was on a dreary night of November that I beheld the accomplishment of my toils. With an anxiety that almost amounted to agony, I collected the instruments of life around me, that I might infuse a spark of being into the lifeless thing that lay at my feet. It was already one in the morning; the rain pattered dismally against the panes, and my candle was nearly burnt out, when, by the glimmer of the half-extinguished light, I saw the dull yellow eye of the creature open; it breathed hard, and a convulsive motion agitated its limbs.

How can I describe my emotions at this catastrophe, or how delineate the wretch whom with such infinite pains and care I had endeavoured to form? His limbs were in proportion, and I had selected his features as beautiful. Beautiful! Great God! His yellow skin scarcely covered the work of muscles and arteries beneath; his hair was of a lustrous black, and flowing; his teeth of a pearly whiteness; but these luxuriances only formed a more horrid contrast with his watery eyes, that seemed almost of the same colour as the dun-white sockets in which they were set, his shrivelled complexion and straight black lips.

The different accidents of life are not so changeable as the feelings of human nature. I had worked hard for nearly two years, for the sole purpose of infusing life into an inanimate body. For this I had deprived myself of rest and health. I had desired it with an ardour that far exceeded moderation; but now that I had finished, the beauty of the dream vanished, and breathless horror and disgust filled my heart. Unable to endure the aspect of the being I had created, I rushed out of the room and continued a long time traversing my bed-chamber, unable to compose my mind to sleep. At length lassitude succeeded to the tumult I had before endured, and I threw myself on the bed in my clothes, endeavouring to seek a few moments of forgetfulness. But it was in vain; I slept, indeed, but I was disturbed by the wildest dreams. I thought I saw Elizabeth, in the bloom of health, walking in the streets of Ingolstadt. Delighted and surprised, I embraced her, but as I imprinted the first kiss on her lips, they became livid with the hue of death; her features appeared to change, and I thought that I held the corpse of my dead mother in my arms; a shroud enveloped her form, and I saw the grave-worms crawling in the folds of the flannel. I started from my sleep with horror; a cold dew covered my forehead, my teeth chattered, and every limb became convulsed; when, by the dim and yellow light of the moon, as it forced its way through the window shutters, I beheld the wretch—the miserable monster whom I had created. He held up the curtain of the bed; and his eyes, if eyes they may be called, were fixed on me. His jaws opened, and he muttered some inarticulate sounds, while a grin wrinkled his cheeks. He might have spoken, but I did not hear; one hand was stretched out, seemingly to detain me, but I escaped and rushed downstairs. I took refuge in the courtyard belonging to the house which I inhabited, where I remained during the rest of the night, walking up and down in the greatest agitation, listening attentively, catching and fearing each sound as if it were to announce the approach of the demoniacal corpse to which I had so miserably given life.

Oh! No mortal could support the horror of that countenance. A mummy again endued with animation could not be so hideous as that wretch. I had gazed on him while unfinished; he was ugly then, but when those muscles and joints were rendered capable of motion, it became a thing such as even Dante could not have conceived.

Source: Mary Wollstonecraft Shelley. *Frankenstein, or the Modern Prometheus.* London: Lackington, Hughes, Harding, Mavor & Jones, 1818. Available online. URL: http://www.gutenberg.org/etext/84. Accessed June 1, 2009.

Louis Pasteur: The Germ Theory and Its Applications to Medicine and Surgery (April 29, 1878)

This paper was initially read before the French Academy of Sciences and was published in French in Comptes Rendus de l'Academie des Sciences. *Pasteur credits two colleagues, Jourbert and Chamberland, as contributors. This translation is by H. C. Ernst, who includes several notes in the text. In the paper, Pasteur determines that anthrax is a bacterial disease and shows how germ theory applies to septicemia (his term "vibrio" is synonymous with bacteria). In doing so, Pasteur solidified his ideas about the role of microscopic organisms in infectious diseases. The idea of germ theory proved pivotal in the medical advances of the next several generations, and Pasteur himself continues to be heralded as one of the intellectual giants in the history of biology.*

The Sciences gain by mutual support. When, as the result of my first communications on the fermentations in 1857–1858, it appeared that the ferments, properly so-called, are living beings, that the germs of microscopic organisms abound in the surface of all objects, in the air and in water; that the theory of spontaneous generation is chimerical; that wines, beer, vinegar, the blood, urine and all the fluids of the body undergo none of their usual changes in pure air, both Medicine and Surgery received fresh stimulation. A French physician, Dr. Davaine, was fortunate in making the first application of these principles to Medicine, in 1863.

Our researches of last year, left the etiology of the putrid disease, or septicemia, in a much less advanced condition than that of anthrax. We had demonstrated the probability that septicemia depends upon the presence and growth of a microscopic body, but the absolute proof of this important conclusion was not reached. To demonstrate experimentally that a microscopic organism actually is the cause of a disease and the agent of contagion, I know no other way, in the present state of Science, than to subject the microbe (the new and happy term introduced by M. Sedillot) to the method of cultivation out of the body. It may be noted that in twelve successive cultures, each one of only ten cubic centimeters volume, the original drop will be diluted as if placed in a volume of fluid equal to the total volume of the earth. It is just this form of test to which M. Joubert and I subjected the anthrax bacteridium. [Note: In making the translation, it seems wiser to adhere to Pasteur's nomenclature. *Bacillus anthracis* would be the term employed to-day. —Translator] Having cultivated it a great number of times in a sterile fluid, each culture being started with a minute drop from the preceding, we then demonstrated that the product of the last culture was capable of further development and of acting in the animal tissues by producing anthrax with all its symptoms. Such is—as we believe—the indisputable proof that ANTHRAX IS A BACTERIAL DISEASE.

Our researches concerning the septic vibrio had not so far been convincing, and it was to fill up this gap that we resumed our experiments. To this end, we attempted the cultivation of the septic vibrio from an animal dead of septicemia. It is worth noting that all of our first experiments failed, despite the variety of culture media we employed—urine, beer yeast water, meat water, etc. Our culture media were not sterile, but we found—most commonly—a microscopic organism showing no relationship to the septic vibrio, and presenting the form, common enough elsewhere, of chains of extremely minute spherical granules possessed of no virulence whatever. [Note: It is quite possible that Pasteur was here dealing with certain septicemic streptococci that are now know to lose their virulence with extreme rapidity under artificial cultivation. —Translator] This was an impurity, introduced, unknown to us, at the same time as the septic vibrio; and the germ undoubtedly passed from the intestines—always inflamed and distended in septicemic animals—into the abdominal fluids from which we took our original cultures of the septic vibrio. If this explanation of the contamination of our cultures was correct, we ought to find a pure culture of the septic vibrio in the heart's blood of an animal recently dead of septicemia. This was what happened, but a new difficulty presented itself; all our cultures remained sterile. Furthermore this sterility was accompanied by loss in the culture media of (the original) virulence.

It occurred to us that the septic vibrio might be an obligatory anaerobe and that the sterility of our inoculated culture fluids might be due to the destruction of the septic vibrio by the atmospheric oxygen dissolved in the fluids. The Academy may remember that I have previously demonstrated facts of this nature in regard to the vibrio of butyric fermentation, which not only lives without air but is killed by the air.

It was necessary therefore to attempt to cultivate the septic vibrio either in a vacuum or in the presence of inert gases—such as carbonic acid. Results justified our attempt; the septic vibrio grew easily in a complete vacuum, and no less easily in the presence of pure carbonic acid.

These results have a necessary corollary. If a fluid containing septic vibrios be exposed to pure air, the vibrios should be killed and all virulence should disappear. This is actually the case. If some drops of septic serum be spread horizontally in a tube and in a very thin layer, the fluid will become absolutely harmless in less than half a day, even if at first it was so virulent as to produce death upon the inoculation of the smallest portion of a drop.

Furthermore all the vibrios, which crowded the liquid as motile threads, are destroyed and disappear. After the action of the air, only fine amorphous granules can be found, unfit for culture as well as for the transmission of any disease whatever. It might be said that the air burned the vibrios.

If it is a terrifying thought that life is at the mercy of the multiplication of these minute bodies, it is a consoling hope that Science will not always remain powerless before such enemies, since for example at the very beginning of the study we find that simple exposure to air is sufficient at times to destroy them.

But, if oxygen destroys the vibrios, how can septicemia exist, since atmospheric air is present everywhere? How can such facts be brought in accord with the germ theory? How can blood, exposed to air, become septic through the dust the air contains?

All things are hidden, obscure and debatable if the cause of the phenomena be unknown, but everything is clear if this cause be known. What we have just said is true only of a septic fluid containing adult vibrios, in active development by fission: conditions are different when the vibrios are transformed into their germs, [Note: By the terms *germ* and *germ corpuscles*, Pasteur undoubtedly means "spores," but the change is not made, in accordance with note above. —Translator] that is into the glistening corpuscles first described and figured in my studies on silk-worm disease, in dealing with worms dead of the disease called "flacherie." Only the adult vibrios disappear, burn up, and lose their virulence in contact with air: the germ corpuscles, under these conditions, remain always ready for new cultures, and for new inoculations.

All this however does not do away with the difficulty of understanding how septic germs can exist on the surface of objects, floating in the air and in water.

Where can these corpuscles originate? Nothing is easier than the production of these germs, in spite of the presence of air in contact with septic fluids.

If abdominal serous exudate containing septic vibrios actively growing by fission be exposed to the air, as we suggested above, but with the precaution of giving a substantial thickness to the layer, even if only one centimeter be used, this curious phenomenon will appear in a few hours. The oxygen is absorbed in the upper layers of the fluid—as is indicated by the change of color. Here the vibrios are dead and disappear. In the deeper layers, on the other hand, towards the bottom of this centimeter of septic fluid we suppose to be under observation, the vibrios continue to multiply by fission—protected from the action of oxygen by those that have perished above them: little by little they pass over to the condition of germ corpuscles with the gradual disappearance of the thread forms. So that instead of moving threads of varying length, sometimes greater than the field of the microscope, there is to be seen only a number of glittering points, lying free or surrounded by a scarcely perceptible amorphous mass. [Note: In our note of July 16th, 1877, it is stated that the septic vibrio is not destroyed by the oxygen of the air nor by oxygen at high tension, but that under these conditions it is transformed into germ-corpuscles. This is, however, an incorrect interpretation of facts. The vibrio is destroyed by oxygen, and it is only where it is in a thick layer that it is transformed to germ-corpuscles in the presence of oxygen and that its virulence is preserved. —Translator] Thus is formed, containing the latent germ life, no longer in danger from the destructive action of oxygen, thus, I repeat, is formed the septic dust, and we are able to understand what has before seemed so obscure; we can see how putrescible fluids can be inoculated by the dust of the air, and how it is that putrid diseases are permanent in the world.

The Academy will permit me, before leaving these interesting results, to refer to one of their main theoretical consequences. At the very beginning of these researches, for they reveal an entirely new field, what must be insistently demanded? The absolute proof that there actually exist transmissible, contagious, infectious diseases of which the cause lies essentially and solely in the presence of microscopic organisms. The proof that for at least some diseases, the conception of spontaneous virulence must be forever abandoned—as well as the idea of contagion and an infectious element suddenly originating in the bodies of men or animals and able to originate diseases which propagate themselves under identical forms: and all of those opinions fatal to medical progress, which have given rise to the gratuitous hypotheses of spontaneous generation, of albuminoid ferments, of hemiorganisms, of archebiosis, and many other conceptions without the least basis in observation. What is to be sought for in this instance is the proof that along with our vibrio there does not exist an independent virulence belonging to the surrounding fluids or solids, in short that the vibrio is not merely an epiphenomenon of the disease of which it is the obligatory accompaniment. What then do we see, in the results that I have just brought out? A septic fluid, taken at the moment that the vibrios are not yet changed into germs, loses its virulence completely upon simple exposure to the air, but preserves this virulence, although exposed to air on the simple condition of being in a thick layer for some hours. In the first case, the virulence once lost by exposure to air, the liquid is incapable of taking it on again upon cultivation: but, in the second case, it preserves its virulence and can propagate, even after exposure to air. It is impossible, then, to assert that there is a separate virulent substance, either fluid or solid, existing, apart from the adult vibrio or its germ. Nor can it be supposed that there is a virus which loses its virulence at the moment that the adult vibrio dies; for such a substance should also lose its virulence when the vibrios, changed to germs, are exposed to the air. Since the virulence persists under these conditions it can only be due to the germ corpuscles—the only thing present. There is only one possible hypothesis as to the existence of a virus in solution, and that is that such a substance, which was present in our experiment in nonfatal amounts, should be continuously furnished by the vibrio itself, during its growth in the body of the living animal. But it is of little importance since the hypothesis supposes the forming and necessary existence of the vibrio.

Source: Louis Pasteur. "The Germ Theory and Its Applications to Medicine and Surgery." *Comptes Rendus de l'Academie des Sciences* 86 (1878): 1,037–1,043. Translated by H. C. Ernst. Available online. URL: http:// biotech.law.lsu.edu/cphl/history/articles/pasteur.htm#paperII. Accessed June 1, 2009.

Francis Galton: "Eugenics: Its Definition, Scope and Aims" (1904)

Galton was a British renaissance man. He was a cousin of Charles Darwin and a noted scientist, mathematician, meteorologist, anthropologist, and psychologist. He was a prolific author, and in his later years he pioneered the study of identifying individuals by their fingerprints. He was fascinated by the question of nature v. nurture, and his observations led to his concept of eugen-

ics. The following was originally delivered as a speech at London University in 1904 and was later published in his collection Essays in Eugenics. *The essay outlines Galton's ideas on eugenics; basically, he advocates the reproduction of individuals who, by parentage, are likely to be healthy, free of disease, intelligent, and able members of society.*

Eugenics is the science which deals with all influences that improve the inborn qualities of a race; also with those that develop them to the utmost advantage. The improvement of the inborn qualities, or stock, of some one human population, will alone be discussed here.

What is meant by improvement? What by the syllable *Eu* in Eugenics, whose English equivalent is *good*? There is considerable difference between goodness in the several qualities and in that of the character as a whole. The character depends largely on the proportion between qualities whose balance may be much influenced by education. We must therefore leave morals as far as possible out of the discussion, not entangling ourselves with the almost hopeless difficulties they raise as to whether a character as a whole is good or bad. Moreover, the goodness or badness of character is not absolute, but relative to the current form of civilisation. A fable will best explain what is meant. Let the scene be the Zoological Gardens in the quiet hours of the night, and suppose that, as in old fables, the animals are able to converse, and that some very wise creature who had easy access to all the cages, say a philosophic sparrow or rat, was engaged in collecting the opinions of all sorts of animals with a view of elaborating a system of absolute morality. It is needless to enlarge on the contrariety of ideals between the beasts that prey and those they prey upon, between those of the animals that have to work hard for their food and the sedentary parasites that cling to their bodies and suck their blood, and so forth. A large number of suffrages in favour of maternal affection would be obtained, but most species of fish would repudiate it, while among the voices of birds would be heard the musical protest of the cuckoo. Though no agreement could be reached as to absolute morality, the essentials of Eugenics may be easily defined. All creatures would agree that it was better to be healthy than sick, vigorous than weak, well fitted than ill-fitted for their part in life. In short that it was better to be good rather than bad specimens of their kind, whatever that kind might be. So with men. There are a vast number of conflicting ideals of alternative characters, of incompatible civilisations; but all are wanted to give fulness and interest to life. Society would be very dull if every man resembled the highly estimable Marcus Aurelius or Adam Bede. The aim of Eugenics is to represent each class or sect by its best specimens; that done, to leave them to work out their common civilisation in their own way.

BIOTECHNOLOGY AND GENETIC ENGINEERING

A considerable list of qualities can be easily compiled that nearly every one except "cranks" would take into account when picking out the best specimens of his class. It would include health, energy, ability, manliness and courteous disposition. Recollect that the natural differences between dogs are highly marked in all these respects, and that men are quite as variable by nature as other animals in their respective species. Special aptitudes would be assessed highly by those who possessed them, as the artistic faculties by artists, fearlessness of inquiry and veracity by scientists, religious absorption by mystics, and so on. There would be self-sacrificers, self-tormentors and other exceptional idealists, but the representatives of these would be better members of a community than the body of their electors. They would have more of those qualities that are needed in a State, more vigour, more ability, and more consistency of purpose. The community might be trusted to refuse representatives of criminals, and of others whom it rates as undesirable.

Let us for a moment suppose that the practice of Eugenics should hereafter raise the average quality of our nation to that of its better moiety at the present day and consider the gain. The general tone of domestic, social and political life would be higher. The race as a whole would be less foolish, less frivolous, less excitable and politically more provident than now. Its demagogues who "played to the gallery" would play to a more sensible gallery than at present. We should be better fitted to fulfil our vast imperial opportunities. Lastly, men of an order of ability which is now very rare, would become more frequent, because the level out of which they rose would itself have risen.

The aim of Eugenics is to bring as many influences as can be reasonably employed, to cause the useful classes in the community to contribute *more* than their proportion to the next generation.

The course of procedure that lies within the functions of a learned and active Society such as the Sociological may become, would be somewhat as follows:—

1. Dissemination of a knowledge of the laws of heredity so far as they are surely known, and promotion of their farther study. Few seem to be aware how greatly the knowledge of what may be termed the *actuarial* side of heredity has advanced in recent years. The *average* closeness of kinship in each degree now admits of exact definition and of being treated mathematically, like birth and death-rates, and the other topics with which actuaries are concerned.

2. Historical inquiry into the rates with which the various classes of society (classified according to civic usefulness) have contributed to the population

at various times, in ancient and modern nations. There is strong reason for believing that national rise and decline is closely connected with this influence. It seems to be the tendency of high civilisation to check fertility in the upper classes, through numerous causes, some of which are well known, others are inferred, and others again are wholly obscure. The latter class are apparently analogous to those which bar the fertility of most species of wild animals in zoological gardens. Out of the hundreds and thousands of species that have been tamed, very few indeed are fertile when their liberty is restricted and their struggles for livelihood are abolished; those which are so and are otherwise useful to man becoming domesticated. There is perhaps some connection between this obscure action and the disappearance of most savage races when brought into contact with high civilisation, though there are other and well-known concomitant causes. But while most barbarous races disappear, some, like the negro, do not. It may therefore be expected that types of our race will be found to exist which can be highly civilised without losing fertility; nay, they may become more fertile under artificial conditions, as is the case with many domestic animals.

3. Systematic collection of facts showing the circumstances under which large and thriving families have most frequently originated; in other words, the *conditions* of Eugenics. The names of the thriving families in England have yet to be learnt, and the conditions under which they have arisen. We cannot hope to make much advance in the science of Eugenics without a careful study of facts that are now accessible with difficulty, if at all. The definition of a thriving family, such as will pass muster for the moment at least is one in which the children have gained distinctly superior positions to those who were their class-mates in early life. Families may be considered "large" that contain not less than three adult male children. It would be no great burden to a Society including many members who had Eugenics at heart, to initiate and to preserve a large collection of such records for the use of statistical students. The committee charged with the task would have to consider very carefully the form of their circular and the persons entrusted to distribute it. The circular should be simple, and as brief as possible, consistent with asking all questions that are likely to be answered truly, and which would be important to the inquiry. They should ask, at least in the first instance, only for as much information as could be easily, and would be readily, supplied by any member of the family appealed to. The point to be ascertained is the *status* of the two parents at the time of their marriage, whence its more or less eugenic character might have been predicted, if the larger knowledge that we now hope to obtain had then existed. Some account would, of course, be wanted of their race, profession, and residence; also of

their own respective parentages, and of their brothers and sisters. Finally, the reasons would be required why the children deserved to be entitled a "thriving" family, to distinguish worthy from unworthy success. This manuscript collection might hereafter develop into a "golden book" of thriving families. The Chinese, whose customs have often much sound sense, make their honours retrospective. We might learn from them to show that respect to the parents of noteworthy children, which the contributors of such valuable assets to the national wealth richly deserve. The act of systematically collecting records of thriving families would have the further advantage of familiarising the public with the fact that Eugenics had at length become a subject of serious scientific study by an energetic Society.

4. Influences affecting Marriage. The remarks of Lord Bacon in his essay on Death may appropriately be quoted here. He says with the view of minimising its terrors:

> "There is no passion in the mind of men so weak but it mates and masters the fear of death. Revenge triumphs over death; love slights it; honour aspireth to it; grief flyeth to it; fear pre-occupateth it."

Exactly the same kind of considerations apply to marriage. The passion of love seems so overpowering that it may be thought folly to try to direct its course. But plain facts do not confirm this view. Social influences of all kinds have immense power in the end, and they are very various. If unsuitable marriages from the Eugenic point of view were banned socially, or even regarded with the unreasonable disfavour which some attach to cousin-marriages, very few would be made. The multitude of marriage restrictions that have proved prohibitive among uncivilised people would require a volume to describe.

5. Persistence in setting forth the national importance of Eugenics. There are three stages to be passed through. *Firstly* it must be made familiar as an academic question, until its exact importance has been understood and accepted as a fact; *Secondly* it must be recognised as a subject whose practical development deserves serious consideration; and *Thirdly* it must be introduced into the national conscience, like a new religion. It has, indeed, strong claims to become an orthodox religious tenet of the future, for Eugenics co-operates with the workings of Nature by securing that humanity shall be represented by the fittest races. What Nature does blindly, slowly, and ruthlessly, man may do providently, quickly, and kindly. As it lies within his power, so it becomes his duty to work in that direction; just as it is his duty to succour neighbours

who suffer misfortune. The improvement of our stock seems to me one of the highest objects that we can reasonably attempt. We are ignorant of the ultimate destinies of humanity, but feel perfectly sure that it is as noble a work to raise its level in the sense already explained, as it would be disgraceful to abase it. I see no impossibility in Eugenics becoming a religious dogma among mankind, but its details must first be worked out sedulously in the study. Over-zeal leading to hasty action would do harm, by holding out expectations of a near golden age, which will certainly be falsified and cause the science to be discredited. The first and main point is to secure the general intellectual acceptance of Eugenics as a hopeful and most important study. Then let its principles work into the heart of the nation, who will gradually give practical effect to them in ways that we may not wholly foresee.

Source: Francis Galton. "Eugenics: Its Definition, Scope and Aims." In *Essays in Eugenics.* London: Eugenics Education Society, 1909. Available online. URL: http://galton.org. Accessed June 1, 2009.

G. K. Chesterton: *Eugenics and Other Evils* (1922) (excerpt)

Chesterton was an influential writer—and a very witty one—of the early 20th century. In Eugenics and Other Evils, *he skewers eugenics as a morally degraded concept and urges others to turn against an idea in which "the baby that does not exist can be considered even before the wife who does." At the time he wrote the book, eugenics organizations were becoming established around the world, along with mandatory sterilization laws. "The First Obstacles" is the book's second chapter.*

The First Obstacles

Now before I set about arguing these things, there is a cloud of skirmishers, of harmless and confused modern sceptics, who ought to be cleared off or calmed down before we come to debate with the real doctors of the heresy. If I sum up my statement thus: "Eugenics, as discussed, evidently means the control of some men over the marriage and unmarriage of others; and probably means the control of the few over the marriage and unmarriage of the many," I shall first of all receive the sort of answers that float like skim on the surface of teacups and talk. I may very roughly and rapidly divide these preliminary objectors into five sects; whom I will call the Euphemists, the Casuists, the Autocrats, the Precedenters, and the Endeavourers. When we have answered the immediate protestation of all these good, shouting, short-sighted people, we can begin to do justice to those intelligences that are really behind the idea.

BIOTECHNOLOGY AND GENETIC ENGINEERING

Most Eugenists are Euphemists. I mean merely that short words startle them, while long words soothe them. And they are utterly incapable of translating the one into the other, however obviously they mean the same thing. Say to them "The persuasive and even coercive powers of the citizen should enable him to make sure that the burden of longevity in the previous generations does not become disproportionate and intolerable, especially to the females?"; say this to them and they sway slightly to and fro like babies sent to sleep in cradles. Say to them "Murder your mother," and they sit up quite suddenly. Yet the two sentences, in cold logic, are exactly the same. Say to them "It is not improbable that a period may arrive when the narrow if once useful distinction between the anthropoid *homo* and the other animals, which has been modified on so many moral points, may be modified also even in regard to the important question of the extension of human diet"; say this to them, and beauty born of murmuring sound will pass into their faces. But say to them, in a simple, manly, hearty way "Let's eat a man!" and their surprise is quite surprising. Yet the sentences say just the same thing. Now, if anyone thinks these two instances extravagant, I will refer to two actual cases from the Eugenic discussions. When Sir Oliver Lodge spoke of the methods "of the stud-farm" many Eugenists exclaimed against the crudity of the suggestion. Yet long before that one of the ablest champions in the other interest had written "What nonsense this education is! Who could educate a racehorse or a greyhound?" Which most certainly either means nothing, or the human stud-farm. Or again, when I spoke of people "being married forcibly by the police," another distinguished Eugenist almost achieved high spirits in his hearty assurance that no such thing had ever come into their heads. Yet a few days after I saw a Eugenist pronouncement, to the effect that the State ought to extend its powers in this area. The State can only be that corporation which men permit to employ compulsion; and this area can only be the area of sexual selection. I mean somewhat more than an idle jest when I say that the policeman will generally be found in that area. But I willingly admit that the policeman who looks after weddings will be like the policeman who looks after wedding-presents. He will be in plain clothes. I do not mean that a man in blue with a helmet will drag the bride and bridegroom to the altar. I do mean that nobody that man in blue is told to arrest will even dare to come near the church. Sir Oliver did not mean that men would be tied up in stables and scrubbed down by grooms. He meant that they would undergo a loss of liberty which to men is even more infamous. He meant that the only formula important to Eugenists would be "by Smith out of Jones." Such a formula is one of the shortest in the world; and is certainly the shortest way with the Euphemists.

The next sect of superficial objectors is even more irritating. I have called them, for immediate purposes, the Casuists. Suppose I say "I dislike this spread of Cannibalism in the West End restaurants." Somebody is sure to say "Well, after all, Queen Eleanor when she sucked blood from her husband's arm was a cannibal." What is one to say to such people? One can only say "Confine yourself to sucking poisoned blood from people's arms? and I permit you to call yourself by the glorious title of Cannibal." In this sense people say of Eugenics, "After all, whenever we discourage a schoolboy from marrying a mad negress with a hump back, we are really Eugenists." Again one can only answer, "Confine yourselves strictly to such schoolboys as are naturally attracted to hump-backed negresses; and you may exult in the title of Eugenist, all the more proudly because that distinction will be rare." But surely anyone's common-sense must tell him that if Eugenics dealt only with such extravagant cases, it would be called common-sense—and not Eugenics. The human race has excluded such absurdities for unknown ages; and has never yet called it Eugenics. You may call it flogging when you hit a choking gentleman on the back; you may call it torture when a man unfreezes his fingers at the fire; but if you talk like that a little longer you will cease to live among living men. If nothing but this mad minimum of accident were involved, there would be no such thing as a Eugenic Congress, and certainly no such thing as this book.

I had thought of calling the next sort of superficial people the Idealists; but I think this implies a humility towards impersonal good they hardly show; so I call them the Autocrats. They are those who give us generally to understand that every modern reform will "work" all right, because they will be there to see. Where they will be, and for how long, they do not explain very clearly. I do not mind their looking forward to numberless lives in succession; for that is the shadow of a human or divine hope. But even a theosophist does not expect to be a vast number of people at once. And these people most certainly propose to be responsible for a whole movement after it has left their hands. Each man promises to be about a thousand policemen. If you ask them how this or that will work, they will answer, "Oh, I would certainly insist on this"; or "I would never go so far as that"; as if they could return to this earth and do what no ghost has ever done quite successfully—force men to forsake their sins. Of these it is enough to say that they do not understand the nature of a law any more than the nature of a dog. If you let loose a law, it will do as a dog does. It will obey its own nature, not yours. Such sense as you have put into the law (or the dog) will be fulfilled. But you will not be able to fulfill a fragment of anything you have forgotten to put into it.

Along with such idealists should go the strange people who seem to think that you can consecrate and purify any campaign for ever by repeating the names of the abstract virtues that its better advocates had in mind. These people will say "So far from aiming at *slavery*, the Eugenists are seeking *true* liberty; liberty from disease and degeneracy, etc." Or they will say "We can assure Mr. Chesterton that the Eugenists have no intention of segregating the harmless; justice and mercy are the very motto of—" etc. To this kind of thing perhaps the shortest answer is this. Many of those who speak thus are agnostic or generally unsympathetic to official religion. Suppose one of them said "The Church of England is full of hypocrisy." What would he think of me if I answered, "I assure you that hypocrisy is condemned by every form of Christianity; and is particularly repudiated in the Prayer Book"? Suppose he said that the Church of Rome had been guilty of great cruelties. What would he think of me if I answered, "The Church is expressly bound to meekness and charity; and therefore cannot be cruel"? This kind of people need not detain us long. Then there are others whom I may call the Precedenters; who flourish particularly in Parliament. They are best represented by the solemn official who said the other day that he could not understand the clamour against the Feeble-Minded Bill as it only extended the "principles" of the old Lunacy Laws. To which again one can only answer "Quite so. It only extends the principles of the Lunacy Laws to persons without a trace of lunacy." This lucid politician finds an old law, let us say, about keeping lepers in quarantine. He simply alters the word "lepers" to "long-nosed people," and says blandly that the principle is the same.

Perhaps the weakest of all are those helpless persons whom I have called the Endeavourers. The prize specimen of them was another M. P. who defended the same Bill as "an honest attempt" to deal with a great evil: as if one had a right to dragoon and enslave one's fellow citizens as a kind of chemical experiment; in a state of reverent agnosticism about what would come of it. But with this fatuous notion that one can deliberately establish the Inquisition or the Terror, and then faintly trust the larger hope, I shall have to deal more seriously in a subsequent chapter. It is enough to say here that the best thing the honest Endeavourer could do would be to make an honest attempt to know what he is doing. And not to do anything else until he has found out. Lastly, there is a class of controversialists so hopeless and futile that I have really failed to find a name for them. But whenever anyone attempts to argue rationally for or against any existent and recognizable *thing*, such as the Eugenic class of legislation, there are always people who begin to chop hay about Socialism and Individualism; and say "*You* object to all State interference; I am in favour of State interference. You are an Individualist; I, on the other hand," etc. To which I can only answer, with heart-broken patience, that I am not an Individualist, but a poor fallen but baptized journalist who is trying to write a book about Eugenists, several of whom he has met; whereas he never met an Individualist and is by no means certain he would recognize him if he did. In short, I do not deny, but strongly affirm, the right of the State to interfere to cure a great evil. I say in this case it would interfere to create a great evil; and I am not going to be turned from the discussion of that direct issue to bottomless botherations about Socialism and Individualism, or the relative advantages of always turning to the right and always turning to the left.

And for the rest, there is undoubtedly an enormous mass of sensible, rather thoughtless people, whose rooted sentiment it is that any deep change in our society must be in some way infinitely distant. They cannot believe that men in hats and coats like themselves can be preparing a revolution; all their Victorian philosophy has taught them that such transformations are always slow. Therefore, when I speak of Eugenic legislation, or the coming of the Eugenic State, they think of it as something like *The Time Machine* or *Looking Backward*: a thing that, good or bad, will have to fit itself to their great-great-great-grandchild, who may be very different and may like it; and who in any case is rather a distant relative. To all this I have, to begin with, a very short and simple answer. The Eugenic State has begun. The first of the Eugenic Laws has already been adopted by the Government of this country; and passed with the applause of both parties through the dominant House of Parliament. This first Eugenic Law clears the ground and may be said to proclaim negative Eugenics; but it cannot be defended, and nobody has attempted to defend it, except on the Eugenic theory. I will call it the Feeble-Minded Bill, both for brevity and because the description is strictly accurate. It is, and quite simply and literally, a Bill for incarcerating as madmen those whom no doctor will consent to call mad. It is enough if some doctor or other may happen to call them weak-minded. Since there is scarcely any human being to whom this term has not been conversationally applied by his own friends and relatives on some occasion or other (unless his friends and relatives have been lamentably lacking in spirit), it can be clearly seen that this law, like the early Christian Church (to which, however, it presents points of dissimilarity), is a net drawing in of all kinds. It must not be supposed that we have a stricter definition incorporated in the Bill. Indeed, the first definition of "feeble-minded" in the Bill was much looser and vaguer than the phrase "feeble-minded" itself. It is a piece of yawning idiocy about "persons who though capable of earning their living under favourable circumstances" (as if anyone could earn his living if circumstances were directly unfavourable to his doing so), are nevertheless "incapable of managing their affairs with proper prudence"; which is exactly what all the world and his wife are saying about their neighbours all over this planet. But as an incapacity for any kind of thought is now regarded as statesmanship, there is nothing so very novel about such slovenly drafting. What is novel and what is vital is this: that the *defence* of this crazy Coercion Act is a Eugenic defence. It is not only openly said, it is eagerly urged, that the aim of the measure is to prevent any person whom these propagandists do not happen to think intelligent from having any wife or children. Every tramp who is sulky, every labourer who is shy, every rustic who is eccentric, can quite easily be brought under such conditions as were designed for homicidal maniacs. That is the situation; and that is the point. England has forgotten the Feudal State; it is in the last anarchy of the Industrial State; there is much in Mr. Belloc's theory that it is approaching the Servile State; it cannot at present get at the Distributive State; it has almost certainly missed the Socialist State. But we are already under the Eugenist State; and nothing remains to us but rebellion.

Source: G. K. Chesterton. "The First Obstacles." From *Eugenics and Other Evils.* London: Cassell, 1922. Available online. URL: http://www.cse.dmu.ac.uk/~mward/gkc/books/Eugenics.html. Accessed June 1, 2009.

Nuremberg Code (1947)

The Nuremberg Code was established by the judges who rendered the guilty verdict in United States of America v. Karl Brandt et al. *at the post-World War II Nuremberg Trials, along with Dr. Leo Alexander, an American neurologist of Austrian and Jewish ancestry who was a medical adviser during the trials. Brandt was Hitler's personal physician and leader of the Nazi program of euthanasia and experiments on concentration camp detainees involving freezing, gassing, bone transplants, sterilization, and infection with malaria and typhus. Alexander drafted six principles for the ethical treatment of medical research involving humans, to which the judges added four, for a total of ten. The ideas expressed in the succinct code—informed consent, beneficence, good experiment design—became the cornerstone of future ethics guidelines, such as the Declaration of Helsinki, the Belmont Report, and the Asilomar Conference precautionary principle.*

1. The voluntary consent of the human subject is absolutely essential. This means that the person involved should have legal capacity to give consent; should be so situated as to be able to exercise free power of choice, without the intervention of any element of force, fraud, deceit, duress, over-reaching, or other ulterior form of constraint or coercion; and should have sufficient knowledge and comprehension of

the elements of the subject matter involved as to enable him to make an understanding and enlightened decision. This latter element requires that before the acceptance of an affirmative decision by the experimental subject there should be made known to him the nature, duration, and purpose of the experiment; the method and means by which it is to be conducted; all inconveniences and hazards reasonable to be expected; and the effects upon his health or person which may possibly come from his participation in the experiment.

The duty and responsibility for ascertaining the quality of the consent rests upon each individual who initiates, directs or engages in the experiment. It is a personal duty and responsibility which may not be delegated to another with impunity.

2. The experiment should be such as to yield fruitful results for the good of society, unprocurable by other methods or means of study, and not random and unnecessary in nature.

3. The experiment should be so designed and based on the results of animal experimentation and a knowledge of the natural history of the disease or other problem under study that the anticipated results will justify the performance of the experiment.

4. The experiment should be so conducted as to avoid all unnecessary physical and mental suffering and injury.

5. No experiment should be conducted where there is an a priori reason to believe that death or disabling injury will occur; except, perhaps, in those experiments where the experimental physicians also serve as subjects.

6. The degree of risk to be taken should never exceed that determined by the humanitarian importance of the problem to be solved by the experiment.

7. Proper preparations should be made and adequate facilities provided to protect the experimental subject against even remote possibilities of injury, disability, or death.

8. The experiment should be conducted only by scientifically qualified persons. The highest degree of skill and care should be required through all stages of the experiment of those who conduct or engage in the experiment.

9. During the course of the experiment the human subject should be at liberty to bring the experiment to an end if he has reached the

physical or mental state where continuation of the experiment seems to him to be impossible.

10. During the course of the experiment the scientist in charge must be prepared to terminate the experiment at any stage, if he has probable cause to believe, in the exercise of the good faith, superior skill and careful judgment required of him that a continuation of the experiment is likely to result in injury, disability, or death to the experimental subject.

Source: "Nuremberg Code." From *Trials of War Criminals before the Nuremberg Military Tribunals under Control Council Law* No. 10, Vol. 2, pp. 181–182. Washington, D.C.: U.S. Government Printing Office, 1949. Available online. URL: http://ohsr.od.nih.gov/guidelines/nuremberg.html. Accessed June 1, 2009.

JAPAN

Japan's Biotechnology Strategy Guidelines (2002) (excerpt)

Japan's Biotechnology Strategy Council was charged with creating a road map for the country's life science endeavors. The result is the Biotechnology Strategy Guidelines: Three Strategies Opening the Way to Vast Improvements in Three Basic Aspects of the Human Experience: Our Health, Our Food, Our Lifestyles. Recognizing that the life sciences are to the 21st century what electronics was to the 20th, government leaders want to ensure that Japan remains a world leader in providing services for its own people and for others around the world.

The 20th Century was the Century of Electronics. Japan had a firm grasp on the science and technology involved, and took appropriate action in all areas. As a result, Japan succeeded in taking a lead position in the application of electronics technologies, the lifestyle of our citizens became an abundant one, and electronics manufacturing became the chief pillar in Japan's industrial backbone.

The 21st Century is the Life Sciences Century—the century of biotechnology ("biotech" below). Biotech represents a truly epochal revolution—one which will bring about fundamental changes in three basic areas of the human experience: our food, our health, and our lifestyles.

The scientific progress that serves as the driving force behind this revolution is now making explosive leaps forward around the world. Since the technological results known as biotech were born from astounding advances made in the scientific treatment of humanity's very foundation (that is, of life itself), there is every likelihood that biotech will create entirely new technologies as well as whole new industries—possibilities of a sort that no one can foresee. In addition, existing industries and the existing technology base will also be subjected to enormous change.

For these reasons, nations throughout the world are doing a great deal to strengthen their positions in the biotech arena. In this context, it would be wrong to boast that Japan has already established a firm lead. On the contrary, one cannot help but note that there are a number of areas in which we are actually lagging quite a ways behind. If we do not immediately make every effort to strengthen our biotech initiatives at a national level, Japan runs the risk of being left behind in the most significant scientific and technological advances of the 21st Century. The danger is that we could fall back in terms of improvements to our citizens' standard of living, and that the foundations of many of our industries might suffer blows large enough for them to start to crumble. If we do not face this possibility head-on, we run the risk of having a major bill to pay sometime in our future.

This Biotechnology Strategy Council feels sure that developments in biotech will be the most significant scientific results of the 21st Century. We are also deeply aware that biotech will give rise to great changes, both in industry and in human life on this planet. Because of these understandings, we want to emphasize that, while we must face biotech-related issues of ethics and safety head-on, it is absolutely necessary for our would-be techno-superpower to link the fruits of the outstanding developments in biotech to both improvements in the lives of our citizenry, and to industry and manufacturing.

The current progress in biotech worldwide is proceeding at an extremely fast pace. It is for this very reason that the next five to ten years are sure to prove decisive. In order for Japan to effect a speedy response to this opportunity, with our public and private sectors both giving their all in the effort to make a major leap forward, our country is setting its sights on the year 2010 and setting out this document, our "Biotechnology Strategy Guidelines." . . .

The following sorts of things are possible through application of biotech:

- Because investigation of individuals' genes shows individual characteristics, such as the susceptibility of that individual to developing cancer or hypertension, it is possible to advise people to exercise or diet in order to maximize their chance of avoiding illness. Additionally, it enables the choice of foods most suitable to individuals' constitutions. In the event someone falls ill, it is possible to receive highly effective treatment suitable to the individual's constitution and causing minimal side effects.

- If regenerative medicine is commercialized and it becomes possible to transplant insulin-secreting cells into a diabetic's body, numerous daily insulin shots will become unnecessary. Also, people using wheelchairs as a result of a serious spinal injury will be able to walk again after receiving transplanted nerve stem cells.

- As the causes and mechanisms become clear for various dementia illnesses, including Alzheimer's disease, immunoallergic diseases, and rheumatoid arthritis, for lifestyle-related illnesses such as cancer and hypertension, as well as for infectious diseases caused by new pathogens, such as O-157, medical technologies and preventive methods, including groundbreaking new drugs and vaccines, are being developed and mortality rates are declining.

- As new medical equipment (e.g. minimally invasive equipment) is developed, and it becomes easier to receive treatment with minimal strain on the body, earlier diagnosis and quicker treatment become possible and treatment technology becomes more advanced, allowing more effective treatment.

Using biotech, "tailor-made" medical treatments are now being used that enable prevention and effective treatment with minimal side effects tailored to each individual based on genetic information, permitting safer and more effective treatment. The reduction in side effects is expected to lead to a reduction of medical expenses. Furthermore, it will be possible to prevent illnesses by understanding each person's constitution at the genetic level. Functional regeneration of, for example, organs and body tissues that are functionally impaired or dysfunctional by using stem cells will allow transplant of organs and cells, such as nerve stem cells, without their being rejected, for use in treating motor paralysis, Parkinson's disease and lifestyle-related illnesses such as diabetes, due to technologies related to functionally regenerative medicines. Through utilizing genomics and protein sciences, we can expect the creation of medical technologies and groundbreaking new drugs that will enable causal treatment of illnesses, such as cancer, strokes, diabetes, asthma, hypertension and dementia, including Parkinson's disease. In addition, it is anticipated that by using these developments in medical treatment centers there will be a reduction in the time people will be bed-ridden and, furthermore, that it will contribute to moderating medical expenses.

By using biotech in responding to auto-immune diseases, immune and allergic diseases, such as hay fever, and infectious diseases caused by new pathogens found in prion diseases, such as [E. coli] 0157 and BSE, methods of treatment and prevention of transmission will be developed through better understanding of the transmission mechanisms, and light will be thrown on immune control mechanisms, which are the starting point for immune system control.

Using biotech in responding to mounting suicide rates and psychological health issues, which are increasingly problematic in modern society, will advance understanding of peoples' brain functions. Utilizing genomics, biotech will help to elucidate peoples' behavior and mental activities at the molecular level, and illuminate the effects of stress on the brain, thus making it possible to play a role in preventing problems.

In addition, there will be significant improvements in peoples' health through advances in effective and safe scientific evaluation that will satisfy the increasingly high degree of health-consciousness, as well as developments in effective and safe functional foods, contributing to health self-management.

From another perspective, a robust safety guarantee system will be established due to advances in measures to guarantee the safety of newly developed medical-related technologies, medical products and equipment and functional foods—as indispensable to the development of biotech as technological developments. While guaranteeing safety, communication of those safety aspects will be promoted between government and citizens as well as between citizens and researchers and corporations, and citizens will be able to choose and make their own judgments on using the new technologies based on an understanding of the risks and benefits involved.

These developments in biotech will permit extensions to healthy life expectancy, and especially for Japan, which is confronted with a declining birthrate and ageing society, will be key in helping to guarantee a healthy, fully active and high-quality life irrespective of age and gender. Additionally, these developments will also provide a starting point to deal with emerging infectious diseases that have been threatening peoples' lives in recent years, issues of various chemical substances, including endocrine-disrupting substances, and, furthermore, peoples' spiritual health issues.

Food (Better Eating)
The following sorts of things are possible through application of biotech:

- Swift detection of pollutants and residual agricultural chemicals in food products will be possible, which have become an issue with imported vegetables, and the food products we eat everyday can be enjoyed more safely.

195

- It will become possible through scientific examinations to spot false labeling of production areas and brand misrepresentation of high-quality Japanese beef, for instance, leading to consumer confidence in labeling.
- Information relating to the safety of food products will become easier to access by providing it over the Internet, enabling consumers to select food products in the satisfaction that their choices are based on scrupulously scientific grounds.
- By being able to create product types with controlled volumes and types of starch and protein, it will be possible to make tastier rice and rice for people with kidney illnesses.
- Diets will be enriched by product improvements that strengthen resistance to diseases and pests and by always and continuously being able to get high-quality food products.

Using biotech, technology for swift, low-cost and easy detection of harmful substances in food products will be established and standardized; scientific confirmation of labels will be possible by the establishment of, for example, type and production-area identification technology that utilizes trace constituents and DNA, and scientific response methods will become available to address the issue of false labeling of food products.

In addition to the establishment of food safety and technology to guarantee safety, people will be able to select their food based on their own evaluation as a result of the availability of safety information and the establishment of a steadfast food safety guarantee system, owing to the introduction of risk analysis methods (comprised of risk evaluation, risk management and risk communication) thus ensuring food safety and peace of mind.

Regarding food development, high value-added foods will be efficiently developed, such as allergen-free foods (devoid of allergy-inducing substances) and high-quality foods with, for example, high-nutrition values and great tastes. For these high value-added foods, varieties will be efficiently developed that are resistant to adverse environments, diseases and pests. Recently, due to a string of issues of falsification of food product labels and the BSE issue, confidence in food has been shaken dramatically, but due to the effective implementation of strategies directed at biotech development, food safety and piece of mind have been regained and the steady supply of safe, high-quality, high value-added food products will enable us to enjoy an enriched diet.

Source: Biotechnology Strategy Guidelines (Draft): Three Strategies Opening the Way to Vast Improvements in Three Basic Aspects of the Human Experience: Our Health, Our Food, Our Lifestyles. Biotechnology Strategy Council, December 6, 2002.

INDIA

Biotechnology: A Vision (September 7, 2001) (excerpt)

India's Department of Biotechnology in the Ministry of Science and Technology is charged with promoting biotechnology within the country, both in the educational sector and the business sector. It strives to make India a hospitable place for research and development in order to create long-term prosperity for its citizens through a robust health infrastructure that includes a world-class life sciences industry. This report marked the 10th anniversary of the country's Department of Biotechnology and includes a road map for development related to medicine, agriculture, genomics, biofuels, and other new technologies.

Our Mission

Realising biotechnology as one of the greatest intellectual enterprises of humankind, to provide the impetus that fulfills this potential of understanding life processes and utilizing them to the advantage of humanity.

- To launch a major well directed effort with significant investment, for harnessing biotechnological tools for generation of products, processes and technologies to enhance the efficiency and productivity and cost effectiveness of agriculture, nutritional security, molecular medicine, environmentally safe technologies for pollution abatement, biodiversity conservation and bioindustrial development.
- Scientific and technological empowerment of India's incomparable human resource.
- Creation of a strong infrastructure both for research and commercialisation, ensuring a steady flow of bioproducts, bioprocesses and new biotechnologies.

. . .

Executive Summary

In order to realise the full potential of biotechnology as a frontline area of research and development with an overwhelming impact on society, the Indian biotechnological enterprise will be systematically nurtured at three distinct levels. The focus would be on: basic research in modern biotechnology, including genomics and bioinformatics; agriculture, plant and animal biotechnology; medical biotechnology; environment and biodiversity; biofuels; product and process development and bioinstrumentation; human resource development; creation and strengthening of infrastructure in existing and new institutions; biotechnology for societal development; biosafety, ethical issues and biotechnology related policy issues; conduct of cutting edge research, large scale demonstrations, partnership with private and public sector industries for commercialisation and marketing of bioproducts.

Enhancing the knowledge base and generating highly skilled human resource:

The national bioscience research endeavour in elucidating the molecular basis of plant, animal and microbial life processes will be honed to the cutting-edge by applying global standards. Necessary informational resources will be systematically developed through data banks, inventories, and germ plasm repositories. Human resource development in biosciences and biotechnology will be enhanced to achieve widespread excellence in both teaching quality and support resources. At least 20 Distinguished Professorships in Biotechnology would be instituted to recognize the excellence and provide opportunities for furthering research. Awards and incentives would be instituted to recognize meritorious efforts. Selected missions in identified areas would be launched.

Nurturing leads of potential utility:

Life technology development leads emerging from the bioscience enterprises will be vigilantly identified and fostered in three major areas; agriculture, health care and the environment. Widely available information resources will be developed for this interface via regulated and comprehensive repositories, systematic biological standardisation, and patent support mechanisms. Industrial transitions will be facilitated with large-scale demonstrations and seed partnerships. Pro-active steps will be taken to address societal concerns by establishing transparent mechanisms of systematic public dissemination of bio-information, and by putting into place comprehensive stringent frameworks for both bioethics and biosafety.

The widely spread bioinformatics network would be maximally utilised to ensure connectivity as also sharing and exchange of information, nationally and internationally, data analysis, software development and for dissemination of information. The required infrastructure facilities would be strengthened and created wherever necessary.

Bringing bio-products to the marketplace:

Innovative policies will be developed and implemented, in conjunction with other government departments and agencies, to enhance the biotechnological landscape for investment, to champion Indian biotechnology in the global marketplace, and to design innovative as well as defensive strategies for global intellectual property rights. Systematic interventions at this level will include pilot-scale production and training units, short- as well as long-term partnerships with the biotechnology industry, and coordination of public investment in societally essential products with low commercial returns. We aim at achieving excellence in this field, indigenous self reliance and international competitiveness.

. . .

The Knowledge Base:
Basic Research in New Biology and Biotechnology

In the identified areas, mechanisms, relationships and pathways would be evolved leading to the strengthening of the knowledge base and potential leads for product and process development. In the areas of genomics and bioinformatics, basic research would be more time targeted and related to the identified products. The genomics and bioinformatics infrastructure and networking would be completed within 2–3 years. After the completion of 5 years, about 1000 trained experts in bioinformatics would be available with a large number of database and with abilities for data mining, data annotation, comparative and functional genomics.

Agriculture

Transgenics of rice, brassica, mongbean, pigeonpea, cotton, potato, tomato, and some vegetables like cabbage, cauliflower etc. would complete field assessment and some of them would be ready for the large scale seed production by 2005. Nutritionally enhanced potato and BT-cotton are among the important ones. Transgenic wheat with more protein content and better quality and also higher lysine content and marker assisted breeding programme is expected to be introduced in farmers' field by 2003–2005. The sequencing of Chromosome 11 in rice would be completed by 2005 with an annual contribution of 2 Mb to international rice genome project. This would ensure that India would have the total information on rice genome; and functional genomics work would start by identification of the important markers and genes.

Edible vaccines, particularly for cholera, rabies and hepatitis B, work on which is already under progress, would be ready for clinical trials by 2003–2004, with an expression gene in tomato, cabbage and banana.

Biofertilizers and Biopesticides

Transgenic biofertilizers and biopesticides, particularly botanicals which already have been developed, would have been field tested for commercial

199

production and at least 20 more technology packages would be ready for testing.

Bioprospecting and Molecular Taxonomy

It is expected to complete the prospecting and molecular characterisation and documentation of the economically and ecologically important hot spots of biodiversity in the country, such as the Western Ghats and the Northeastern Region, by 2004, in complementation of classical approaches. The molecules thus identified would be simultaneously taken up for product development: drugs, vaccines, biofertilizers, biopesticides and therapeutics. The characterisation and inventorisation of much of the biological resources in Andaman and Nicobar islands would be completed by 2005. A bio-monitoring system for fragile ecosystems rich in biodiversity would be put in place by 2003.

Plant Tissue Culture

Complete packages, different tissue culture protocols for coffee, tea, spices and apple would be ready for commercialisation by 2002–2003. The regional hardening facilities to provide the benefit of the tissue culture technology at the grassroot level would be expanded to cover the most plant rich regions of the country which need massive afforestation and wasteland recovery.

Medicinal and Aromatic Plants

A number of herbal products are expected to be in market starting from 2001 onwards. These would be in the form of new formulations, immuno-modulators and drugs. The diseases addressed are septic shock, diabetes, malaria and cancer. At least 20 agro-biotechnological packages for many herbal products would be introduced for higher and better yield.

Animals

Vaccine research for major livestock diseases such as haemorrhagic septi-caemia, recombinant vaccine for Anthrax would be ready for evaluation. Vero cell based rabies and Zona pellucida peptide for fertility control in dogs would complete the evaluation by 2002. Diagnostics for PPR (Peste-des-petis-ruminants) and Blue Tongue are ready for evaluation which would be completed by 2002. Bovine Tuberculosis diagnostic kit is under development. A diagnostic kit for White Spot Virus in prawns would be completing its evaluation by the end of 2001. Animal feed through con-version of lignocellulasic material and its enrichment would be a major research priority. As soon as it is ready for evaluation, it would go for field testing. Genomic studies would be taken up for major livestock species e.g., buffalo and would reach an advanced stage by the end of the Tenth Plan.

Transgenic fish expressing growth hormone gene are expected for further evaluation towards the end of 2001.

. . .

Medical Biotechnology

Diagnostic kits for major infectious diseases like tuberculosis, malaria, Japanese Encephalitis, HIV, dengue, hepatitis as well as non-communicable diseases like hormonal disorders (several of which have already been licensed to industry) would be in the market by 2002. Upscaling and commercialisation of PCR-based diagnostics would be completed by 2002 onwards. The DNA vaccines for rabies in dogs would be ready for manufacture in 2002–2003. The cholera vaccine would complete its trials by 2004–2005. Vaccines for HIV, TB and malaria are expected to enter Phase-I & II trials by 2004–2005. Rotaviral diarrhoea vaccine would enter Phase-I trial in 2001 and is expected to obtain the approvals in 2 years. A vaccine for Hepatitis C would enter Phase-I clinical trials by 2003. Gene therapy trials against cancer will be initiated in 2001–2002.

Upscaling and probably commercialisation of the newer genomics-based technologies (e.g., microarrays) would commence from 2002 onwards and some would be in market by the end of the Tenth Plan period.

Reproductive health and contraceptive research would continue to be major priority.

Bioengineering of Crops for Biofuels and Bioenergy

Identification and development of crops for bioengineering for biofuels and bioenergy would take at least 2–5 years for completion of various objectives. Field testing of such crops may begin during the Tenth Plan.

. . .

Genomics: Structural and Functional

- To establish GEN-NET INDIA to identify and control genetic disorders prevalent in the country.
- Exploit the knowledge created by Human Genome Sequencing and also that of some pathogenic organisms and parasites so as to generate diagnostic and therapeutic products of special relevance for the country mostly for dreadful diseases like malaria, HIV, tuberculosis, cancer and brain disorders.
- Identifying genomic factors responsible for genetic disorders, development of molecular diagnostics and personalised drugs for the

201

treatment, understanding of the biochemical pathways of the diseases leading to a safe and powerful treatment regime. Comparative genomics, functional and structural genomics, studies of single nucleotide polymorphism, proteomics, data annotation, integration and analysis.

- Creation of DNA polymorphism maps and databases of the peoples of India for predictive and preventive health care.
- Creation of microarray facilities for defining the expression and functions of genes. For important crops like rice, wheat, brassica, chickpea, a map-based marker assisted technology development for precision breeding, as well as gene identification through in situ molecular hybridization.

Functional Genomics

- To exploit the sequence information we have to understand the specific biological functions encoded by a sequence through detailed genetic and phenotypic analysis. For this purpose, genetic resources, e.g. mutants, isogenic lines, elite breeding lines, and high throughput facilities such as microarrays and proteomics would be developed. The programme would initially focus on selected high-priority traits such as tolerance to biotic and abiotic stresses. Bioinformatics capability for analytical and computational ability to infer gene function based on sequence information is equally essential. To enhance scientific knowledge and to discover new genes for crop improvement, a national functional genomics program is needed to make information from functional genomic studies broadly available to address practical problems.
- Mapping of the buffalo and the silk worm genomes.
- Comparative genomics of pathogenic microbes.

Source: Department of Biotechnology, Indian Ministry of Science and Technology. *Biotechnology: A Vision* (9/7/01). Available online. URL: http://dbtindia.nic.in/uniquepage.asp?Id_Pk=102. Accessed June 1, 2009.

GERMANY

Commission of the European Communities: *Life Sciences and Biotechnology: A Strategy for Europe* (2002) (excerpt)

As a member of the European Union (EU), Germany cooperates with the other member states in forming guidelines to regulate the life sciences and biotechnology. Germany has a long history of conservative views on genetically modified food and agriculture, and when it comes to genetics it is unique in Europe for its conservative stance despite its well-developed biotech industry. The following document was written in conjunction with German delegates.

The Potential of Life Sciences and Biotechnology

Life sciences and biotechnology are widely regarded as one of the most promising frontier technologies for the coming decades. Life sciences and biotechnology are enabling technologies—like information technology, they may be applied for a wide range of purposes for private and public benefits. On the basis of scientific breakthroughs in recent years, the explosion in the knowledge on living systems is set to deliver a continuous stream of new applications. There is a huge need in global health care for novel and innovative approaches to meet the needs of ageing populations and poor countries. There are still no known cures for half of the world's diseases, and even existing cures such as antibiotics are becoming less effective due to resistance to treatments. Biotechnology already enables cheaper, safer and more ethical production of a growing number of traditional as well as new drugs and medical services (e.g., human growth hormone without risk of Creutzfeldt-Jakobs disease, treatment for haemophiliacs with unlimited sources of coagulation factors free from AIDS and hepatitis C virus, human insulin, and vaccines against hepatitis B and rabies). Biotechnology is behind the paradigm shift in disease management towards both personalised and preventive medicine based on genetic predisposition, targeted screening, diagnosis, and innovative drug treatments. Pharmacogenomics, which applies information about the human genome to drug design, discovery and development, will further support this radical change. Stem cell research and xenotransplantation offer the prospect of replacement tissues and organs to treat degenerative diseases and injury resulting from stroke, Alzheimer's and Parkinson's diseases, burns and spinal-cord injuries.

In the agro-food area, biotechnology has the potential to deliver improved food quality and environmental benefits through agronomically improved crops. Since 1998, the area cultivated with genetically modified crops world-wide has nearly doubled to reach some 50 million hectares in 2001 (in comparison with about 12,000 hectares in Europe). Food and feed quality may be linked to disease prevention and reduced health risks. Foods with enhanced qualities ("functional foods") are likely to become increasingly important as part of life-style and nutritional benefits. Plant genome analysis, supported by a FAIR research project, has already led to the genetic improvement of a traditional European cereal crop (called Spelt) with an increased protein yield (18%) which may be used as an alternative source of protein for animal feed. Considerable reductions in pesticide use have been recorded in crops with modified resistance. The enhancement of natural resistance to disease or stress in plants and animals can lead to reduced use of chemical pesticides, fertilisers and drugs, and increased use of conservation tillage—and hence more sustainable agricultural practices, reducing soil erosion and benefiting the environment. Life sciences and biotechnology are likely to be one of the important tools in fighting hunger and malnutrition and feeding an increasing human population on the currently cultivated land area, with reduced environmental impact.

Biotechnology also has the potential to improve non-food uses of crops as sources of industrial feedstocks or new materials such as biodegradable plastics. Plant-based materials can provide both molecular building blocks and more complex molecules for the manufacturing, and energy and pharmaceutical industries. Modifications under development include alterations to carbohydrates, oils, fats and proteins, fibre and new polymer production. Under the appropriate economic and fiscal conditions, biomass could contribute to alternative energy with both liquid and solid biofuels such as biodiesel and bioethanol as well as processes such as bio-desulphurisation.

Plant genomics also contributes to conventional improvements through the use of marker-assisted breeding.

New ways to protect and improve the environment are offered by biotechnology including bioremediation of polluted air, soil, water and waste as well as development of cleaner industrial products and processes, e.g. based on use of enzymes (biocatalysis). . . .

A Key Element for Responsible Policy:
Governing Life Sciences and Biotechnology

The public debate on life sciences and biotechnology and the fundamental values highlight the need for responsible and coherent policies to govern these fastmoving technologies. All key stakeholders have stressed the importance of governance, i.e. attention to the way public authorities prepare, decide, implement and explain policies and actions.

The Commission proposes to apply the highest standards of governance of life sciences and biotechnology along 5 main action lines:

- Societal dialogue and scrutiny should accompany and guide the development of life sciences and biotechnology
- Life sciences and biotechnology should be developed in a responsible way in harmony with ethical values and societal goals
- Informed choice should facilitate demand-driven applications
- Science-based regulatory oversight should enhance public confidence

- Basic regulatory principles and legal obligations should be respected to safeguard the Community single market and international obligations.

Societal scrutiny and dialogue

Life sciences and biotechnology have given rise to significant public attention and debate. The Commission welcomes this public debate as a sign of civic responsibility and involvement. Life sciences and biotechnology should continue to be accompanied and guided by societal dialogue. Dialogue in our democratic societies should be **inclusive, comprehensive, well informed and structured.** Constructive dialogue requires mutual respect between participants, innovative approaches, and time. It should be structured in agreement with stakeholders to allow progress, for example in the provision of better information and mutual understanding. Experience also shows how important it is that dialogue takes place at the local and national levels, as well as internationally, and the Commission invites Member States and local actors to take relevant initiatives.

Dialogue should be open for **all stakeholders.** Public authorities should help to ensure participation by stakeholders with limited resources. Economic operators, industry and users, who have economic interests at stake, as well as the scientific community, bear a particular responsibility for active participation. The Commission invites these parties to respond to public concerns, for example through transparency of their visions, policies and ethical standards.

Relevant public **information** is essential for meaningful dialogue. Providing it requires focused and pro-active efforts. It is especially important that the information needs formulated by the broad public are taken seriously and responded to. We shall also strive for a balanced and rational approach, distinguishing between real issues, on which we must act, and false claims.

Developing life sciences and biotechnology in harmony with ethical values and societal goals

Without broad public acceptance and support, the development and use of life sciences and biotechnology in Europe will be contentious, benefits will be delayed and competitiveness will be likely to suffer.

The debate and the public consultation carried out by the Commission indicate that the European public is quite prepared and capable to enter into complex weighting of benefits against disadvantages, guided by fundamental values. Although sometimes polarised, the public debate demonstrates many points of converging views. Public opinion depends crucially on the perceived benefits of life sciences and biotechnology. Eurobarometer surveys reveal that public expectations of biotechnology, apart from medical advances, are moderate. And there is also considerable public uncertainty about some applications, and aversion towards their distributional impacts and the risks involved.

There is broad support for many guiding values and goals. Some of these, such as the freedom of research, intrinsic value of new knowledge and the moral obligations to help alleviate illness or hunger, tend to favour the development and application of these new technologies. Others help to clarify the criteria and conditions for the development and applications of life sciences and biotechnology, in particular the need to take into account the ethical and societal implications, and the importance of transparency and accountability in decision-making, minimising risk, and freedom of choice.

It is therefore of key importance to support information and dialogue to help the public and stakeholders better understand and appreciate these complex issues and to develop methods and criteria for assessing benefits against disadvantages or risks, including the distribution of impacts among different parts of society.

Our democratic societies should offer the necessary safeguards to ensure that the development and application of life sciences and biotechnology take place respecting the fundamental values recognised by the EU in the **Charter of Fundamental Rights,** in particular by confirming the respect for human life and dignity. The Community has also banned funding of research into human reproductive cloning.

Support should be given to the Franco-German initiative, addressed to the UN, for a world-wide convention on the prohibition of human reproductive cloning. Other issues such as stem cell research clearly require attention and further debate. Europe has also taken clear positions on the importance of freedom of choice for consumers as well as for economic operators with respect to GM foods, and we have established broad societal agreement on the need to safeguard European agricultural practices.

However, scientific and technological progress will continue to give rise to new ethical or societal implications. The Commission considers that these issues should be addressed pro-actively and with a broad perspective, taking into account the moral obligations towards present and future generations and the rest of the world. We should not content ourselves with acting defensively only when our core values are being transgressed.

These issues cannot be adequately addressed within the narrow context of regulatory product approvals but require more flexible and forward-looking approaches. Europe needs an active and on-going public dialogue, accompanied by focused fact-finding on both benefits and disadvantages to allow the public to contribute to the complex process of setting priorities. In the context of its Science & Society initiative, the Commission has already proposed a series of actions intended to strengthen the ethical dimension in sciences and new technologies.

To be at the front of developments, Europe should have the capacity for foresight/prospective analysis and the necessary expertise to help clarify the often complex issues for policy makers and the public, and to place them in their scientific and socio-economic context.

The Commission welcomes the key role played by the **European Group on Ethics in Science and New Technologies** since its creation in the early 1990s and proposes, as part of the present strategy, to enhance its role and to reinforce the networking with and between national ethical bodies. To this end, an additional targeted consultation of the other Community institutions is envisaged.

Moreover, transparency, accountability and participatory approaches in public policy-making need to be reinforced. These objectives coincide with those of the Commission's White Paper on European Governance and will be pursued through the actions proposed therein. . . .

Europe's Responsibilities towards the Developing World

Life sciences and biotechnology hold the promise of meeting some of the fundamental needs for food and health facing the developing world. The UNDP, in its 2001 Human Development Report, highlights the potential of biotechnology for the developing world. Some emerging economies such as China, India and Mexico have already initiated ambitious national development programmes.

Life sciences and biotechnology are not a panacea and will not resolve the distributional problems of the developing world—but they will be one of the important tools. New capacities should help developing countries reconcile yield increases, sustainable use of natural resources, economic efficiency and social acceptability. Potential applications must be adequately researched and assessed, taking full account of both the environmental safety issues and the needs expressed by the populations concerned to reduce poverty and strengthen food security and nutritional quality.

As a major actor in life sciences and technologies, Europe has a particular responsibility to help the developing world deal with the risks, challenges and opportunities, and to facilitate the safe and orderly development of these technologies at the global level. Europe already holds an influential position in international deliberations on life sciences and biotechnology. This needs to be taken forward with responsible policies to achieve our strategic objectives and to allow the safe and efficient use of life sciences and biotechnology in developing countries.

207

- Europe should continue to promote protection of biodiversity and the implementation of the Biosafety Protocol for international trade in living modified organisms. Moreover, Europe should continue to support negotiated multilateral frameworks such as the Convention on Biological Diversity and the FAO International Undertaking on Plant Genetic Resources. These international instruments regulate access to genetic resources and the sharing of the benefits arising from their use, in view of providing compensation to the centres of origin of genetic resources and the holders of traditional knowledge used in biotechnological inventions. The EC should contribute to ensure that the benefits generated by biotechnological inventions, including intellectual property income, are properly shared with the providers of genetic resources or traditional knowledge.
- Europe should contribute to technical assistance, capacity-building and technology transfer to allow developing countries to participate in negotiating and implementing international agreements and standards, notably on risk governance, and to safely develop and apply these new technologies if they so wish. Europe should support local initiatives for dialogue on biotechnology among public and private stakeholders and civil society in partner countries.
- Europe should encourage equitable and balanced North-South partnerships and public research for demand-driven applications of life sciences and biotechnology.
- Domestic European policies with regard to life sciences and biotechnology are bound to have major impacts on developing countries. Whilst not compromising EU food safety requirements or consumer information policies, we should provide technical assistance and capacity building to ensure that our policies do not, unwittingly, prevent developing countries from harvesting desired benefits. In particular, we should guard against regulatory requirements that may be manageable only in the industrial world but are unachievable by developing countries, thereby either upsetting existing trade or effectively blocking developing countries from developing life sciences and biotechnology at their own wish and pace.

Implementation and Coherence across Policies, Sectors and Actors

Europe does not have a single policy for life sciences and biotechnology but a patchwork of specific regulation, overlaid by many sectoral and horizontal policies at international, Community, Member State and local levels. If, with so many actors and policies involved, Europe is to successfully manage life sciences and biotechnology and reap the benefits for society, we should proceed on the basis of a shared vision for a co-operative approach and with effective implementing mechanisms to compensate for absence of overall responsibility and control. Without such mechanisms, life sciences and biotechnology risk to continue to suffer indecision or short-sighted and local solutions.

Source: Commission of the European Communities. "The Potential of Life Sciences and Biotechnology." In *Life Sciences and Biotechnology—A Strategy for Europe.* Brussels: 2002, pp. 5–20.

SOUTH AFRICA

A National Biotechnology Strategy for South Africa: Executive Summary and Foreword (2001)

South Africa's biotechnology strategy, as drafted by the Ministry of Arts, Culture, Science and Technology, outlines what the government hopes to accomplish in developing the life sciences industries—namely, to provide food, health care, and other services in accordance with first world standards. It is understood that this will require a multipronged approach that begins with education and includes good business practices in order to attract companies that will help attain these goals.

South Africa has a solid history of engagement with traditional biotechnology. It has produced one of the largest brewing companies in the world; it makes wines that compare with the best; it has created many new animal breeds and plant varieties, some of which are used commercially all over the world and it has competitive industries in the manufacture of dairy products such as cheese, yoghurt and maas and baker's yeast and other fermentation products.

However, South Africa has failed to extract value from the more recent advances in biotechnology, particularly over the last 25 years with the emergence of genetics and genomic sciences (the so-called 3rd generation). Already many companies and public institutions elsewhere in the world are offering products and services that have arisen from the new biotechnology. In the USA alone, there are 300 public biotechnology companies with a market capitalisation of $353 billion and an annual turnover of $22 billion p.a. Moreover, the growth of biotechnology industries is not restricted to the developed countries. Developing countries such as Cuba, Brazil and China have been quick to identify the potential benefits of the technology and have established measures both to develop such industries and to extract value where possible and relevant.

BIOTECHNOLOGY AND GENETIC ENGINEERING

The strategy outlined in this document is designed to make up for lost ground and to stimulate the growth of similar activities in South Africa. Biotechnology can make an important contribution to our national priorities, particularly in the area of human health (including HIV/AIDS, malaria and TB), food security and environmental sustainability. In the pursuit of these priorities, we are fortunate in that we can be guided by the experiences of other countries. For instance, we know that to achieve success a country requires a government agency to champion biotechnology, to build human resources proactively, and to develop scientific and technological capabilities. In addition, successful commercialisation of public sector-supported research and development (R&D) requires strong linkages between institutions within the National System of Innovation and a vibrant culture of innovation and entrepreneurship, assisted by incubators, supply-side measures and other supporting programmes and institutions.

Some of these components of a successful biotechnology sector are already in place in South Africa. However, a number of gaps are identified in this document and certain interventions are suggested to address these problems. The recommendations are divided into two categories, namely new institutional arrangements and specific actions for Government departments. In the case of the former, the Panel has recommended the establishment of a Biotechnology Advisory Committee (BAC), under the auspices of the Cabinet's Economics Cluster, the responsibilities of which will include the implementation of this strategy, co-ordination of biotechnology R&D and alignment with national priorities.

A key component of the strategy is the creation of several regional innovation centres (RICs) to act as nuclei for the development of biotechnology platforms, from which a range of businesses offering new products and services can be developed. The RICs will be required to work in close collaboration with academia and business in order for the centres to become active nodes for the growth of the biotechnology sector. Using both existing funds and new allocations specifically designated for biotechnology, and employing well-trained scientists, engineers and technologists in a multi-disciplinary environment, the centres will stimulate the creation of new intellectual property (IP). The successful protection and exploitation of this IP will be made possible by a new venture capital fund and an array of new and existing support structures. It is emphasised that the main focus of the RICs will be the creation of economic growth and employment through innovation.

A number of recommendations are made to Government, including support, both financial and at a policy level, for the formation of the BAC, which will be responsible for the implementation of this strategy. The proposed actions will require an annual budget of R182 million, of which R135 million is required for the funding of the RICs and the associated R&D programmes, R20 million for the venture capital fund, R25 million for additional funding to strengthen the link between academia and industry and R2 million to run the BAC, plus a once-off establishment cost of R45 million for the RICs. This document also urges the Government to complete a number of important revisions to the legislative and regulatory environment, including the extension of the activities of the Bioethics Committee and the revision of the Patents Act, in order for the strategy to be successful.

Finally, careful attention must be given to the development of the appropriate human resources and to the public understanding of biotechnology. It is Government's responsibility to ensure that new biotechnology products or services do not threaten the environment or human life, or undermine ethics and human rights. . . .

Foreword

The first century of the new millennium will belong not only to communications, or information technologies, but also to biotechnology, which will bring unprecedented advances in human and animal health, agriculture and food production, manufacturing and sustainable environmental management.

To embrace biotechnology is to further embrace our commitment to the realization of our national imperatives and specifically:

- To improve access to and affordability of health care.
- To provide sufficient nutrition at low cost.
- To create jobs in manufacturing.
- To protect and cherish our rich environment.

To achieve our objectives, we will be required to assimilate biotechnology skills rapidly in order to commercialise country-specific applications and reduce the economic gap between developed and developing countries.

Without doubt, we will need to exercise caution and judgement in the application of biotechnology.

We will need to ensure that the potential risks to human health and the environment arising from the commercial use of genetically modified organisms in food production are properly managed.

We will need to continuously assess our biotechnology programmes within the framework of the constitution, which ensures our rights to safety, to choice and to information.

We will need to establish suitable regulatory systems in order to participate as exporters and importers in the international trade in biotechnology products.

We will need to increase the level of public awareness and acceptance of these products.

In many respects we are fortunate: new advances in biotechnology promise to make the path of progress a great deal easier and shorter. We stand at the crossroads and our response to this opportunity will shape our future.

Minister of Arts, Culture, Science and Technology, Dr Ben Ngubane
11 June 2001

Source: South Africa's Ministry of Arts, Culture, Science and Technology. *A National Biotechnology Strategy for South Africa.* (June 2001). Available online. URL: http://www.dst.gov.za/publications-policies/strategies-reports/reports/dst_biotechnology_strategy.PDF. Accessed June 1, 2009.

PART III

Research Tools

6

How to Research Biotechnology

GETTING STARTED

No matter what your topic, start by considering your deadline. Work backward from that date, allotting sufficient time and making interim deadlines for each step: research, outline, rough draft, and revision. The more organization you do up front, the better. It will save you time and energy down the road.

Select a Topic

What issue resonates most with you—cloning, eugenics, finding a cure for AIDS? It all depends on where your interests lie and finding a way to connect biotechnology to what is going on in your own life. If your family is involved in agriculture, for example, you may have firsthand knowledge of how GM seed affects your family's bottom line. Or perhaps you are interested in how genetic engineering may help epidemiologists contain an influenza pandemic similar to the one your great-great grandparents experienced. Maybe you play video games that simulate war and you have become interested in how the U.S. Department of Defense is preparing for possible biowarfare. If you do not have much choice in what you write about, then try to find a connection that turns an unfamiliar topic into something you can relate to. Make your job easier by selecting a topic that will sustain your interest in the research and writing process from start to finish.

Determine Your Thesis Statement

Be specific and, if asked to do so, take a position. Do not write a paper about DDT; write a paper about how DDT should (or should not) be used to combat malaria in developing countries. Take a position; have a point of view. Maybe you think cloning is bad. Narrow that down a bit; why is it bad? Is all cloning bad? What if scientists clone only certain cells that can be used to replace an individual's damaged cells? Keep in mind the issue is not whether or not your position is right or wrong, but how you defend it. Equally good papers can be written in favor of cloning and against it.

Consider the following thesis statements.

Ineffective thesis statement: "Cloning should be illegal."
Effective thesis statement: "All nations should sign a treaty forbidding the cloning of a human being's entire DNA sequence."

Ineffective thesis statement: "Genetically modified organisms are dangerous."
Effective thesis statement: "Genetically modified organisms promote mono-culture, which results in unhealthy land use and a loss of biodiversity."

Ineffective thesis statement: "The government must make sure designer babies don't become a reality."
Effective thesis statement: "Genetic techniques that may lead to 'designer babies' are an updated form of eugenics and should be limited by law."

Read and Watch the News

You can never be too well informed. Make a point of visiting news Web sites regularly, even just to peruse the headlines. Channel-surf the news networks every now and then to see if they are reporting on your topic. The radio, and particularly public radio, still has worthy information.

Diversify your news sources: Bookmark a few key Web sites. Local, national, and international outlets should all be in your mix. Knowing the international picture may give you a different perspective on the national one, and both will provide a context for the local situation. If you speak or read a language in addition to English, brush up by reading foreign-language newspapers or Web sites. You might be surprised at the different perspectives this provides.

Diversify your media sources: Internet, print, television, and radio sources are all important, so capitalize on the strengths of each. The Internet is great for breaking news and finding specific items, but not so great for local news if you live outside a major metropolitan area. Newspapers and magazines often include articles on subjects you wouldn't have thought to look up on the Internet. News programs on television (such as *20/20* or *60 Minutes*) often present an in-depth look at a topic and include interviews with experts. Radio stations often host local call-in shows that air the viewpoints of people on the street.

How to Research Biotechnology

Learn the History of Your Topic

Josef Mengele's heinous medical experiments on Nazi concentration camp inmates remain a vivid historical event that informs current discussions of genetics. For issues that lack such a sordid history it is still important to know the details. Consider that biowarfare is nearly as old as warfare itself. The Greeks dipped their spears in poison, the Romans contaminated their enemy's water supply with the carcasses of rotting animals, and soldiers in medieval times sent diseased individuals into enemy territory in hopes of igniting an epidemic. In this light, modern concerns about weaponized anthrax are a variation on a theme rather than a new development. Thus, rather than tackling a new problem, the 1972 Biological Weapons Convention represented a new political attempt to address a very old problem.

Seek Out Different Opinions

Acknowledging the validity of differing opinions is a strength. To properly defend your position, start first by defining the parameters of the issue, then narrow your focus methodically. This grounds your argument along a continuum and lets your reader know exactly where you stand and that you are well versed in your topic.

Take cloning as an example. Here are two extreme perspectives on the issue:

For: Cloning represents the apex of scientific advancement and will result in valuable medical applications, such as the ability to generate new organs for those who would die without them. At the very worst, it creates an identical twin of a living (or once living) organism. Since twins occur in nature, the only difference here is that a clone would be born at a different time than its twin. Moreover, what would be so terrible about creating another Einstein or resurrecting a species like the Dodo bird, which humans rendered extinct?

Against: Cloning is unnatural and should be prohibited under all circumstances. It is a crime against nature. Tampering with the building blocks of life may lead to unknown consequences that may not be able to be stopped, such as unleashing damaged or ill-formed genes that could contaminate a species' genome.

By locating your own thesis somewhere between these points, you identify both as extreme and thereby ground your position in a sense of reasoned credibility. Not acknowledging the more extreme views on a topic can weaken your argument, which may appear not to be anchored in a solid foundation.

BIOTECHNOLOGY AND GENETIC ENGINEERING

Find Points That Both Sides Agree On

Common ground can provide the springboard from which to launch your argument. Establish what those on both sides of an issue have in common. In the case of genetically modified seed, both those who advocate for its use as a way to produce more food for the world and those who advocate for organic farming, integrated pest management, and protecting biodiversity would know that genetically modified organisms are manufactured by inserting genes from one organism into the DNA of another organism, or by "turning on" or "turning off" specific genes within an organism. Both sides would also agree that the first example of modern-day genetic engineering took place in 1973 when scientists used recombinant DNA technology to create *E. coli* bacteria that contained a gene from the *salmonella* bacteria. Whether or not everyone believes this experiment should have taken place is another matter.

Understand the Connectedness of Issues

Finding the common ground along the spectrum of your issue also allows you to connect the dots to other issues. For example, what does poverty on the Lower East Side of Manhattan in the early 20th century have to do with the rise in organic farming in the early 21st century? Not much at first glance, but consider the following.

- Poverty on Manhattan's Lower East Side spurred Margaret Sanger to advocate for birth control in order to reduce the size of families. She believed that smaller families would help ease poverty by freeing up parents' time and resources to provide better care for their existing children.

- Sanger, in an effort to promote birth control internationally, organized the first World Population Conference in Geneva, Switzerland, in 1927. One of the main sessions at the conference focused on food and population.

- The World Population Conference led to the formation of the Union for the Scientific Study of Population in 1928. It established three research committees, one of which studied population and food.

- The connection between population and food came to a head after World War II, when a substantial increase in food production was required to feed the world's surging population. The problem was addressed through the efforts of Norman Borlaug, the Ford Foundation, the Rockefeller Foundation, and the United States Agency for International Development (USAID), among others. The Green Revolution was made possible

218

through their initiatives. Agricultural production was raised significantly in the countries in danger of mass starvation, thus averting disaster.

- The Green Revolution achieved many of its gains through the heavy use of pesticides, fertilizers, and monoculture.
- The gains of the Green Revolution, by the 21st century, were eroded by pesticide-resistant bugs and land degradation due to monoculture. Organic farming came into its own as a reaction against these and other practices.

Global village or six degrees of separation—no matter what you call it, being able to connect the dots between your topic and others speaks highly of your ability to process information.

Know Your Experts

In 1616, Galileo Galilei was convicted of heresy by the Roman Catholic Church for asserting that the Earth revolved around the Sun. The church had been an established, influential authority for a thousand years; how could one scientist's idea possibly overturn the combined intellectual legacy of generations of society's most learned men, all of whom believed that the Sun revolved around the Earth? The moral of the story is that knowledge evolves. Scientists routinely find new evidence that changes prevailing opinions. A contemporary example of this principle is Dr. Manto Tshabalala-Msimang, who was the minister of health in South Africa from 1999 until she was forced to resign in 2008. Throughout her tenure, she refused to acknowledge that HIV causes AIDS. While guiding the public health policies of the nation, she advocated a diet of beetroot, garlic, olive oil, and lemon for those suffering from AIDS instead of a regimen of antiretroviral medicines that had proven quite effective for other AIDS patients throughout the world. Her policies resulted in the deaths of thousands who under a different health minister may have received the medical treatment they needed. All this took place under the presidency of Thabo Mbeki, who appointed Tshabalala-Msimang and also publicly stated that he did not believe that HIV causes AIDS. Though both officials had received university degrees from well-respected institutions, their education made no difference when it came to the issue of HIV/AIDS in South Africa, and both were eventually forced to resign.

On the other end of the spectrum is Paul Berg. He is an example of a cautious expert who, despite being a leader in the field of biochemistry, put his research on hold until he could convene the Asilomar Conference and gain a consensus from his colleagues on how to proceed with research that could have led to unintended negative consequences. Berg was conducting experiments using Simian Virus 40 cells as a vector to introduce new genes into mammalian cells. Simian Virus 40 was known to cause tumors in rodents. Berg was fairly certain—but not positive—that the cancer-causing parts of the Simian Virus would not escape the transgenic process. Other scientists voiced their concern that the Simian Virus might inadvertently be transmitted to people and cause cancer. The conference created a working set of guidelines to ensure as much as possible that Berg's research and that of others would adhere to safety standards designed to prevent negative outcomes.

Make sure to test the claims of experts against those of others to see how they match. Also beware of those who claim certainty or couch their speech in absolutes, with words such as *always* and *never*. And be sure to avoid the trap of using such words in your own writing.

Know Your Own Bias

Rare is the individual who can remain completely unbiased on a controversial topic. Chances are you have some biases, even if you accept facts that contradict your views. Acknowledging such biases as you undertake your research will enable you to compensate by seeking out opposing information. During this process, you might even find that your bias changes. For example, if your religious views are such that you oppose stem cell research, you would still need to look into the possible benefits of such research and ensure that you explore them sufficiently.

Note that biases can be quite subtle. Perhaps, when looking over your research, you realize that most of your information comes from a few sources: the *New York Times*, the *New England Journal of Medicine*, and *Nature*. What is wrong with that? Maybe nothing, but it is possible that all your sources have a liberal or conservative bent. To reduce this risk, look far and wide. If you cite the *New York Times*, look at what the *Wall Street Journal* has to say about the same issue. Cite several peer-reviewed journals and a variety of other periodicals. Give your research breadth and depth to avoid the pitfalls of bias, unintended as they may be.

Understand the Science

Opinions are fine and good, but they are only as strong as your grasp on the underlying science. It is one thing to say that genetic screening is wrong because it will lead to discrimination and another to understand the process of genetic screening, its limitations, and its promises.

Become conversant in basic terminology. Know the difference between DNA and a gene. Understand what happens during the recombinant DNA process. Know the difference between DNA and RNA. You do not have to be able to diagram protein chains and memorize all the genes on a given chromosome, but you should be comfortable enough with your topic to explain it to someone who is unfamiliar with it.

Follow Your Hunch

Sometimes your instincts will guide you to useful information. For example, you suspect that genetic testing will become routine—possibly even mandatory—within the next few years. Following your hunch, you undertake research into how genetic testing is performed and find that a simple swab of saliva is all that is required. The cost is relatively low in comparison to other medical procedures. Genetic testing of various sorts—such as prenatal testing for Down syndrome—has been undertaken for years. Then you find out that the Genetic Information Nondiscrimination Act was signed into law in 2008, making it illegal for health insurers to deny coverage or charge higher premiums to individuals based on the likelihood of their developing a disease. You also discover that some companies already offer mail-in genetic tests to assess a person's genetic predisposition for a number of diseases. Your hunch, you discover, was well grounded in fact.

Taking Notes and Keeping Track of Sources

Put your pen to paper and write it down or type up notes. Do not trust yourself to remember something. Often the act of writing something down makes it more likely you will remember it. Paraphrase information rather than copying it word for word. Doing so ensures you avoid plagiarism and you really understand what you are reading. If you do want to quote someone, make sure you keep track of source citation information. It is no fun to track this information down after the fact, when you are in the process of compiling endnotes or a bibliography. For books, this information includes the name of the book, the author, and the place and date of publication, and the relevant page numbers. For articles, it includes the author, the name of the publication, the volume number or month and year of publication, and the relevant page numbers. For a Web site, make sure you have the complete URL as well as the root URL if the data is likely to be moved. Also note the date on which you accessed a site or downloaded information.

Seek Out Primary Documents

A primary document is the real thing. It is a bridge to the past that allows you a firsthand glimpse of a previous generation's thinking. For example, Margaret Sanger stated in 1924 that "There are definite reasons when and why parents should not have children, which will be conceded by most thoughtful people. First—Children should not be born when either parent has an inheritable disease, such as insanity, feeble-mindedness, epilepsy, or syphilis."[1] What pops out in this statement is Sanger's language. The term *feeble-mindedness* is seldom used today, yet it was common in 1924. It had no legal definition but was a catchall term for slight mental retardation, learning disabilities, or psychological illnesses such as depression. Taking note of the language used in primary documents and taking the time to discover whether such imprecise terms can really be considered scientific allows you to uncover biases that you should try to avoid in your own work. While feebleminded was once a favorite label of eugenicists, today individuals would be said to suffer from, for example, ADD, Asperger's syndrome, dyslexia, or any other condition that in no major way would limit one's ability to be a productive member of society.

Know the Law

Analyzing legal decisions that relate to your topic is a good way to discover how conventional wisdom has evolved. From the Comstock laws of the 1870s, which prohibited all forms of birth control for all women, to *Griswold v. Connecticut* in 1965, which allowed married couples to obtain birth control, to *Eisenstadt v. Baird* in 1972, which extended the right to birth control for unmarried couples, you can trace the changing social mores of the United States. At the time of *Buck v. Bell* (1926), birth control was illegal for most people, yet sterilization—as a form of birth control—was compulsory for some. Consider what such double standards reveal.

Develop an International Perspective

Perspectives on a topic differ among countries and also among individuals. Government officials, business executives, consumer advocates, and farmers all have different interests. For example, the U.K.'s Environment Minister Phil Woolas claimed "It was the government's 'moral responsibility' to investigate whether genetically modified crops could help provide a solution to hunger in the developing world."[2] In the United States, an article about GMOs quoted Monsanto CEO Hugh Grant framing the hunger issue differently: It "isn't a feel good thing. . . . Satisfying the demand curve is a great business opportunity."[3] In India, one of the countries that supposedly benefits from GMOs, the community leader Vandana Shiva says that GMOs are ruining the country and causing farmers who have lost their livelihoods to commit suicide: "Corporate seeds aren't about increasing productivity—they are about increasing debt. . . . When moneylenders come to repossess the land the farmer cannot bear it, and [he] consumes pesticides to end his life."[4] Thus, biotechnology is seen by some as an ethical imperative, by others as a business opportunity, and by still others as an ethical aberration. These differences may become apparent only when you seek out information originating beyond the borders of Western nations.

USING SOURCES EFFECTIVELY

When to Trust the Internet

Reputable Web sites are usually run by an individual or group that has a credible offline presence. Reputable Web sources do not use pseudonyms, but they do provide links to other sources to support the claims they make and exhibit sufficient transparency into who they are, what their goals are, and how they are funded.

That said, there is plenty of solid information on the Internet. Some of it is available only from subscription sites or requires that you pay an access fee. The good news is that many of these sites can be accessed through your local public or university libraries for free.

Wikipedia: Chances are you can find an introduction to your topic on Wikipedia. A good entry will include footnotes to other Web sites that provide more information, links to related topics, lists of prominent individuals associated with the topic, and photographs and charts that illustrate the topic. On the subject of biotechnology and genetics, many Wikipedia entries are quite detailed and accurate. Be sure to cross-check all facts using independent sources. Also, never quote directly from Wikipedia. The entries are anonymous and there is no way to verify the author's qualifications or biases.

Government Web sites: Many governments host reputable Web sites that provide a wealth of information. In the United States, Web site addresses that end with .gov are those of legitimate government agencies. All three branches of the federal government—executive, legislative, and judicial—have a Web presence. You can find information on the Internet from many of the federal agencies involved in biotechnology, including the Food and Drug Administration (FDA), the Centers for Disease Control (CDC), and the National Institutes of Health (NIH). Many of those sites include pages designed for educators and students.

The National Center for Biotechnology, a division of the NIH, has an extensive online presence and hosts GenBank, which houses the genetic sequence database derived from the Human Genome Project. The site contains software and tools to enable working scientists to share their research results with GenBank and other international initiatives. Much of the data is quite technical, however, which makes it less than ideal for those new to the topic.

The U.S. Congress hosts Web sites for all the committees of the legislative branch of government. These include:

- House Committee on Agriculture: http://agriculture.house.gov/index. shtml. This committee includes several subcommittees involved with biotechnology and genetic engineering issues, including the Subcommittee on Horticulture and Organic Agriculture; the Subcommittee on Livestock, Dairy and Poultry; and the Subcommittee on Specialty Crops, Rural Development, and Foreign Agriculture.
- House Committee on Science and Technology: http://science.house.gov. This site includes the subcommittees on Technology and Innovation and Research and Science Education, both of which deal with issues of biotechnology and life sciences.

Topics pertaining to executive branch functions can be found at http://www.whitehouse.gov, and information regarding the judicial branch—that is the U.S. Supreme Court—can be found at http://www.supremecourtus.gov.

Foundations and Academic Institutions: In evaluating an organization, take stock of its Web site. Does it list a street address? How long has it been established? What are its program areas? Does it have a board of directors? Does it accept members? What is its mission statement? While there are no right answers to these questions, they will help you determine the organization's intention and level of expertise. The membership and purpose of a grassroots organization such as the Sierra Club, which holds specific opinions on genetic engineering, are much different than the goals of a conservative think tank such as the Heritage Foundation, which advocates capitalist free enterprise above all else. Both are valid organizations, but the former is geared toward educating the public and the latter is more concerned with influencing government policy.

Many institutes are associated with universities. The Bio-X research center at Stanford University and the Bioethics Institute at Johns Hopkins University are just two examples.

International Web Sites: The United Nations and the European Union both have Web portals through which you can find information on their biotechnology initiatives. Most individual countries have at least partial information on their programs in English. The BBC News Web site (http://www.bbc.

co.uk) serves as a major news outlet for English speakers in Europe. The *International Tribune Herald* is an English-language daily newspaper that reports news from around the world.

Nonprofit Organizations: Numerous nonprofit organizations deal with biotechnology and genetics. The Foundation for Biotechnology Awareness and Education (http://www.fbae.org), the Council for Biotechnology Information (http://www.whybiotech.com), the Bill and Melinda Gates Foundation (http://www.gatesfoundation.org), and the Council for Responsible Genetics (http://www.gene-watch.org) are just a few places you can obtain information on various topics. Nonprofits typically demonstrate a more cautious, grassroots approach to topics than do industry. All reputable organizations will state their goals up front. Citizens' groups tend to be more consumer-oriented than groups comprised of business or academic entities.

Think Tanks: A think tank is a nonprofit organization comprised of researchers and consultants that develops and advocates policy positions. Think tanks run the gamut of the political spectrum, from liberal to conservative to libertarian, and many aim to influence Washington lawmakers. Some well-known think tanks are the RAND Corporation (http://www.rand.org), which addresses biotechnology through its health and health care and science and technology research areas; the Brookings Institution (http://www. brookings.edu), which addresses issues such as malaria control and agricultural practices through its Africa Growth Initiative; and the Cato Institute (http://www.cato.org), which addresses FDA and drug regulation and genetic engineering in its Regulatory Studies research program.

Blogs: Blogs are useful for gaining a "person on the street" perspective. Some are affiliated with mainstream news organizations; these include TierneyLab (http://tierneylab.blogs.nytimes.com/) and Well (http://well. blogs.nytimes.com/), both of which are written by staff writers at the *New York Times* and cover science and health issues. Biotech Blog (http://www. biotechblog.com) and GMO Africa Blog (http://www.gmoafrica.org) are written by biology professionals. Biotech Blog is a group blog that includes articles on a wide range of biotechnology topics. Blogs can be a great way to become familiar with opposing viewpoints. Again, however, do not quote from blogs (other than in an informal, anecdotal way), do not describe a blogger as an expert unless that person is noted in his or her field beyond the realm of the Internet, and always independently verify claims made by a blogger.

Offline Resources

Offline resources include books, magazines, documentary films, television news reports, radio reports, and personal interviews. Many offline resources have online counterparts or are also published on the Internet.

NEWSPAPERS

The *New York Times, USA Today, Wall Street Journal, Washington Post, Chicago Tribune, Los Angeles Times*, and the *Cleveland Plain Dealer* are some of the most respected newspapers in the country. News wire services, especially the Associated Press and Reuters, are valuable for breaking news that serves as the basis for longer reports in the daily newspapers. Most newspapers have online archives that are searchable by date and topic; many have separate science and technology sections. Old articles are often available for a small fee.

GENERAL INTEREST PERIODICALS

Time, Newsweek, U.S. News & World Report, New Yorker, Atlantic, Harper's, Economist, and *National Geographic* are respected magazines that routinely investigate biotech topics. Your library may have back issues from which you can photocopy articles, or you may be able to purchase individual articles from the magazines' online archives.

SCIENCE MAGAZINES

Nature, Science, Discover, Science News, Scientific American, and *Popular Science* are a few of the many well-respected magazines that cover biotech topics in more detail than is available elsewhere. All of these have a long track record of following major science breakthroughs and interpreting them for the average reader. Typical subjects include what politicians have to say about gene therapy, the latest organisms to have their genome mapped, and drugs that may soon be available. Articles are usually illustrated and easier to understand than the hard research found in peer-reviewed journals. All of the magazines listed above have Web sites that contain articles for free. Older articles can be downloaded for a small fee or may be available at your library free of charge.

PEER-REVIEWED JOURNALS

A peer-reviewed journal is one that publishes original research after the studies have undergone rigorous scrutiny by experts to ensure they adhere to strict scientific standards. This gives the data a degree of validity that is sometimes absent from less scientific reporting published elsewhere. Articles in peer-reviewed journals are available to other scientists to quote, use in their own research, or serve as the basis of further research, with the goal of either replicating the original study's findings or refuting them. In this way, scientific knowledge is advanced.

How to Research Biotechnology

Peer-reviewed biotechnology journals are numerous. They include *African Journal of Biotechnology, Journal of Biomedicine and Biotechnology, International Journal of Biotechnology, Genetic Engineering News, Human Gene Therapy,* and *Genetic Testing.* Journals that specialize in topics such as stem cell research, cloning, or bioethics are also published. It can be a good idea to glance at the abstract (the summary of the study's findings) at the beginning of an article to see if it fits your needs. Sometimes a quote from the abstract or conclusion will suit your purposes.

PUBLIC RADIO AND PUBLIC TELEVISION

Both the Public Broadcasting Service (PBS) and National Public Radio (NPR), which receive little corporate funding, broadcast programs that take an in-depth look at various topics. Radio shows are often archived on the NPR Web site (http://www.npr.org), and PBS (http://www.pbs.org) shows are frequently available on DVD. Extended interviews with authors and other experts are a hallmark of many long-running PBS and NPR series. Several science journalists associated with public radio and television have online podcasts that can be accessed for free.

READ THE BOOK, SEE THE MOVIE

Stories dealing with genetic tampering are as old as literature itself. Mythology features beasts that are half-human, half-animal. Mary Shelley's *Frankenstein* evokes the danger of creating a being from dead body parts, and H. G. Wells's *The Island of Dr. Moreau* concerns a mad scientist who performs gruesome experiments on humans and animals. Aldous Huxley's 1931 novel *Brave New World* features a dystopic society that has rid itself of pathogens and taken reproductive technology to a new level. Science fiction writers have long woven biotechnology themes into their fiction; William Gibson's 1984 novel *Neuromancer* combines the realms of biotechnology and artificial intelligence in the tale of a man whose nervous system is sabotaged by poisonous mycotoxins—real-life toxins derived from mushrooms, mold, and yeast.

Other influential novels with biotech themes include Ira Levin's *The Boys from Brazil* (1976) and *The Stepford Wives* (1975), Michael Crichton's *The Andromeda Strain* (1969), and Robin Cook's *Coma* (1977). Cook and Crichton were medical doctors concerned with the intersection of technology and bioethics; their novels are based in medical fact and extrapolate situations that are made to seem entirely plausible. Margaret Atwood's award-winning novels *The Handmaid's Tale* (1985) and *Oryx and Crake* (2003) both concern the dark side of reproductive biotechnology. Many of these novels have been adapted into highly regarded movies.

Keep It Original

Unless you are quoting a source verbatim, always paraphrase the information or you may be guilty of plagiarism, a serious offense in which you present another person's words as your own. Also be aware of fair use copyright laws that restrict how much you can quote from a source without having to obtain permission or pay a fee. The general rule of thumb is 10 percent, or 100 words, whichever is more. That means in a short article of 100 words, you may quote no more than 10 words in your paper. For articles of a thousand words or more, you should quote no more than 100 words at a time, and the total words quoted should not exceed 10 percent of the entire article. A 2,000-word article, therefore, should have no more than 200 words total included in your paper.

Most sources published prior to 1920 are considered in the public domain, so are no longer covered under copyright law. You may quote freely from public domain sources, but you must still provide source citations.

CONCLUSION

Above all, your success will be driven by your curiosity. As long as you can relate your topic in some way to your own experiences, you will be off to a good start on what could be a compelling project. Prepare an outline and a schedule, tackle each task in turn, and make sure you keep asking and answering the crucial question—why?

---

[1] Sanger, Margaret. "The Case for Birth Control." *Woman Citizen* 8 (2/23/24): 17–18.

[2] BBC News. "Prince 'Must Prove Anti-GM Claim.'" (8/17/08). Available online. URL: http://news.bbc.co.uk/go/pr/fr/-/1/hi/uk/7566012.stm. Accessed May 19, 2009.

[3] Hindo, Brian. "Monsanto on the Menu." *Business Week* (6/23/08).

[4] Davis, Rowenna. "Interview with Vandana Shiva: Environmentalist Extraordinaire." *New Internationalist* (April 2008).

7

Facts and Figures

1. Number of Biotechnology Companies per Country, 2006

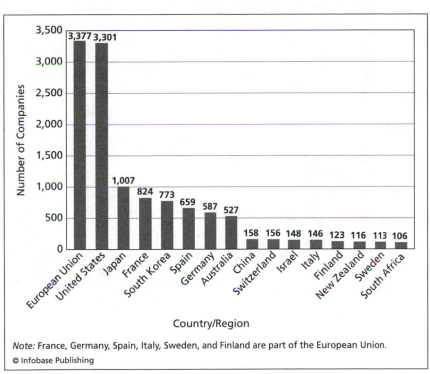

Note: France, Germany, Spain, Italy, Sweden, and Finland are part of the European Union.

© Infobase Publishing

Shows the top 15 nations involved in commercial biotechnology, along with the combined number of biotech companies in the European Union.

Source: Brigitte van Beuzekom and Anthony Arundel. OECD Biotechnology Statistics 2009. Available online. URL: http://www.oecd.org/dataoecd/4/23/42833898.pdf. Accessed June 5, 2009.

2. Biotech Crops in Millions of Hectares per Country, 2007

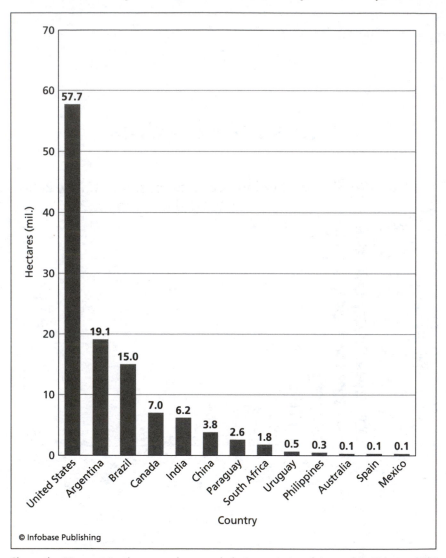

Shows the 13 countries that grow the overwhelming majority of the world's GM crops. GM crops include soybeans, maize, cotton, canola, squash, papaya, alfalfa, tomatoes, poplar, petunias, and sweet peppers.

Source: International Service for the Acquisition of Agri-Biotech Applications, Global Status of Commercialized Biotech/GM Crops: 2007. Available online. URL: http://www.isaaa.org/ resources/publications/briefs/37/executivesummary/default.html. Accessed June 5, 2009.

3. Adoption of Genetically Engineered Crops in the United States

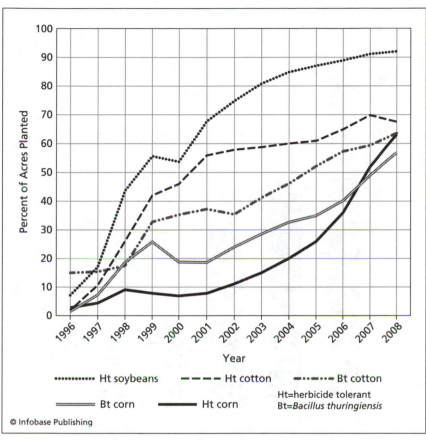

Note that GM seed first became available in 1996, and within 10 years it was firmly established in soybeans, cotton, and corn crops.

Source: United States Department of Agriculture, Economic Research Service. Updated July 2, 2008. Available online. URL: http://www.ers.usda.gov/Data/BiotechCrops. Accessed June 11, 2009.

4. Organic Agriculture Worldwide, 2007

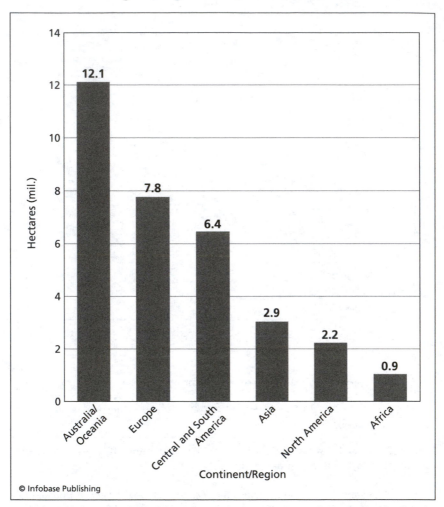

© Infobase Publishing

Australia and the surrounding islands boast the most hectares of organic farmland in the world. Europe, which has little farmland compared to many other continents, comes in second. When it comes to industrial regions, North America is last, at 2.2 million hectares of organic farmland. Much of the world's remaining organic farmland is the result of subsistence agriculture among people in developing countries.

Source: The Research Institute of Organic Agriculture (FiBL) and The International Federation of Organic Agriculture Movements (IFOAM). The World of Organic Agriculture: Statistics and Emerging Trends 2009. Available online. URL: www.organic-world.net. Accessed June 11, 2009.

5. Percentage of Biotech Patent Applications by Country, 2006

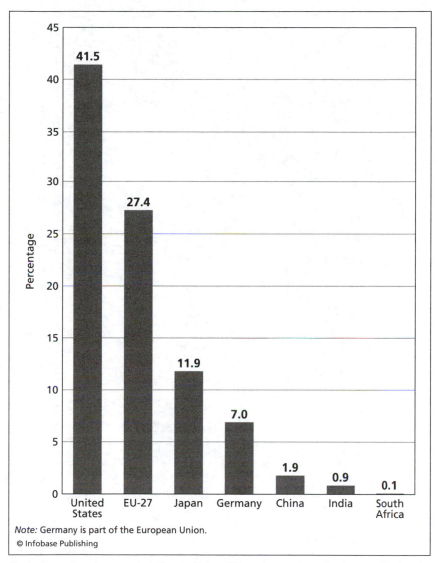

Note: Germany is part of the European Union.
© Infobase Publishing

Chart shows that the overwhelming majority of biotech patent applications are filed in the United States and the European Union. Germany, with 7 percent of biotech patent applications, has the highest rate within the European Union, but it is still below Japan's rate.

Source: Brigitte van Beuzekom and Anthony Arundel. OECD Biotechnology Statistics 2009. Available online. URL: http://www.oecd.org/dataoecd/4/23/42833898.pdf. Accessed June 5, 2009.

6. National Policies on Embryonic Stem Cell Research, ca. 2009

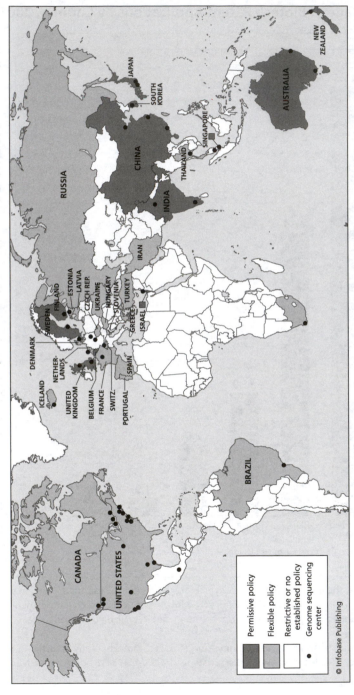

© Infobase Publishing

Countries with permissive or flexible national policies toward stem cell research represent more than half the world's population. A permissive policy means that various embryonic stem cell derivation techniques are permitted by law; a flexible policy means that only embryonic stem cell derivations from fertility clinic donations are permitted by law; restrictive policies range from outright prohibition of human embryo research to permitting research on a limited number of previously established stem cell lines.

Source: William Hoffman. Minnesota Biomedical and Bioscience Net (MBBNet). Available online. URL: http://www.mbbnet.umn.edu/scmap.html. Accessed June 11, 2009.

7. Operations for Eugenic Sterilization Performed in State Institutions under State Laws up to January 1, 1933

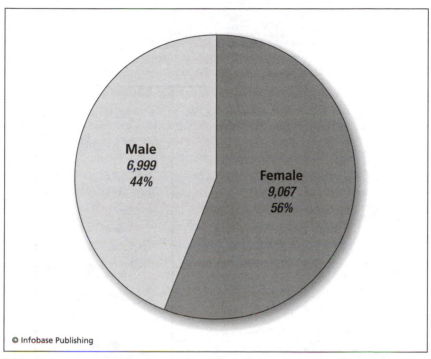

Male
6,999
44%

Female
9,067
56%

© Infobase Publishing

Of the 16,066 eugenic sterilizations performed in U.S. state institutions under state eugenics laws, 56 percent were performed on females. California led the nation, with a total of 8,504 sterilizations through 1932, or 63 percent of the U.S. total. Not included in these totals are sterilizations performed for "therapeutic" purposes or those undertaken by institutions not under the aegis of the law.

Source: Human Betterment Foundation. "Operations for Eugenic Sterilization Performed in State Institutions under State Laws up to January 1, 1933." The Cold Spring Harbor Laboratory Eugenics Archive. Available online. URL: http://www.eugenicsarchive.org. Accessed June 11, 2009.

8.1 Agencies of the U.S. Department of Health and Human Services Involved in Genetics and Biotechnology, 2009

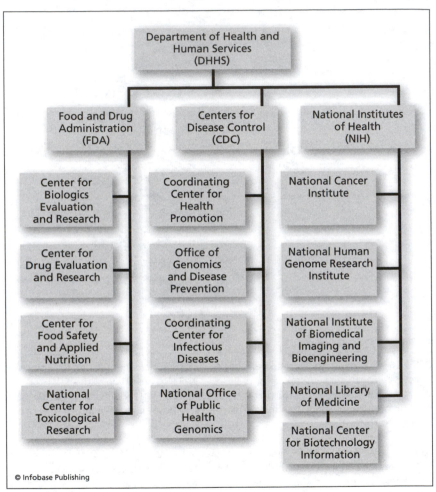

© Infobase Publishing

The Department of Health and Human Services encompasses many agencies that drive biotechnology research and policy in the United States. The secretary of health and human services is a cabinet position and reports directly to the president of the United States.

8.2 Other U.S. Agencies Involved in Biotechnology and Genetic Engineering, 2009

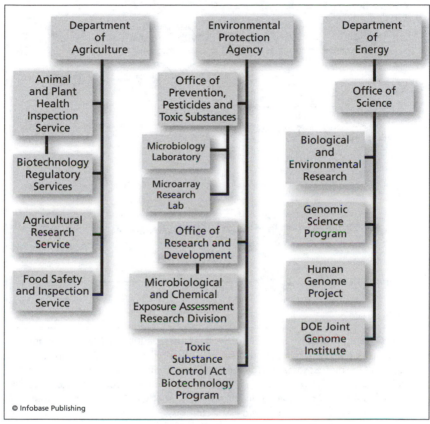

© Infobase Publishing

The U. S. Department of Agriculture (USDA), the Environmental Protection Agency (EPA), and the Department of Energy (DOE) all have important biotechnology programs for research and regulation. Note that the Human Genome Project is overseen by the DOE, and livestock and crop issues are the purview of the USDA. The EPA is largely responsible for overseeing the safety of food as it pertains to environmental pollutants.

9. Somatic Cell Nuclear Transfer Cloning

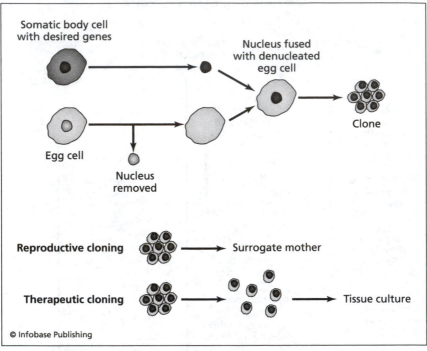

Diagram shows the process of reproductive and therapeutic cloning via somatic cell nuclear transfer (SCNT). In both, the nucleus of an egg cell is removed and replaced with the nucleus of the desired somatic body cell. The egg cell with the new nucleus is prompted to divide until it reaches the blastocyst stage. At that point, reproductive cloning takes place when the cells are implanted into a surrogate mother. Therapeutic cloning takes place when the cells are used for in vitro tissue cultures.

8

Key Players A to Z

PERRY ADKISSON (1929–) Received the 1997 World Food Prize, along with RAY F. SMITH, for popularizing integrated pest management (IPM), a combination of natural techniques that is an ecologically sound way to control pests in agriculture crops.

OSWALD AVERY (1877–1955) Canadian-born American molecular biologist and pioneer of immunochemistry. In 1944, along with fellow researchers Colin MacLeod and Maclyn McCarty, he discovered that DNA is the structure in genes and chromosomes that contains genetic information.

CHRISTIAAN BARNARD (1922–2001) South African surgeon who performed the world's first successful human heart transplant, in 1967.

WOUTER BASSON (1950–) South African cardiologist who spearheaded the country's secret biowarfare program, Project Coast, during the 1980s under the apartheid regime. Basson created dummy corporations that fronted his research into pathogens and genetically engineered weapons.

GEORGE BEADLE (1903–1989) American scientist who received the 1958 Nobel Prize in physiology or medicine with Edwin Tatum for discovering the role of genes within cells. He also helped found the field of molecular biology at Caltech and wrote the 1966 book *The Language of Life*, a primer on genes and cells.

WILLIAM J. BEAL (1833–1924) American botanist who pioneered hybrid corn varieties at Michigan State University, where he also founded the country's oldest experimental botanical garden.

PAUL BERG (1926–) Stanford University biochemist who is sometimes called the father of genetic engineering. He was a pioneer of recombinant

DNA procedures and an advocate of the precautionary principle as outlined at the Asilomar Conference in 1975.

WENDELL BERRY (1934–) American farmer and writer who promotes organic farming, sustainable agriculture, and environmental awareness through his nonfiction writing. He decries factory farming and animal husbandry practices that he feels are detrimental to animal and human health.

NORMAN BORLAUG (1914–2009) American plant geneticist and agronomist, father of the Green Revolution, winner of the Nobel Peace Prize (1970), and proponent of genetically modified crops as a solution to feed the world's surging population. His work in improving crop yields across the globe beginning in the 1970s earned him accolades from many world leaders, who credit him with saving more than a billion lives.

HERBERT BOYER (1936–) American biochemist who pioneered genetic engineering with STANLEY N. COHEN in the 1970s, for which he received the 1990 National Medal of Science. He cofounded Genentech in 1976, the first private company involved in genetic engineering for profit.

SYDNEY BRENNER (1927–) South African molecular biologist who worked with Francis Crick on experiments that resulted in the discovery of messenger RNA, and recipient of the 2002 Nobel Prize in physiology or medicine. Founded the Molecular Sciences Institute in California in 1996.

LOUISE BROWN (1978–) British woman who was the first person conceived and born through in vitro fertilization, a process that has since become routine.

CARRIE BUCK (1906–1983) Plaintiff in *Buck v. Bell,* the U.S. Supreme Court decision that ruled in favor of forced sterilization for the "feeble-minded." In 1924, Buck was committed by her foster parents to the Virginia Colony for Epileptics and Feebleminded, where she gave birth to a daughter. Chief Justice OLIVER WENDELL HOLMES, declaring that "three generations of imbeciles are enough," ordered that Buck comply with the state eugenics laws that called for her sterilization.

EARL BUTZ (1909–2008) American agronomist and secretary of agriculture under Nixon and Ford. Butz spearheaded new policies that encouraged farmers to "get big or get out," which included a transition to commodity farming that led to a surplus of crops (mainly corn) and a drop in food prices. His policies discouraged small family farms and paved the way for the rise of large agricultural corporations and their investment in genetically engineered seeds.

KEITH CAMPBELL (1954–)   British microbiologist who, along with IAN WILMUT, headed the team that cloned the first mammal, Dolly the sheep, at Scotland's Roslin Institute in 1996.

RACHEL CARSON (1907–1964)   American marine biologist and writer, whose best-selling *Silent Spring* (1962) outlined the dangers of widespread pesticide use on the environment. Carson is widely credited with being one of the originators of the modern environmental movement, and her concerns led to the formation of the Environmental Protection Agency in 1970.

GEORGE WASHINGTON CARVER (1864–1943)   American botanist and inventor who revitalized agriculture in the American South by advocating sustainable practices and eliminating monoculture cotton plantations. After a bollworm infestation devastated the economy and land of the South, Carver encouraged farmers to plant other crops, such as peanuts and sweet potatoes, in order to replenish the soil and diversify the economy.

ANANDA CHAKRABARTY (1938–)   Indian-born American microbiologist and defendant in *Diamond v. Chakrabarty* (1980), in which the U.S. Supreme Court ruled in his favor, allowing him to patent a genetically engineered oil-eating bacteria he invented while employed by General Electric. In a 5-4 decision, the Supreme Court wrote that "the fact that micro-organisms are alive is without legal significance for the purpose of patent law." The decision paved the way for much corporate investment in agriculture biotechnology.

IGNACIO CHAPELA (dates unknown)   Mexican microbiologist and mycologist who, along with David Quist, published a controversial research study in *Nature* in 2001 claiming that transgenic maize (genetically engineered corn) had contaminated non–GM maize varieties over a wide area in Mexico. The journal later retracted the story, and Chapela became a controversial figure in academic circles for his outspoken opinions regarding what he believes is an unhealthy alliance between academia and the biotechnology industry.

ERWIN CHARGAFF (1905–2002)   Austrian-born American biochemist whose research assisted JAMES D. WATSON and FRANCIS CRICK in discovering the double-helix structure of DNA. He wrote Chargaff's rules, which explain the relationship between guanine, cytosine, adenine, and thymine within an organism's DNA. In later years he became an outspoken opponent of genetic engineering research.

MARTHA CHASE (1927–2003)   American geneticist whose 1952 experiment with ALFRED HERSHEY proved that DNA, not protein, supplies the genetic material for living organisms. The experiment led JAMES D. WATSON and FRANCIS CRICK shortly thereafter to discover DNA's double-helix structure. Hershey, but not Chase, was awarded the 1969 Nobel Prize in physiology or medicine for the Hershey Chase experiment.

STANLEY N. COHEN (1935–   )   American pioneer of genetic engineering with HERBERT BOYER. In 1972 they invented a method of transferring genes between organisms—recombinant DNA—a revolutionary technique that earned Cohen the 1988 National Medal of Science.

FRANCIS COLLINS (1950–   )   American geneticist whose research led to the discovery of the genes that cause Huntington's and other diseases. Under his direction the Human Genome Project completed its mission to map the human genome ahead of schedule and under budget in 2000.

ROBIN COOK (1940–   )   American medical doctor and best-selling novelist. He writes medical thrillers, such as *Coma* (1977), which deal with biomedicine and the darker side of public health.

MICHAEL CRICHTON (1942–2008)   American medical doctor and best-selling novelist. He wrote medical thrillers that examine the "what if" component of biotechnology gone awry. His novels include *The Andromeda Strain* (1969), *Congo* (1980), and *Jurassic Park* (1990). He also developed and wrote episodes of the long-running television series *ER* (1994–2009), which frequently tackled issues of biomedical ethics.

FRANCIS HARRY COMPTON CRICK (1916–2004)   British molecular biologist who discovered the double-helix structure of DNA in 1953 with his research partner JAMES D. WATSON. He received the 1962 Nobel Prize in physiology or medicine for his work.

CHARLES DARWIN (1809–1882)   British naturalist who originated the idea of natural selection to explain the evolution of plant and animal life on earth. Author of several landmark books, including *On the Origin of Species* (1859).

CHARLES DAVENPORT (1866–1944)   American biologist and eugenicist, founder of the Eugenics Record Office at Cold Spring Harbor Laboratory in 1910. His 1911 book *Heredity in Relation to Eugenics* remained a standard text in support of eugenics for years.

MAX DELBRÜCK (1906–1981)   German-born American geneticist and founder of modern molecular biology through his work with *Drosophila* *melanogaster* and bacteriophages. He received the 1969 Nobel Prize in physiology or medicine for his research on bacterial resistance to viruses.

ASHANTHI DESILVA (1986–)   The first patient successfully treated with gene therapy for severe combined immunodeficiency (SCID), a genetic disease that made her immune system unable to fight even mild infections. In 1990, geneticists inserted a missing gene into her white blood cells, which significantly strengthened her immune system but did not cure the disease.

ROBERT EDWARDS (1925–)   British expert in genetics and reproduction who, along with PATRICK STEPTOE, originated the process of in vitro fertilization that resulted in the birth of the world's first "test-tube baby," LOUISE BROWN, in England in 1978.

EMPEDOCLES (490–430 B.C.E.)   Greek philosopher and one of the first Western scientific thinkers to theorize about how elements of nature combined to create different organisms. His work can be seen as an early musing on what CHARLES DARWIN later called natural selection.

KARL EREKY (1878–1952)   Hungarian scientist who coined the term *biotechnology* in 1919 to describe the process by which biological substances are transformed into useful products.

ALEXANDER FLEMING (1881–1955)   Scottish biologist whose work with bacteria led to the discovery of the curative properties of penicillin in 1928, for which he received the 1945 Nobel Prize in physiology or medicine.

WALTHER FLEMMING (1843–1905)   German biologist who discovered chromosomes in 1875, although he was unaware of their role in heredity.

ROSALIND FRANKLIN (1920–1958)   British scientist and X-ray crystallographer whose images of DNA led to JAMES D. WATSON and FRANCIS CRICK's discovery of its double-helix structure, although her contribution was not acknowledged at the time. Franklin also conducted notable research on the tobacco mosaic virus.

ANDREW FRENCH (dates unknown)   Chief science officer of Stemagen, a California-based biotechnology corporation. In 2008, French and SAMUEL H. WOOD announced they had cloned five human embryos from adult skin cells in a process known as somatic cell nuclear transfer (SCNT), which bypassed the need to use embryonic stem cells. French and Wood concluded that using SCNT technology for human reproductive cloning would be unethical, and the embryos were destroyed.

FRANCIS FUKUYAMA (1952–)   American philosopher and theorist. His book *Our Posthuman Future: Consequences of the Biotechnology Revolution*

(2002) warns that when it becomes possible for scientists to alter human nature, they will do so. This will throw the world's tenuous grasp on liberal democracy off balance and result in perpetual inequality.

FRANCIS GALTON (1822–1911) British scientist, mathematician, anthropologist, and father of eugenics. In books such as *Hereditary Genius* (1869), he outlined the principles of eugenics and investigated the debate over nature versus nurture as it related to human characteristics. He refuted CHARLES DARWIN's theory of pangenesis, after his experiments with rabbits failed to confirm his theory that blood-borne gene-carrying gemmules are responsible for inherited traits.

JESSE GELSINGER (1981–1999) American patient who became the first to die as a result of gene therapy. Doctors treated Gelsinger's liver disease by injecting him with an adenovirus with a modified gene they hoped would replace his malfunctioning gene. Gelsinger's immune system reacted negatively to the virus that transported the corrected gene into his body, and he died days later. The incident put a halt to many other gene therapy experiments.

JOHN R. HELLER (dates unknown) Head of "The Tuskegee Study of Untreated Syphilis in the Negro Male" in its later years, until it was terminated in 1972. During his tenure, Heller forbade subjects to take penicillin, which was standard medical treatment for syphilis at the time, even though the Nuremberg Code, designed to protect patients' rights, was already in place. Heller later told investigators: "The men's status did not warrant ethical debate. They were subjects, not patients; clinical material, not sick people."

ALFRED HERSHEY (1908–1997) American geneticist who proved in 1952 in the Hershey Chase experiment that DNA contains an organism's genetic information, not protein. Received the 1962 Nobel Prize in physiology or medicine for his work on the genetic structure of viruses.

OLIVER WENDELL HOLMES, JR. (1841–1935) United States Supreme Court Justice known for his influential, well-written court decisions. Holmes authored *Buck v. Bell*, which upheld state eugenics laws, noting that CARRIE BUCK, the supposedly "feebleminded" daughter of a "feebleminded" mother, should be forcibly sterilized following the birth of her presumed-to-be "feebleminded" daughter, because "three generations of imbeciles are enough."

ROBERT HOOKE (1635–1703) British scientist called the father of microscopy. Through his microscope, he was the first to view a *cell*—a term he coined—and recognize that it was the basic unit of life.

HWANG WOO-SUK (1953–　)　South Korean professor of biology who claimed in 2005 to have cloned a human embryo from a stem cell, an assertion that was later found to be false. He headed the team that did successfully create the first cloned dog.

ALEC JEFFREYS (1950–　)　British geneticist who developed the process of DNA fingerprinting, or profiling, which is used in forensics to positively identify individuals based on their genetic material.

EDWARD JENNER (1749–1823)　British scientist and physician who refined and popularized the smallpox vaccine.

MICHAEL KABACK (dates unknown)　American geneticist who developed a screening process that identifies carriers of the gene for Tay-Sachs disease. Screening began in the 1970s. It was the first widespread genetic testing to be available in the United States and resulted in nearly eliminating the disease within the Ashkenazi Jewish population.

KATŌ SHIDZUE (1897–2001)　Early Japanese feminist and colleague of MARGARET SANGER, with whom she campaigned for birth control reform in the 1920s in Japan. She founded the Japan Family Planning Association in 1948 and served in the Diet of Japan from 1946–1974.

PHILIP KITCHER (1947–　)　British science philosopher, professor, and author. He frequently writes on biotechnology issues and morality; his books include *The Lives to Come: The Genetic Revolution and Human Possibilities* (1996), and *In Mendel's Mirror: Philosophical Reflections on Biology* (2003).

ROBERT KOCH (1843–1910)　German scientist who first isolated *Bacillus anthracis, Mycobacterium tuberculosis,* and *Vibrio cholerae.* He received the 1905 Nobel Prize in physiology or medicine for his research to identify and treat tuberculosis.

HARRY H. LAUGHLIN (1880–1943)　American eugenicist and director of the Eugenics Record Office, a position he used to advocate mandatory sterilization of all persons "unfit" to be parents.

JOSHUA LEDERBERG (1925–2008)　American molecular biologist and winner of the 1958 Nobel Prize in physiology or medicine for discovering that bacteria can exchange genetic information.

ANTONI VAN LEEUWENHOEK (1632–1723)　Dutch scientist dubbed the father of microbiology for his improvements to the microscope, which allowed him to be the first to view bacteria, spermatozoa, and muscle fibers.

PHOEBUS LEVENE (1869–1940)   Russian-born American chemist who studied nucleic acid and was the first to identify DNA and RNA; he discovered that DNA is comprised of adenine, guanine, thymine, and cytosine.

STEVEN LINDOW (ca. 1951–)   Biologist at University of California at Berkeley who developed the "ice-minus" bacteria, which was the first genetically altered organism to be released into the environment in 1987.

JOSEPH LISTER (1827–1912)   British surgeon who promoted the need for sterility in the medical environment. His research and development of antiseptics vastly reduced infections during surgery and hospital stays.

THABO MBEKI (1942–)   Former president of South Africa who was forced to resign in 2008 in part because of his views on AIDS. He contended that the disease was caused by poverty and did little to enact policies that would allow patients to receive proper medical care in the form of antiretroviral therapy or to educate people on how to prevent transmission of the disease.

BARBARA MCCLINTOCK (1902–1992)   American geneticist who discovered the process of genetic transposition within maize chromosomes, for which she received the 1983 Nobel Prize in physiology or medicine.

VICTOR MCKUSICK (1921–2008)   American geneticist known as the father of clinical medical genetics. In 1966, he published the first catalogue of all known genes and genetic disorders, *Mendelian Inheritance in Man*, which was updated 12 times through 1988 and is freely available as the *Online Mendelian Inheritance in Man*. McKusick became the founding president of the Human Genome Organization (HUGO) in 1989.

GREGOR MENDEL (1822–1884)   Austrian scientist and monk who studied the inheritance of traits in pea plants. His groundbreaking paper "Experiments on Plant Hybridization" (1866) outlined how dominant and recessive genes alter a plant's phenotype, but it was largely ignored by the scientific establishment of the day.

JOSEF MENGELE (1911–1979)   German physician and Nazi officer, who was known as the Angel of Death at the Auschwitz-Birkenau concentration camps during World War II, where he sent thousands of people to their deaths in the gas chambers and subjected many others (especially twins) to grotesque medical experiments to test his theories of genetics, heredity, and eugenics.

FRIEDRICH MIESCHER (1844–1895)   Swiss biologist who discovered nucleic acids.

THOMAS HUNT MORGAN (1866–1945) American scientist and founder of modern genetics who studied genetic mutations in *Drosophila melanogaster*, the fruit fly. He received the 1933 Nobel Prize in physiology or medicine for discovering how chromosomes are involved in heredity.

KARY MULLIS (1944–) American molecular biologist who received the Nobel Prize in chemistry in 1993 for pioneering the polymerase chain reaction (PCR), which has become a crucial technique for sequencing DNA.

CHRISTIANE NUSSLEIN-VOLHARD (1942–) German biologist whose work with *Drosophila melanogaster*, in which she manipulated the genes of embryos, led to her winning the 1995 Nobel Prize in physiology or medicine.

LOUIS PASTEUR (1822–1895) French chemist and a founding father of microbiology, whose research led to germ theory (the idea that fermentation, not spontaneous generation, leads to the growth of microorganisms), several important vaccines, and the process of heating liquids to kill bacteria, or pasteurization.

KARL PEARSON (1857–1936) British-born, German-educated mathematician and staunch eugenicist who advocated race war and wrote that "Mankind as a whole, like the individual man, advances through pain and suffering only. The path of progress is strewn with the wreck of nations; traces are everywhere to be seen of the hecatombs of inferior races."

INGO POTRYKUS (1933–) German plant biologist and creator of Golden Rice, which is genetically engineered to include beta carotene and intended to reduce malnutrition in the developing world.

ÁRPÁD PUSZTAI (1930–) Hungarian-born scientist of plant lectins who announced in 1998 that his research showed that potatoes that were genetically modified with the snowdrop lectin caused harm in laboratory rats. He was fired from the Rowett Research Institute in Scotland, where he had worked for 30 years, and denied access to his research.

REINHOLD RAU (1932–2006) German-born South African naturalist, historian, and founder of the Quagga Project, which uses selective breeding techniques to recreate a zebra that looks like the extinct quagga, a species related to the zebra.

V. K. RAVICHANDRAN (dates unknown) Indian professor of agronomy and proponent of SRI—the system of rice intensification, which aims to increase rice yields in India and other developing countries by promoting natural practices such as spacing plants further apart, watering less vigorously, and using organic fertilizer.

JEREMY RIFKIN (1943–   ) American economist, author, and commentator on science and technology issues, especially biotechnology through his books *Who Should Play God?* (1977) and *The Biotech Century* (1998). As an activist, he is critical of genetic engineering and often organizes demonstrations and files lawsuits with an eye to making the public more aware of the dangers involved in biotechnology.

JONAS SALK (1914–1995) American medical doctor and scientist who developed the Salk vaccine for polio in 1952, which led to the near eradication of polio in the United States and around the world within just a few years.

FREDERICK SANGER (1918–   ) British biochemist who was the first to sequence the amino acids of insulin in 1955, for which he received the Nobel Prize in chemistry in 1958. In 1975, he developed the chain termination method of sequencing DNA, known as the Sanger Method, which he used to sequence the first complete DNA genome. That research earned him a second Nobel Prize in chemistry in 1980 and led to the success of the Human Genome Project.

MARGARET SANGER (1879–1966) American birth control activist and proponent of eugenics, who believed that the poor should limit the size of their families. She organized international conferences, published many books, and founded the organization that became Planned Parenthood.

MATTHIAS SCHLEIDEN (1804–1881) German biologist and founder of cell theory in 1839, with THEODOR SCHWANN. The cell theory states that "all living things are composed of cells and cell products."

THEODOR SCHWANN (1810–1882) German zoologist who discovered pepsin, an enzyme in yeast that causes fermentation, and the process of metabolism. With MATTHIAS SCHLEIDEN in 1939, he formulated cell theory, which states that "all living things are composed of cells and cell products."

VANDANA SHIVA (1952–   ) Indian-born physicist, author, community organizer, and environmental activist who campaigns against GMOs in her native country and promotes the concept of Vedic ecology, in which long-proven natural practices in agriculture are followed to ensure preservation of biodiversity. She received the 1993 Right Livelihood Award, the alternative

Nobel Prize, for promoting women's involvement in eliminating poverty among farmers in India.

LEE M. SILVER (1952–)   American molecular biologist and author whose books, including *Remaking Eden: How Genetic Engineering and Cloning Will Transform the American Family* (1998) and *Challenging Nature: The Clash of Science and Spirituality at the New Frontiers of Life* (2006), promote a positive view of biotechnology breakthroughs, including genetically modified organisms and cloning.

RAY F. SMITH (1919–1999)   American entomologist who, along with PERRY ADKISSON, won the 1997 World Food Prize for his development of integrated pest management (IPM), which uses natural means to improve crop yields and discourages the use of chemical pesticides and fertilizers.

PATRICK STEPTOE (1913–1988)   British obstetrician and gynecologist who, along with ROBERT EDWARDS in 1978, performed the world's first successful in vitro fertilization process resulting in the birth of LOUISE BROWN.

NETTIE STEVENS (1861–1912)   American cytologist who discovered in 1905 that females of a species have two X chromosomes, and males have an X and a Y chromosome. It was the first observable trait linked to chromosomes.

MARIE STOPES (1880–1958)   Scottish paleobotanist, eugenicist, author, and birth control advocate. Stopes promoted forced sterilization of people deemed unfit to be parents and to maintain racial purity and in 1921 founded a birth control clinic that later became Marie Stopes International.

ALFRED STURTEVANT (1891–1970)   American geneticist who in 1913 created the first genetic map of a chromosome. A student of THOMAS HUNT MORGAN, Sturtevant worked with *Drosophila melanogaster* and pioneered the processes used in later genetic mapping initiatives.

ROBERT SWANSON (1947–1999)   American venture capitalist and cofounder with HERBERT BOYER of Genentech in 1976. Swanson is widely considered a founding father of the biotech revolution.

JAMES THOMSON (1958–)   American cell biologist and director of the Morgridge Institute for Research at the University of Wisconsin–Madison. In 1998, he led the first team to successfully isolate human embryonic stem cells, and in 2007 he isolated induced pluripotent stem cells (iPS), human skin cells that have the characteristics of embryonic stem cells and thus remove the need to destroy human embryos for research purposes.

MANTO TSHABALALA-MSIMANG (1940–)   South Africa's minister of health from 1999 until 2008. As minister of health, Tshabalala-Msimang denied that HIV causes AIDS and recommended a diet of garlic, beetroot, olive oil, and lemon as a cure for the disease instead of a regimen of antiretroviral drugs.

NORMAN UPHOFF (ca. 1941–)   Former director of the Cornell International Institute for Food, Agriculture and Development (CIIFAD) and promoter of the system of rice intensification (SRI), which aims to solve hunger in the developing world by introducing low-tech, non-GMO-based agricultural practices that will be less damaging to land and better for local economies.

HAROLD VARMUS (1939–)   American scientist who received the 1989 Nobel Prize in physiology or medicine for discovering the link between genetics and cancer and the cellular nature of retroviral oncogenes. Varmus was the director of the National Institutes of Health from 1993 to 1999.

J. CRAIG VENTER (1946–)   American biologist and founder of the Institute for Genomic Research. In the late 1980s, he spearheaded a private initiative to map the human genome, a parallel effort of the U.S. government's Human Genome Project that used a different technique.

RUDOLF VIRCHOW (1821–1902)   German biologist known as the father of pathology for having identified the biological symptoms of leukemia, lung cancer, and other diseases. He also rejected spontaneous generation and promoted cell theory, which states that every living cell originates from another living cell and that cells are the basic unit of organization for all living things. Virchow also standardized autopsy procedures that are still in use more than 100 years later.

JAMES D. WATSON (1928–)   American molecular biologist who shared the 1962 Nobel Prize in physiology or medicine with FRANCIS CRICK and MAURICE WILKINS for discovering the double-helix structure of DNA. In 1988 he was named head of the U.S. Department of Energy's Human Genome Project.

OLIVER C. WENGER (dates unknown)   American physician and director of "The Tuskegee Study of Untreated Syphilis in the Negro Male" during its inception. He misled subjects regarding their diagnosis and withheld treatment for years.

NANCY WEXLER (1945–)   Columbia University researcher whose work on Huntington's disease resulted in finding the gene responsible for the disease. She was also the first head of the Ethical, Legal, and Social Issues Working Group of the National Center for Human Genome Research.

MAURICE WILKINS (1916–2004)   New Zealand–born molecular biologist who shared the 1962 Nobel Prize in physiology or medicine with JAMES D. WATSON and FRANCIS CRICK for his contribution to discovering the double-helix structure of DNA through X-ray diffraction.

IAN WILMUT (1944–)   Scottish embryologist and supervisor of the research team that cloned Dolly the sheep in 1996.

EDMUND BEECHER WILSON (1856–1939)   Pioneering American cell biologist who, along with Nettie Stevens, discovered that gender corresponds to an organism's chromosomes.

SAMUEL WOOD (dates unknown)   American reproductive endocrinologist who donated his DNA for Stemagen's somatic cell nuclear transfer (SCNT) process, which resulted in the creation of five cloned embryos in 2008. Wood is a founder and CEO of Stemagen; his cloned embryos were later destroyed.

YAMANAKA SHINYA (dates unknown)   Japanese stem cell researcher and professor at Kyoto University who generated human-induced pluripotent stem cells (iPS) in 2007, a process that eliminates the need to destroy human embryos to obtain stem cells.

9

Organizations and Agencies

Access Excellence @ the National Health Museum
URL: http://www.accessexcellence.org
1350 Connecticut Avenue NW, 5th floor
Washington, DC 20036
Phone: (650) 712-1723

This Web site for life science offers students and educators a wide variety of free resources, including articles about the history of biotechnology and key individuals and dates.

AfricaBio
http://www.africabio.com
Phone: (27-12) 667-1844

A nonprofit association of research and policy professionals that advocates for the responsible use of research and application of biotechnology, mostly in agriculture. The Web site presents fact sheets, reports, the full text of government laws, and lists of key contacts in Africa.

American Genetic Association
URL: http://www.theaga.org
2030 SE Marine Science Drive
Newport, OR 97365
Phone: (541) 867-0334

A professional organization that began in 1903 as the American Breeders Association; the American Genetic Association took its current name in 1914. The organization publishes primary research in its *Journal of Heredity*.

American Society for Reproductive Medicine
URL: http://www.asrm.org
1209 Montgomery Highway
Birmingham, AL 35216-2809

A nonprofit organization that focuses on issues related to infertility, menopause, and contraception. It seeks to educate the public and provides continuing education for medical professionals.

American Society of Gene Therapy
URL: http://www.asgt.org
555 East Wells Street
Milwaukee, WI 53202
Phone: (414) 278-1341

A membership organization devoted to public and professional education and promotion of gene, cell, and nucleic acid therapies.

American Society of Human Genetics
URL: http://www.ashg.org
9650 Rockville Pike
Bethesda, MD 20814-3998
Phone: (301) 634-7300

A membership organization formed in 1948 for professionals in the field of human genetics. It publishes the *American Journal of Human Genetics,* promotes education for students and teachers, and sponsors the annual DNA Day Essay Contest.

American Society of Law, Medicine and Ethics
URL: http://www.aslme.org
765 Commonwealth Drive, Suite 1634
Boston, MA 02215
Phone: (617) 262-4990

Professional organization founded in 1911 that provides interdisciplinary education to its members, who are attorneys, nurses, ethicists, and health workers. One major program area is biomedical science and research, where it monitors the intersection of science and research.

Bill & Melinda Gates Foundation
URL: http://www.gatesfoundation.org/Pages/home.aspx
PO Box 23350
Seattle, WA 98102
Phone: (206) 709-3100

The world's largest philanthropic foundation; seeks to improve lives around the world by eliminating disease and improving health. The agriculture program gives grants to help farmers in the developing world use biotechnology to increase their harvests. One of its major health initiatives is developing an AIDS vaccine.

Biowatch South Africa
URL: http://www.biowatch.org.za
PO Box 13477
Mowbray 7705
South Africa
Phone: +27 (0) 82 435 5812

A grassroots organization that "strives to prevent biological diversity from being privatised for corporate gain." This Web site includes information on GM crops in South Africa and links to articles related to the food supply.

Biotechnology Industry Organization (BIO)
URL: http://www.bio.org
1201 Maryland Avenue SW, Suite 900
Washington, DC 20024
Phone: (202) 962-9200

A pro-industry lobbying group formed in 1993, with individual and corporate membership. It is dedicated to promoting policies on GM crop regulation, national health care, and FDA reform.

Celera Genomics
https://www.celera.com
1401 Harbor Bay Parkway
Alameda, CA 94502-7070
Phone: (510) 749-4200

Formed in 1998 by J. Craig Venter for the purpose of sequencing the human genome using its own "shotgun sequencing" process, Celera is now dedicated to personalizing disease management by conducting research and developing products such as the Cystic Fibrosis Genotyping Assay. The Web site includes news and information on products and research.

Center for Biologics Evaluation and Research
http://www.fda.gov/cber
1401 Rockville Pike, Suite 200N
Rockville, MD 20852-1448
Phone: (800) 835-4709

An agency of the FDA that regulates biologics—drugs and substances derived from living sources, which includes many pharmaceuticals, blood products, vaccines, allergenics, tissues, and cell and gene therapies. The Web site has up-to-date information on research and consumer and product information.

Center for Food Safety
URL: http://www.centerforfoodsafety.org
660 Pennsylvania Avenue SE, Suite 302
Washington, DC 20003
Phone: (202) 547-9359

A nonprofit public interest group devoted to campaigning against GMOs, both plant crops and livestock. The group is also concerned about the use of bovine growth hormone, mad cow disease, food irradiation, and the cloning of animals. It was established in 1997 by the International Center for Technology Assessment.

Center for Genetics and Society
URL: http://www.geneticsandsociety.org
436 14th Street, Suite 700
Oakland, CA 94612
Phone: (510) 625-0819

A nonprofit group that promotes the responsible use of human genetics and reproductive technologies and seeks to provide access to its beneficial applications to all needy people. It publishes the blog *Biopolitical Times*.

Centers for Disease Control and Prevention
URL: http://www.cdc.gov
1600 Clifton Road
Atlanta, GA 30333
Phone: (800) 232-4636

The center's Web site contains information on a variety of topics for students, teachers, and citizens, including statistics on mortality, morbidity, infectious disease, and chronic disease. The search function enables users to find information about the agency's work in genetics and biotechnology.

Codex Alimentarius Commission
URL: http://www.codexalimentarius.net
Viale delle Terme di Caracalla
00153 Rome
Italy
Phone: (39-6) 5705-1

Web site on food standards developed by the Food and Agriculture Organization and the World Health Organization, both agencies of the United Nations. Published in English, French, and Spanish, the Web site offers the text of the 1963 *Codex Alimentarius* and links to other reports.

Cold Spring Harbor Laboratory
URL: http://www.cshl.edu
1 Bungtown Road
Cold Spring Harbor, NY 11724
Phone: (516) 367-8800

Founded in 1890, the laboratory was once the epicenter of the American eugenics movement. Today it is a private, nonprofit institution that conducts research in the areas of plant biology, genomics, and bioinformatics, among others. The Web site contains a wealth of information for students and teachers, as well as links to the CSH-sponsored Dolan DNA Learning Center and EugenicsArchives.org.

Council for Biotechnology Information
URL: http://www.whybiotech.com
1201 Maryland Avenue SW, Suite 900
Washington, DC 20024
Phone: (202) 962-9200

A lobbying organization for the agricultural biotechnology industry that promotes sustainable development internationally through GE crops, biofuels, and other advances. Member companies include BASF PlantScience, Bayer CropScience, Dow AgroSciences, DuPont, Monsanto, and Syngenta. The Web site includes fact sheets and issue briefs.

Council for Responsible Genetics
URL: http://www.gene-watch.org
5 Upland Road, Suite 3
Cambridge, MA 02140
Phone: (617) 868-0870

A nonprofit organization that fosters discussion on the social, ethical, and environmental consequences of biotechnology. It publishes *GeneWatch* magazine.

CropLife America
URL: http://www.croplifeamerica.org
1156 15th Street NW, Suite 400
Washington, DC 20005
Phone: (202) 833-4480

A lobbying organization for the plant science industry. Members include BASF PlantScience, Bayer CropScience, Dow AgroSciences, DuPont, Monsanto, and Syngenta. It focuses on promoting biotechnology and pesticides internationally.

Department of Agriculture Animal and Plant Health Inspection Service (APHIS)
URL: http://www.aphis.usda.gov
Biotechnology Regulatory Service
4700 River Road, Unit 147
Riverdale, MD 20737
Phone: (301) 734-7324

This is the government agency charged with protecting American agriculture and regulating GMOs.

European Federation of Biotechnology
URL: http://www.efb-central.org
Lluis Companys 23
08010 Barcelona
Spain
Phone: (34 93) 268 77 03

A European organization with members from 56 countries that promotes safe, sustainable use of biotechnology and promotes biotech research. The Web site includes many briefing papers on genetics and biotechnology issues as they relate to both medicine and agriculture.

European Society of Human Genetics
URL: http://www.eshg.org
c/o Vienna Medical Academy
Alser Strasse 4
1090 Vienna
Austria
Phone: (43-1) 405 13 83 20

A professional organization of researchers involved in biochemical genetics, cytogenetics, and genomics. The Web site includes many recent news items about biotechnology.

Gene Therapy Advisory Committee
URL: http://www.advisorybodies.doh.gov.uk/genetics/gtac
GTAC Secretariat
Department of Health
Area 604
Wellington House
135-155 Waterloo Road
London SE1 8UG
United Kingdom
Phone: (44-020) 7972-3057

Oversees the clinical trials of gene and stem cell therapies in the United Kingdom to verify that they meet ethics standards. The Web site contains information on the country's Department of Health definitions of and requirements for gene and stem cell therapy.

Genetic Interest Group
URL: http://www.gig.org.uk
Unit 4D, Leroy House
436 Essex Road
London N1 3QP
Phone: (44-020) 7704-3141

A nonprofit organization that works for the rights of patients afflicted with genetic disorders. It promotes education, research, technology transfer, and patient rights. The Web site contains good general information on specific genetic diseases and current issues.

Genetics and Public Policy Center
URL: http://www.dnapolicy.org

1717 Massachusetts Avenue NW, Suite 530
Washington, DC 20036
Phone: (202) 663-5971

Affiliated with Johns Hopkins University and formed in response to the scientific advances forged by the Human Genome Project, the center surveys public attitudes toward genetics, monitors technology transfer, and advises policy makers on issues relating to genetics.

Genetics Society of America
URL: http://www.genetics-gsa.org
9650 Rockville Pike
Bethesda, MD 20814-3998
Phone: (301) 634-7300

A professional organization that advocates for responsible genetics research, fosters communication between researchers, and educates students and the public on recent developments in genetics. It publishes the journal *Genetics* and produces the ongoing DVD series *Conversations in Genetics.*

Genome Programs of the U.S. Department of Energy
URL: http://genomics.energy.gov

A Web portal for students, teachers, and citizens that contains educational materials, research databases, online libraries, and databanks on many topics related to the Human Genome Project, microbial genome research, genomics, biofuels, medicine, and ethics issues.

Germany Federal Office of Consumer Protection and Food Safety (BVL)
URL: http://www.bvl.bund.de
Bundesallee 50, Building 247
38116 Braunschweig
Phone: (05-31) 214-970

Web site, available in English, that provides information on the German government's position on GMOs, genetic engineering, and other biotechnology issues.

GMO Compass
URL: http://www.gmo-compass.org/eng/home
Genius Biotechnologie Gmbh

Robert-Bosch-Strasse 7
64293 Darmstadt
Germany
Phone: (49-6151) 872-4040

Financed by the European Union, GMO Compass is a consumer-oriented Web portal that provides information on issues related to genetic engineering in EU member countries. It includes EU country reports, news stories on GMOs from around the world, and links to EU agencies.

Greenpeace
URL: http://www.greenpeaceusa.org
702 H Street NW, Suite 300
Washington, DC 20001
Phone: (202) 462-1177

Greenpeace's genetic engineering program calls for thorough testing and labeling of GE food and seed. The Web site contains information for the general public about how the issue affects them.

Hastings Center
URL: http://www.thehastingscenter.org
21 Malcolm Gordon Road
Garrison, NY 10524
Phone: (845) 424-4040

"A nonpartisan research institution dedicated to bioethics and the public interest since 1969." It publishes reports, periodicals, monographs, and *The Hastings Center Report,* which provides in-depth analysis of ethics topics in medicine, including those related to genetics.

Howard Hughes Medical Institute
URL: http://www.hhmi.org
4000 Jones Bridge Road
Chevy Chase, MD 20815-6789
Phone: (301) 215-8500

Founded in 1953 by aviator Howard Hughes, the HHMI is a leading non-profit research organization that also boasts a strong educational component for students and teachers. It publishes the *HHMI Bulletin* and funds cutting-edge biomedical research at 64 laboratories in universities across the country.

Indian Department of Biotechnology
URL: http://dbtindia.nic.in/index.asp
Block 2, C.G.O. Complex
Lodhi Road
New Delhi 110 003
India

Available in English; this is the Web site for the department of biotechnology, which operates under the Ministry of Science and Technology. Its mandate is to promote large-scale use of biotechnology in India, research and development, education, and commerce, and to establish policy guidelines. The department includes many institutions, such as the Institute of Immunology, the Centre for DNA Fingerprinting, and the Rajiv Gandhi Centre for Biotechnology.

International Center for Technology Assessment
URL: http://www.icta.org
660 Pennsylvania Avenue SE, Suite 302
Washington, DC 20003
Phone: (202) 547-9359

A nonprofit watchdog organization with programs in nanotechnology, corporate accountability, and human biotechnology. It warns of eugenic tendencies in modern biotechnology practices and lobbies for strong regulations. It is also concerned with the patenting of human cloning techniques and natural resources. Its sister organization is the Center for Food Safety.

International Centre for Genetic Engineering and Biotechnology
URL: http://www.icgeb.trieste.it/home.html
AREA Science Park
Padriciano 99
34012 Trieste
Italy
Phone: (39-040) 37571

Established by the General Assembly of the United Nations in 1983, the ICGEB has offices in Trieste, New Delhi, and Cape Town. It is an "international organisation dedicated to advanced research and training in molecular biology and biotechnology, with special regard to the needs of the developing world."

International Federation of Human Genetics Societies
URL: http://www.ifhgs.org

9650 Rockville Pike
Bethesda, MD 20814
Phone: (301) 634-7300

An organization founded in 1996 that sponsors the International Congress of Human Genetics, which takes place every five years. The next one will be in Montreal, Canada, in 2011. The IFHGS's membership is comprised of professional membership organizations from different geographical regions, including the African Society of Human Genetics, the American Society of Human Genetics, the German Society of Human Genetics, the Indian Society of Human Genetics, and the Japan Society of Human Genetics.

International Food Information Council
URL: http://www.ific.org
1100 Connecticut Avenue NW, Suite 430
Washington, DC 20036
Phone: (202) 296-6540

The Web site, available in English and Spanish, includes a glossary of food biotechnology terms and a link to the organization's 2008 report, *Food Biotechnology: A Study of U.S. Consumer Attitudinal Trends.*

International Service for the Acquisition of Agri-Biotech Applications (ISAAA)
URL: http://www.isaaa.org
417 Bradfield Hall
Cornell University
Ithaca, NY 14853
Phone: +1 607 255-1724

A nonprofit organization that brings agricultural biotechnology to poor people in developing countries, with an emphasis on technology transfer. The Web site is multilingual and includes statistics, videos, and other teaching tools. The organization has offices in Nairobi, Kenya, and Ithaca, New York.

International Society for Stem Cell Research
URL: http://www.isscr.org
111 Deer Lake Road
Suite 100

Deerfield, IL 60015
Phone: (847) 509-1944

An independent, nonprofit organization that promotes stem cell research and education about that research.

J. Craig Venter Institute
URL: http://www.jcvi.org
9704 Medical Center Drive
Rockville, MD 20850
Phone: (301) 795-7000

A nonprofit research institute with more than 400 scientists working in areas including human genomic medicine, plant genomics, bioenergy, and bioinformatics. It was the first organization to publish the complete human diploid genome and produces the DiscoverGenomics! Science Education Program for K–12 students.

Molecular Biology Society of Japan
URL: http://wwwsoc.nii.ac.jp/mbsj/en
20 Sankyo Building 11F
3-11-5 Iida bashi
Chiyoda-ku
Tokyo 102-0072
(81-3) 3556-9600

A professional organization for biotechnology researchers in Japan. Some Web pages are in English.

Monsanto
URL: http://www.monsanto.com
800 N. Lindbergh Boulevard
St. Louis, MO 63167
Phone: (314) 694-1000

Web site of the agroscience corporation. It provides detailed explanations of the company's position on many issues, including "terminator technology" seeds, suicides among Indian farmers, and issues related to the labeling of GMOs and gene patenting.

National Center for Biotechnology Information (NCBI)
URL: http://www.ncbi.nlm.nih.gov

8600 Rockville Pike
Bethesda, MD 20894
Phone: (301) 496-2475

A data-rich portal Web site sponsored by the U.S. National Library of Medicine. It provides links to all public databases that collect and publish research information, from GenBank to dbGaP, the database of Genotype and Phenotype, which provides data from studies conducted to discover the link between genes and disease.

National Human Genome Research Institute
URL: http://www.genome.gov
Building 31, Room 4B09
31 Center Drive, MSC 2152
9000 Rockville Pike
Bethesda, MD 20892-2152
Phone: (301) 402-0911

The institute was established in 1989 to carry out the mandate of the Human Genome Project and has expanded since that project's completion in 2003 with research into the human genome's role in disease. One of the 27 research institutes that comprise the National Institutes of Health.

National Society of Genetic Counselors
URL: http://www.nsgc.org
401 N. Michigan Avenue
Chicago, IL 60611
Phone: (312) 321-6834

A professional organization for genetic counselors; the Web site allows users to find counselors in their area. It includes an extensive FAQ section about how to become a genetic counselor, what counselors do, and their code of ethics.

Nature Publishing Group
URL: http://www.nature.com
75 Varick Street, 9th floor
New York, NY 10013-1917
Phone: (212) 726-9200

Web site of the journal *Nature* (founded in 1869), as well as many other scientific journals. It contains the full text of many articles relating to biotechnology, genetics, and the Human Genome Project that are printed in

Nature, as well as abstracts of research studies published in the group's other journals.

Organic Consumers Association
URL: http://www.organicconsumers.org
6771 S. Silver Hill Drive
Finland, MN 55603
Phone: (218) 226-4164

A grassroots nonprofit organization dedicated to promoting the interests of consumers and producers of organic food. It is interested in food safety, industrial agriculture, and genetic engineering. The Web site contains many articles on biodynamics, cloning, nanotechnology, Mad Cow disease, and local farming.

President's Council on Bioethics
URL: http://www.bioethics.gov
1425 New York Avenue NW, Suite C100
Washington, DC 20005
Phone: (202) 296-4669

Formed by President George W. Bush in 2001, the President's Council on Bioethics is a panel of appointed experts that advises the president on advancements in biotechnology. The council publishes numerous reports, all of which are available on the Web site, on issues ranging from human cloning and stem cell research to bioethics.

Society for Biotechnology, Japan
URL: http://www.sbj.or.jp/e
c/o Faculty of Engineering
Osaka University
2-1 Yamadaoka, Suita
Osaka
Japan 565-0871
Phone: (81-6) 6876-2731

Web site for a professional organization of biotechnology researchers available in English. The society publishes the *Journal of Biosciences and Bioengineering.*

Union of Concerned Scientists
URL: http://www.ucsusa.org
2 Brattle Square

Cambridge, MA 02238-9105
Phone: (617) 547-5552

A nonprofit organization founded at Massachusetts Institute of Technology in 1969; now has a membership of 250,000 scientists and laypeople. Food and agriculture is a major program area. The Web site includes the full text of the organization's reports on concentrated animal feeding operations (CAFOs), pharmaceutical crops, antibiotic abuse in relation to livestock, transgenic contamination, and other biotechnology topics.

United States Food and Drug Administration
URL: http://www.fda.gov
5600 Fishers Lance
Rockville, MD 20857-0001
Phone: (888) 463-6332

The federal agency that assures the safety of the nation's drugs, biological products, and other substances in order to protect public health. The Web site presents a wealth of information for students and educators, including the text of speeches and transcripts of congressional testimony.

10

Annotated Bibliography

The following annotated bibliography focuses on biotechnology and genetic engineering issues in the United States and in the countries examined in chapter 3. Entries are grouped into the following five categories:

Bioethics

Biotechnology and Agriculture

Biotechnology, Medicine, Stem Cell Research, and Biowarfare

Genetic Engineering/Human Genome Project

Biotechnology History and Biography

Each category is subdivided into four sections: *Books, Articles and Papers, Web Documents,* and *Other Media.*

BIOETHICS

Books

Brannigan, Michael C., ed. *Cross-Cultural Biotechnology.* Lanham, Md.: Rowman & Littlefield, 2004. Collection of essays explaining how biotechnology is viewed in various cultures. Includes the chapters "Islamic Perspectives on Biotechnology" and "Agricultural Biotechnology in African Countries" and an extensive bibliography.

Dewar, Elaine. *The Second Tree: Stem Cells, Clones, Chimeras, and Quests for Immortality.* New York: Carroll & Graf, 2004. Dewar is a Canadian investigative journalist whose book presents stories from the fringes of biotechnology and examines the moral and political issues they raise. She interviews key figures and lets them tell their own stories.

Fukuyama, Francis. *Our Posthuman Future: Consequences of the Biotechnology Revolution.* New York: Farrar, Straus & Giroux, 2002. Noted political economist Fukuyama was a member of the President's Council on Bioethics and believes that biotechnology is the last vulnerability of humankind, capable of launching us into a "post-human" world. He advocates strict regulations when it comes to research and development in biotech fields.

Glover, Jonathan. *Choosing Children: Genes, Disability, and Design.* New York: Oxford University Press, 2008. Proposes that genetics has led humankind to a turning point and offers ethical arguments for allowing babies to continue being born with disabilities, despite medicine's ability to prevent them. Challenges the ideas of eugenics and offers the example of a deaf couple who chose to have a deaf child.

Guillemin, Jeanne. *Biological Weapons: From the Invention of State-Sponsored Programs to Contemporary Bioterrorism.* New York: Columbia University Press, 2005. Guillemin is a professor of sociology and a security studies expert whose book traces the origins of weaponized biological agents in the 20th century through the new fears caused by the 9/11 attacks.

Human Dignity and Bioethics: Essays Commissioned by the President's Council on Bioethics. Washington, D.C.: President's Council on Bioethics, 2008. Collection of scholarly position papers from members of the council that examine the history of human dignity and how it manifests today, whether or not it is in sync with modern-day medicine.

Kitcher, Philip. *The Lives to Come: The Genetic Revolution and Human Possibilities.* New York: Free Press, 1997. A philosophy professor, Kitcher outlines the moral quandaries Americans are likely to face in coming years as biotechnology begins to seep into new corners of the culture such as health insurance, job security, genetic screening, and abortion.

———. *In Mendel's Mirror: Philosophical Reflections on Biology.* New York: Oxford University Press, 2003. Collection of previously published essays on biology as it relates to everyday life.

Kurzweil, Ray. *The Singularity Is Near: When Humans Transcend Biology.* New York: Viking, 2005. Kurzweil is a scientist, inventor, artificial intelligence expert, and futurist who claims that a combination of genetics, nanotechnology, and robotics will soon lead to the "singularity," a condition of limitless intelligence and immortality, which will thrust humankind into a new, highly advantageous epoch. Such a philosophy is known as "transhumanism," which allows for people to use artificial means to extend their lives indefinitely.

Nelson, Gerald C., ed. *Genetically Modified Organisms in Agriculture: Economics and Politics.* San Diego, Calif.: Academic Press, 2002. College-level textbook that explores GMOs in depth from a global economic viewpoint.

Rifkin, Jeremy. *The Biotech Century.* New York: Putnam, 1998. Rifkin takes a cautionary approach to the promises of biotechnology in the 21st century. He theorizes that the risks outweigh the benefits when it comes to recombinant DNA, gene mapping, and cloning. He believes that loss of biodiversity and increased security risks from bioweapons are just two problems people need to consider.

Silver, Lee M. *Challenging Nature: The Clash of Science and Spirituality at the New Frontiers of Life.* New York: Ecco, 2006. Silver is a molecular biologist at

Princeton University and a proponent of biotechnology advances such as cloning and GM food. The opposition to biotechnology in the West often centers on Christianity and a spiritualized idealism of Mother Nature, he says, while Eastern religions are much more accepting of the new technology.

Spar, Debora L. *The Baby Business: How Money, Science, and Politics Drive the Commerce of Conception.* Cambridge, Mass.: Harvard Business School Press, 2006. Spar is a Harvard Business School professor who looks at the ethical and business aspects of reproductive technology, including extremely profitable fertility clinics and the possible exploitation of egg donors and surrogates worldwide. Spar wonders how close fertility clinics come to the practice of actually selling babies.

Sulston, John, and Georgina Farry. *The Common Thread: A Story of Science, Politics, Ethics and the Human Genome.* Washington, D.C.: Joseph Henry Press, 2002. Sulston, who won a Nobel Prize for his work in genome sequencing, tells the inside story of the quest to keep the results of the Human Genome Project in the public domain. A tale of scientific discovery, the book is also a chronicle of the happy accidents, outsized personalities, and power struggles behind the scenes of the project, which could have had much different results.

Takahashi, Takao, ed. *Taking Life and Death Seriously: Bioethics from Japan.* San Diego, Calif.: Elsevier, 2005. A college-level study from researchers affiliated with the Kumamoto University Bioethics Research Group, which maintains that bioethics in Japan differs qualitatively from that in America. Bioethics started in Japan 10 years later than in the United States, and still only a fraction of academics are involved in the discipline.

Articles, Reports, and Papers

Berg, Paul, and David Ewing Duncan. "Bio Brain Backs Stem Cells: 'I'm an Experimentalist and the Only Way I Can Tell You If It Works Is to Try It.'" *Discover* 26.4 (April 2005): 18. Interview with Berg, in which he talks about lobbying for stem cell research, which he believes is necessary for medical advancement. He considers political attempts to limit such research unethical, and believes that GM food is both safe and necessary.

Brown, Eric. "The Dilemmas of German Bioethics." *New Atlantis* (Spring 2004): 37–53. Long article about the history of bioethics in Germany, from pre–World War II eugenics policies to current public opinion. Explores the writings of latter-day German philosophers, including Peter Sloterdijk, author of *Rules for the Human Zoo.*

Collier, Paul. "The Politics of Hunger: How Illusion and Greed Fan the Food Crisis." *Foreign Affairs* (November/December 2008). Article from the Council on Foreign Relations about the global food crisis, advocating for reform from politicians and policymakers to alleviate hunger in developing countries, particularly among poor urban dwellers. Collier believes that GM foods are essential to solving the food shortage and that all bans against them should be lifted.

Dixon, Darrin P. "Informed Consent or Institutional Eugenics? How the Medical Profession Encourages Abortion of Fetuses with Down Syndrome." *Issues in Law and Medicine* 24.1 (summer 2008): 3. Long article that claims the medical establishment gives pregnant women misleading, erroneous, and incomplete information when it comes to genetic testing for Down syndrome, which constitutes a violation of informed consent and results in a very high abortion rate for Down syndrome fetuses.

Dugger, Celia W. "Study Cites Toll of AIDS Policy in South Africa." *New York Times* (11/26/08). Article tells of a Harvard study that estimates that the deaths of 365,000 South Africans between 2000 and 2005 could have been prevented if the government had provided antiretroviral drugs to AIDS patients. Failure to make treatment available to AIDS patients stemmed from President Mbeki's belief that poverty—not the HIV virus—causes AIDS.

En-Chang, Li. "Bioethics in China." *Bioethics* 22.8 (October 2008): 448–454. Overview of the development of a framework for bioethics in China, from the Hanghong euthanasia case, to the establishment of the Chinese Medical Association and the Chinese Society of Medical Ethics, culminating in the Eighth World Congress of Bioethics, held in Beijing in 2006.

Lawton, Kim. "How Far Is Too Far in Enhancing Kids' Genetic Traits?" *National Catholic Reporter* 43.38 (9/21/07). Explores the ethics of genetic enhancement: Is it qualitatively different from cosmetic enhancement, and who should pay for it?

Mishra, Pankaj. "How India Reconciles Hindu Values and Biotech." *New York Times* (8/21/05). An Indian writer notes that India's system of traditional medicine, Ayurveda, and the country's national myth, the *Mahabharata*, encapsulate ideas that appear to condone modern-day biotechnology. Nevertheless, he warns that India's huge population of poor people could be exploited in biotech efforts to develop new drugs and procedures.

Morioka, Masahiro. "Bioethics and Japanese Culture: Brain Death, Patients' Rights, and Cultural Factors." *Eubios Journal of Asian and International Bioethics* 5 (1995): 87–90. Morioka is a leading bioethicist in Japan and has written several books on the topic. In this paper, he explores how modern biotechnology is in conflict with traditional culture and religion, particularly when it comes to brain death and organ transplants, which are central ethical issues in Japan.

Pinker, Steven. "The Stupidity of Ethics." *New Republic* 238.4836 (5/28/08): 28. Pinker, a leading linguist and social thinker, criticizes the book *Human Dignity and Bioethics*, a compilation of essays written by members of the President's Council on Bioethics. Pinker believes the authors fail to adequately define "dignity" and are delaying advances in biotechnology and medicine by playing hardcore religious politics.

Pollack, Andrew. "Fighting Diseases with Checkbooks." *New York Times* (7/8/06). Article about private medical philanthropy, including the Ellison Medical

Foundation, the Bill & Melinda Gates Foundation, and others that use private wealth to tackle diseases that may not receive significant public funding.

Pusztai, Arpad, and Stanley W. B. Ewen. "Effect of Diets Containing Genetically Modified Potatoes Expressing Galanthus Nivalis Lectin on Rat Small Intestine." *Lancet* 354.9187 (10/16/99): 1,353–1,354. The research study that ended Pusztai's career at the Roslin Institute in Scotland. After stating that he would not eat food genetically modified with the lectin, he lost his job and an international furor over his comments ensued.

Sandel, Michael. "Designer Babies: The Problem with Genetic Engineering." *Tikkun* 22.5 (Sept.–Oct. 2007): 40. Sandel is a professor of philosophy at Harvard, where he teaches the course Ethics and Biotechnology; he was also a member of George W. Bush's President's Council on Bioethics. Sandel links the idea of designer babies to privatized eugenics, which he calls "morally troubling," and believes that children are gifts who should be welcomed for who they are, not who their parents want them to be.

Tierney, John. "Are Scientists Playing God? It Depends on Your Religion." *New York Times* (11/20/07). Exploration of the East–West religious divide and its impact on biotechnology from the *Times's* biotechnology beat reporter. Generally, Tierney states that Eastern-based religions have few taboos against cloning and technologies that cause consternation in the West.

Web Documents

Brooks, Jamie D., and Meredith L. King. "Geneticizing Disease: Implications for Racial Health Disparities." Center for Genetics and Society (1/15/08). Available online: URL: http://geneticsandsociety.org/downloads/2008_geneticizing_disease.pdf. Accessed June 9, 2009. Joint report of the Center for American Progress and the Center for Genetics and Society that raises concerns about the increasing number of genetic links to disease that correspond with race. The authors fear a new era of scientific racism will obstruct many people's access to health care, resulting in health care disparities that correspond to race.

de Grey, Aubrey. "Why We Age and How We Can Avoid It." (July 2005). 23-minute video of TED conference lecture. Available online. URL: http://www.ted.com/index.php/talks/aubrey_de_grey_says_we_can_avoid_aging.html. Accessed June 9, 2009. De Grey, a Cambridge University biomedical gerontologist, believes aging is a disease that can be cured and that within a few decades it will be possible through biotechnology to extend human life past 1,000 years. His thesis can be summarized as, "Why should we cure aging? Because it kills people."

Kimura, Rihito. "Jurisprudence in Genetics." (1991). Available online. URL: http://www.bioethics.jp/licht_genetics.html. Accessed June 9, 2009. Essay first published in *Ethical Issues of Molecular Genetics in Psychiatry*, 1991. Outlines the history of genetics and eugenics in Japan, with regard to the current Eugenics Protection Law. Kimura is the president of Keisen University in Tokyo.

Macer, Darryl R. J. "Regional Perspectives in Bioethics: Japan." *Annals of Bioethics: Foundational Volume on Regional Perspectives,* edited by J. Peppin. The Netherlands: Swets & Zeitlinger, 2003, pp. 321–337. Available online. URL: http://www.eubios.info/Papers/japbe.htm. Accessed June 9, 2009. Macer analyzes Japan's unique take on bioethics, which he says is a combination of traditional Buddhist, Confucian, and Shinto beliefs mixed with Western beliefs and intended to be pragmatic. The country is transitioning from a paternalistic attitude toward medical care, in which doctors were assumed to know all, to a more individualistic one that values the rights of the patient.

Mandela, Nelson. "Science and Technology in a Democratic South Africa: An ANC Perspective." (August 1993). Available online. URL: http://www.anc.org.za/ancdocs/history/mandela/1993/sp930830.html. Accessed June 9, 2009. Text of a speech Mandela gave to the South African Institute of Mechanical Engineers before he became president, in which he states that the organization's first 100 years was devoted to applying technology to uphold the apartheid system and that in its next 100 years must be used to raise the standard of living of impoverished citizens who were oppressed by apartheid.

O'Mathúna, Dónal P. "Human Dignity in the Nazi Era: Implications for Contemporary Bioethics." *BioMed Central* (3/14/06). Available online. URL: http://www.biomedcentral.com/1472-6939/7/2. Accessed June 9, 2009. Long, scholarly article by a health care ethicist who states that Nazi medical atrocities were a reflection of warped beliefs about human dignity.

Parrington, John. "Why We Should Defend Stem Cell Research." *Science for the People* (1/21/07). Available online. URL:http://www.socialistworker.org.uk/article.php?article_id=10534. Accessed June 9, 2009. Pro–stem cell research article on a Web site maintained by the *Socialist Worker* newspaper, the organ of the International Socialist Organization.

Safian, Mohd, and Yasmin Hanani. "Islam and Biotechnology: With Reference to Genetically Modified Foods." Available online. URL: http://www.metanexus.net/conference2005/pdf/mohd_safian.pdf. Accessed June 9, 2009. Paper given at the Science and Religion: Global Perspectives Conference in Philadelphia, PA, June 4–8, 2005 in which the authors examine the concept of *maslahah* (public interest) as it pertains to GM food in determining if a corporation's attempt to patent and copyright GMOs is allowable under shariah.

Sofair, André N., and Lauris C. Kaldjian. "Eugenic Sterilization and a Qualified Nazi Analogy: The United States and Germany, 1930–1945." *Annals of Internal Medicine* 132.4 (2/15/00): 312–319. Available online. URL: http://www.annals.org/cgi/reprint/132/4/312.pdf. Accessed June 9, 2009. The authors look at historical documents and determine that eugenics programs in the United States and Nazi Germany shared many similarities, but also some important differences that resulted in the U.S. program losing much of its power by World War II.

World Medical Association. *Declaration of Helsinki: Ethical Principles for Medical Research Involving Human Subjects* (June 1964). Available online. URL: http://www.wma.net/e/policy/b3.htm. Accessed June 9, 2009. First adopted in Helsinki, Finland, in June 1964, and amended since, the declaration is a major international treaty outlining the ethics of human experimentation and establishing the rights and dignity that belong to all subjects.

Other Media

Charlie Rose interview with Leon R. Kass. (7/1/03) 38 minutes. Available online. URL: http://www.charlierose.com/guest/view/1881. Accessed June 9, 2009. Interview with Kass, a long-time opponent of embryonic stem cell research and cloning, and also the former head of President George W. Bush's President's Council on Bioethics.

Charlie Rose interview with Michael Crichton. (2/19/07) 52 minutes. Available online. URL: http://www.charlierose.com/view/interview/1. Accessed June 9, 2009. Rose interviews Crichton about his book *Next,* which looks at the issue of disease ownership, in which corporations patent genes and take legal ownership of certain diseases and their cures.

"Memory Pill." *60 Minutes.* (11/16/06). Leslie Stahl examines the new drug propranolol, which weakens traumatic memories. Should those who are perpetually traumatized by memories of rape, violence, war, or death be able to "forget" them with the help of pharmaceuticals?

BIOTECHNOLOGY AND AGRICULTURE
Books

Charles, Daniel. *Lords of the Harvest: Biotech, Big Money, and the Future of Food.* Cambridge, Mass.: Perseus, 2001. Charles is a science reporter and radio correspondent who looks at the agri-biotech industry. In this book, he is particularly interested in Monsanto's role in agribusiness and the controversy surrounding some of the company's products and business decisions.

Eisnitz, Gail A. *Slaughterhouse: The Shocking Story of Greed, Neglect, and Inhuman Treatment Inside the U.S. Meat Industry.* Revised edition. Amherst, N.Y.: Prometheus Books, 2006. Eisnitz outlines the changes in the meatpacking industry since the 1980s, with deregulation and industry conglomeration leading to larger and more polluting factory farms.

Fedoroff, Nina, and Nancy Marie Brown. *Mendel in the Kitchen: A Scientist's View of Genetically Modified Foods.* Washington, D.C.: Joseph Henry Press, 2004. The authors take a positive view of GMOs, explaining that they differ little from the hybrid crosses that biologists and horticulturalists have created for generations. The book explains the basics of genetic engineering and examines many current issues, such as Golden Rice, which the authors maintain will save millions of lives and therefore should not be branded Frankenfood, as GM food so often is by its detractors.

Fukuda-Parr, Sakiko, ed. *The Gene Revolution: GM Crops and Unequal Development.* London, England: Earthscan, 2007. Collection of essays exploring the problems of establishing GM crops in Argentina, Brazil, India, China, and South Africa.

Lambrecht, Bill. *Dinner at the New Gene Café.* New York: Thomas Dunne, 2001. Lambrecht, a journalist, investigates GM food and agribusiness, uncovering information that is not well known to the public regarding the food that the bulk of Americans currently eat.

Martineau, Belinda. *First Fruit: The Creation of the Flavr Savr Tomato and the Birth of Genetically Engineered Food.* New York: McGraw-Hill, 2001. Martineau is a geneticist who worked on the Flavr Savr tomato at Calgene, and here she provides an insider's view of its rise and fall. The science that brought about GMOs was ultimately not in sync with the marketing and public relations aspects of GMOs, she claims, which presented a conflict that ultimately doomed the Flavr Savr.

Mellon, Margaret, and Jane Rissler. *The Ecological Risks of Engineered Crops.* Cambridge, Mass.: MIT Press, 1996. Published in conjunction with the Union of Concerned Scientists, the book cautiously analyzes GM crops and does not offer a wholesale rejection of them. Instead, the authors analyze the methodology of research into transgenic crops and raise questions that need to be answered in order to ensure their safety.

Mgbeoji, Ikechi. *Global Biopiracy: Patents Plants, and Indigenous Knowledge.* Ithaca, N.Y.: Cornell University, 2006. Defines biopiracy as "the appropriation of plants and traditional knowledge by corporations and other entities," usually by Western interests against non-Western cultures. The author advocates for the protection of the developing world's interests against those of the industrialized world.

Midkiff, Ken. *The Meat You Eat: How Corporate Farming Has Endangered America's Food Supply.* New York: St. Martin's Press, 2004. Midkiff is a director of the Sierra Club and concentrates on the issue of sustainable agriculture. He believes that corporate domination of the U.S. food supply is detrimental to human health, the environment, and the financial well-being of the country. He advocates a return to smaller farms and a strong local network of growers, vendors, and consumers.

Mushita, Andrew, and Carol B. Thompson. *Biopiracy of Biodiversity: Global Exchange as Enclosure.* Lawrenceville, N.J.: Africa World Press, 2007. Focuses on the effect of trading seed on African nations; the authors state that privatizing seed, a resource that had been free for thousands of years, is a threat to sustainable biodiversity. Both authors are specialists in the economics of small-scale farming in Africa.

Nestle, Marion. *Safe Food: Bacteria, Biotechnology, and Bioterrorism.* Berkeley: University of California Press, 2004. Nestle analyzes issues surrounding the

U.S. food supply, from byzantine government regulation, industry resistance to food safety, GM foods, irradiation, and bioterrorism. Written from a grass-roots perspective.

Nottingham, Stephen. *Eat Your Genes: How Genetically Modified Food Is Entering Our Diet,* revised edition. London, England: Zed Books, 2003. Nottingham writes about GM corn, soybeans, fruit, and vegetables and the possibility of GM fish, meat, and poultry in the near future. The dangers of widespread GMOs may include an increase in food allergies and increased resistance to antibiotics, but the benefits may be food with a longer shelf-life and higher yields. He also delves into the pharming industry, which uses genetic modi-fication to combine food and pharmaceuticals, an area that promises to be enormously profitable in the coming years.

Paarlberg, Robert. *Starved for Science: How Biotechnology Is Being Kept Out of Africa.* Cambridge, Mass.: Harvard University Press, 2008. President Jimmy Carter and Green Revolution pioneer Norman Borlaug both contributed forewords to Paarlberg's book, in which he warns that Africa's reluctance to adopt GM crops is contributing to a global food crisis. Part of the problem, he believes, is that African nations rely on international aid for much of their food, and many of the contributing countries—especially those in the European Union—are foisting their preference for organic food on a part of the world for which it is an inappropriate solution.

Pringle, Peter. *Food, Inc.: Mendel to Monsanto: The Promise and Perils of the Bio-tech Harvest.* New York: Simon & Schuster, 2005. Pringle is a journalist who examines the hyperbole on both sides of the GMO debate, from Africans who think GM crops will poison them, to corporations who seek to patent seeds that farmers have cultivated for hundreds of years. He argues that the reality is somewhere in the middle; GM crops have the capacity to help millions of people at risk from hunger and starvation in the developing world, but corpo-rations need to act responsibly and respect their consumers.

Rampton, Sheldon, and John Stauber. *Mad Cow USA: Could the Nightmare Hap-pen Here?* Monroe, Maine: Common Courage Press, 2002. The authors trace what they consider to be an unhealthy symbiosis between the government and the food industry, relating how mad cow disease and other maladies could make their way into the food supply. They argue that both the U.S. and British governments have suppressed information about mad cow disease that could have negative consequences for public health.

Ronald, Pamela C., and R. W. Adamchak. *Tomorrow's Table: Organic Farming, Ge-netics, and the Future of Food.* New York: Oxford University Press, 2008. The authors are a married couple who also happen to be an organic farmer and a plant geneticist. They argue that organic farming and GM foods are not dia-metrically opposed and that both technologies should be used together for the benefit of humankind and to achieve true agricultural sustainability.

Shiva, Vandana. *Biopiracy: The Plunder of Nature and Knowledge.* Cambridge, Mass.: South End Press, 1999. Shiva, long an anti–GM activist from India,

argues that GMOs and cloning represent a commercialization of natural resources that is harmful to people in the developing world.

Smith, Jeffrey M. *Seeds of Deception: Exposing Industry and Government Lies
about the Safety of the Genetically Altered Foods You're Eating.* Fairfield, Iowa:
Yes! Books, 2003. Smith does not pretend to give a balanced view but rather
argues that GM foods are foisted on an unsuspecting public with little testing
and possibly catastrophic long-term consequences for agriculture and human
health.

———. *Genetic Roulette: The Documented Health Risks of Genetically Engineered
Food.* White River Junction, Vt.: Chelsea Green, 2007. Smith continues his
exploration of GM food, demonstrating that safety assessments are misleading
and/or incomplete.

Winston, Mark L. *Travels in the Genetically Modified Zone.* Cambridge, Mass.:
Harvard University Press, 2002. Winston traveled the world, speaking with
scientists, policy makers, farmers, business people, and government officials
to understand the politics of GMOs. He relates each position and tries to find
common ground among those who are unwavering in their support for or
their opposition to GMOs.

Articles, Reports, and Papers

Barboza, David. "As Biotech Crops Multiply, Consumers Get Little Choice." *New
York Times* (6/10/01). Although dated, this article outlines the rise of GM
crops, which took place independent of consumer knowledge and/or approval. He notes that GM corn and soybeans quickly established themselves as
mainstay crops and a reversal to non–GM crops is most unlikely. The Starlink
controversy is covered.

Bauer, Martin W. "Controversial Medical and Agri-Food Biotechnology: A Cultivation Analysis." *Public Understanding of Science* 11 (2002): 93–111. Article
that contemplates the differences between the red/green divide of biotechnology, and how in Great Britain, red biotechnology—medical advances—are
perceived as desirable, but green biotechnology—GM food—is perceived as
undesirable.

"Better Living Through Chemurgy." *Economist* (6/26/08). Announces a resurgence
in chemurgy, the practice of using agricultural feedstock for consumer products. Corporations such as BASF and Novozymes have begun manufacturing
bio-plastics that are not derived from petroleum.

Broad, William J. "Food Revolution That Starts with Rice." *New York Times*
(6/17/08). Story about Norman Uphoff, a proponent of the system of rice intensification, or SRI, which he believes will help solve the world food crisis.

Brown, David. "The 'Recipe for Disaster' That Killed 80 and Left a £5bn Bill." *Telegraph* (6/19/01). Traces the mad cow disease outbreak back to the first sick
cow in 1984, whose condition went unrecognized for two years after its death.
Links the incident to the deaths of 80 people from variant Creutzfeldt-Jakob

disease and the agricultural practices that are destined to lead to additional deaths.

Charman, Karen. "Spinning Science into Gold." *Sierra* 86.4 (July 2001): 40. Relates the story of Arpad Pusztai and other scientists whose careers have been ruined by voicing opinions that are contradictory to the goals of the bio-agriculture industry.

Davies, W. Paul. "An Historical Perspective from the Green Revolution to the Gene Revolution." *Nutrition Reviews* 61.6 (June 2003): S124. Traces the advances in agricultural biotechnology from the Green Revolution of the 1960s to 2003, stating that genetic diversity and biotechnology are becoming increasingly important in feeding the world's people.

Freidberg, Susanne, and Leah Horowitz. "Converging Networks and Clashing Stories: South Africa's Agricultural Biotechnology Debate." *Africa Today* 51.1 (fall 2004): 2. Lengthy article about the polarization of biotechnology in South Africa around the year 2002, when GM food became more common and more controversial, and acknowledges the country's role in determining the future of GM crops for the rest of the continent.

Hindo, Brian. "Monsanto on the Menu." *Business Week* 4089 (6/23/08): 32. Short article on the company's plans to boost food production worldwide by distributing royalty-free seeds to Africa in order to take advantage of the business opportunity presented by the world food crisis.

Journal of Environmental Law and Practice 12.3 (2003). Special issue on GMOs and environmental law.

Kershen, Drew L. "Of Straying Crops and Patent Rights." *Washburn Law Journal* 43 (2004): 575–610. The author notes the legal precedence of favoring patent holders in GM seed rights cases and concludes that the best solution for future lawsuits is to apply the law of stray animals. Under these established rules, both the farmer and the patent holder will be protected from inadvertent stray seeds and plants.

Lee, Maria, and Robert Burrell. "Liability for the Escape of GM Seeds: Pursuing the 'Victim'?" *Modern Law Review* 65 (2002): 517–537. British legal experts foresee that the rise in GM seed in Great Britain will bring with it a legal quagmire caused by unwanted cross-pollination between GM and non–GM seed.

Losey, John, Linda S. Rayor, and Maureen E. Carter. "Transgenic Pollen Harms Monarch Larvae." *Nature* 399 (5/20/99): 214. Brief study that concludes that Bt corn harms butterfly larvae when the pollen becomes airborne and lands on nearby milkweed, which is home to the larvae.

Miller, Henry I. "It's Frankenfood v. the Killer Tomatoes." *New York Post* (6/11/08). Opinion piece by the former director of the FDA's Office of Biotechnology that advocates recombinant DNA technology to protect the U.S. food supply from *Salmonella, E. coli,* and other harmful bacteria.

O'Boyle, Michael. "Breaking with Tradition: Mexico Mulls the Implications of Genetically Modified Corn." *Business Mexico* 12.6 (June 2002): 52. Recounts the findings of Ignacio Chapela of the University of California, who found that transgenic corn had contaminated native varieties in rural Mexico, despite the country's ban on GE seed.

Oguamanam, Chidi. "Tension on the Farm Fields: The Death of Traditional Agriculture?" *Bulletin of Science, Technology and Society* 27.4 (August 2007): 260–273. Written by an expert in Canadian law, this technical article examines the issue of "gene wandering" and its implications for agricultural communities. The author argues for better regulations in Canada to deal with the socioeconomic impact of agro-biotechnology.

Pollack, Andrew. "In Lean Times, Biotech Grains Are Less Taboo." *New York Times* (4/21/08). Pollack, who writes frequently on the agriculture industry, notes that rising global food prices are causing many countries who typically shun GMOs to reconsider their position. Japan and South Korea now allow GM corn to be used in processed food, and although Europe continues to shy away from GM products, proponents of agriculture biotechnology see the food crisis as an opportunity to expand globally.

Robinson, Simon. "Grains of Hope: Genetically Engineered Crops Could Revolutionize Farming." *Time* 156.5 (7/31/00): 38. Long, detailed cover story on the efforts of plant geneticists to limit starvation and health problems such as blindness in developing countries with beta-carotene infused, or Golden Rice. Creating the grain was only half the battle; the other half was navigating a complicated web of patent rights and international public relations crises.

Sayre, Laura. "Protecting Milk from Monsanto." *Mother Earth News* (June–July 2008). Article about increasing consumer demand for milk that does not contain artificial bovine growth hormone (rBGH), and Monsanto's begrudging acceptance of the fact in light of major retailers, including Starbucks and Wal-Mart, now offering their customers rBGH-free milk.

Silver, Lee. "Why GM Is Good for Us." *Newsweek International* (3/20/06). Pro–GMO article by leading biotechnology proponent. Silver believes that transgenic foods are safe for human consumption and can even solve health problems, and that genetically modified livestock could be even safer than non–GM livestock.

Smith, Peter. "'Hormone-Free' Milk Spurs Labeling Debate." *Christian Science Monitor* (4/21/08). Article on the controversy over Monsanto's Posilac, or rBGH, used to increase milk yields, and state-led initiatives to change labeling laws. Notes that many mainstream retailers, including Starbucks and Wal-Mart, are now offering customers rBGH-free milk.

Web Documents

Becker, Geoffrey S., and Tadlock Cowan. "Agricultural Biotechnology: Background and Recent Issues." Congressional Research Service (CRS) Reports and Issue Briefs (September 2006). Available online. URL: http://digital.library.unt.edu/govdocs/crs/permalink/meta-crs-9392. Accessed June 9, 2009. In-depth analysis of GM agricultural practices in the United States through 2006. Acknowledges the controversy surrounding GM seed and how that controversy plays out on the international trade scene. Looks at which government agencies should be responsible for overseeing GM seed, especially as biotechnology moves further into the biopharmaceutical realm. Notes that labeling procedures are stricter in EU countries than in the United States and that Congress continues to monitor oversight procedures.

Bessin, Ric. "Bt-Corn: What It Is and How It Works." University of Kentucky College of Agriculture. Available online. URL: http://www.ca.uky.edu/entomology/entfacts/ef130.asp. Accessed June 9, 2009. Short overview for the general reader on how Bt seed varieties work and details of their FDA approval.

Canadian Health Coalition. "GM Foods and Denial of Rights and Choices: Interview with Arpad Pusztai." (11/10/00). Available online. URL: http://www.healthcoalition.ca/pusztai.html. Accessed June 9, 2009. Long interview in which Pusztai speaks frankly and specifically about his research on GM potatoes, which he stands by.

Cartagena Protocol on Biosafety to the Convention on Biological Diversity. Secretariat of the Convention on Biological Diversity (Montreal), 2000. Available online. URL: http://www.cbd.int/biosafety/protocol.shtml. Accessed June 9, 2009. Complete text in several languages of the official FAO policy regarding what it calls "living modified organisms."

Case, Christine. "Microbial Fermentations: Changed the Course of Human History." Access Excellence @ the National Museum of Health Web site. Available online. URL: http://www.accessexcellence.org/LC/SS/ferm_background.php. Accessed June 9, 2009. Overview of how the field of microbiology arose from the study of fermentation, and how fermentation processes are used in biotechnology.

Clark, E. Ann. "On the Implications of the Schmeiser Decision." Genetics Society of Canada (June 2001). Available online. URL: http://www.plant.uoguelph.ca/research/homepages/eclark/percy.htm. Accessed June 9, 2009. Examines the court decision against Saskatchewan farmer Percy Schmeiser, whose crops were found to contain patented Roundup Ready seed that he did not buy, and concludes "The harm that has been done to Percy and Louise Schmeiser, now in their 70s, is grievous. But of even greater concern is how this incomprehensible decision will affect all western Canadian farmers—regardless of whether they even grow canola, let alone GM canola."

Coalition Provisional Authority Order Number 81: Patent, Industrial Design, Undisclosed Information, Integrated Circuits and Plant Variety Law (June 2004). Available online. URL: http://www.cpa-iraq.org/regulations/20040426_CPAORD_81_Patents_Law.pdf. Accessed June 9, 2009. Text of Paul Bremer's Order 81, which provides intellectual property rights for patent holders of GM seed in Iraq. Some believe this policy will lead to lawsuits from seed companies seeking to enforce their rights, even among farmers whose crops have been unknowingly affected by transgenic contamination.

"Conversations about Plant Biotechnology." Available online. URL: http://www.monsanto.com/biotech-gmo/asp/default.asp. Accessed June 9, 2009. Article on Monsanto's Web site that contains firsthand accounts from farmers world-wide who have used Bt cotton and other GM seed with great success.

Cowan, Tadlock, and Geoffrey S. Becker. "Biotechnology in Animal Agriculture: Status and Current Issues." CRS Report for Congress (3/27/06). Available online. URL: http://digital.library.unt.edu/govdocs/crs/permalink/meta-crs-8961. Accessed June 9, 2009. Summary of biotechnology practices as they relate to U.S. livestock, including genetic engineering, recombinant DNA, development of biopharmaceuticals, and cloning. Illustrates the regulatory oversight responsibilities of the FDA and the USDA, as well as various policy concerns.

Delgado, Christopher L., and Claire A. Narrod. "Impact of Changing Market Forces and Policies on Structural Change in the Livestock Industries of Selected Fast-Growing Developing Countries." Report submitted to the UN Food and Agriculture Organization (10/25/01). Available online. URL: http://www.fao.org/wairdocs/lead/x6115e/x6115e00.htm#contents. Accessed June 9, 2009. Lengthy report on the growth of the livestock industry worldwide, with special emphasis on developing countries that are swiftly increasing their production of meat and dairy products as their economic fortunes rise.

Environmental Protection Agency. "What Is a CAFO?" (2/27/08). Available online. URL: http://www.epa.gov/region7/water/cafo/index.htm. Accessed June 9, 2009. EPA Web page about CAFOs, or concentrated animal feeding operations, which are to livestock what monoculture is to agriculture. CAFOs play a major role in the spread of disease among livestock, as well as soil and water pollution.

"FAO Statement on Biotechnology." (March 2000). Available online. URL: http://www.fao.org/biotech/stat.asp. Accessed June 9, 2009. Presents the UN Food and Agriculture Organization's official position on biotechnology as presented at the Codex Alimentarius Ad Hoc Intergovernmental Task Force on Foods Derived from Biotechnology meeting in Japan in March 2000. States that the "FAO recognizes that genetic engineering has the potential to help increase production and productivity in agriculture, forestry and fisheries. It could lead to higher yields on marginal lands in countries that today cannot grow enough food to feed their people. There are already examples where genetic engineering is helping to reduce the transmission of human and animal diseases through new vaccines. Rice has been genetically engineered to contain pro-vitamin A (beta carotene) and iron, which could improve the health of many low-income communities."

Food Biotechnology: A Study of U.S. Consumer Attitudinal Trends. (October 2008). Available online. URL: http://www.ific.org/research/biotechres.cfm. Accessed June 9, 2009. Report from the International Food Information Council that finds most Americans are confident in the safety of the U.S. food supply, a majority are not overly concerned about consuming GM food, and most are unfamiliar with the concept of sustainable agriculture—although the number of those who are familiar with it is rising. When it comes to biotechnology and animals, public opinion is more varied. Nevertheless, a sizable number of people would eat food derived from cloned animals if the FDA determined it was safe, and many people view favorably biotechnology aimed at reducing the toxicity of animal waste.

Fresco, Louise O. "A New Social Contract on Biotechnology." *FAO Magazine* (2003). Available online. URL: http://www.fao.org/ag/magazine/0305sp1.htm. Accessed June 9, 2009. Advocates for more discussion and political will in establishing the benefits of GM crops in order to eradicate the "molecular divide" that separates developed and developing countries.

Genetically Modified Pest-Protected Plants: Science and Regulation. Board on Agriculture and Natural Resources. National Resources Council, Washington, D.C.: National Academy Press, 2000. Available online. URL: http://www.nap.edu/openbook.php?isbn=0309069300. Accessed June 9, 2009. Government study that explores the risks related to GM crops and suggests a regulatory framework for dealing with these risks responsibly.

"GM in India: The Battle over Bt Cotton." SciDevNet (12/20/06). Available online. URL: http://www.scidev.net/en/features/gm-in-india-the-battle-over-bt-cotton.html. Accessed June 9, 2009. In-depth article about the controversy over Bt cotton in India, which began in 1995 when Monsanto began testing the seed under local conditions. Research efforts were hampered by lax regulations that allowed illegal seed to flow to farmers without having been tested properly for local conditions. Subsequent cotton crops failed, and many farmers committed suicide. Meanwhile, the bollworm, which was supposed to be killed by the Bt toxin, has proven resistant to it.

International Convention for the Protection of New Varieties of Plants. December 2, 1961, revised October 23, 1978, and March 19, 1991. Available online. URL: http://www.upov.int/en/publications/conventions/1991/act1991.htm. Accessed June 9, 2009. Revised text of the convention that outlines the rights of plant breeders who create new varieties of plants. The document predates most work on GMOs.

International Treaty on Plant Genetic Resources for Food and Agriculture. (2001). Available online. URL: ftp://ftp.fao.org/ag/cgrfa/it/ITPGRe.pdf. Accessed June 9, 2009. Web page with links to the text of the Food and Agriculture Organization's International Treaty, as well as studies on its compliance and a FAQ regarding what the treaty entails.

Ismael, Yousouf, Richard Bennett, and Stephen Morse. "Benefits from Bt Cotton Use by Smallholder Farmers in South Africa." *AgBioForum* 5.1 (2002): 1–5. Available online. URL: http://www.agbioforum.org/v5n1/v5n1a01-morse.htm. Accessed June 9, 2009. Scholarly study of small farms in the Makhathini region in Kwazulu-Natal that harvested Bt cotton. Study shows that Bt cotton had higher yields than non-Bt cotton, which more than made up for the higher cost of the seed.

Monsanto Canada Inc. v. Schmeiser. (5/21/04). Available online. URL: http://www.canlii.org/en/ca/scc/doc/2004/2004scc34/2004scc34.html. Accessed June 9,

2009. Decision by the Supreme Court of Canada, which ruled that Schmeiser violated Monsanto's patent on Roundup Ready canola when it was found in his fields in 1998, even though the seed had appeared in his fields without his knowledge or consent.

Pardey, Philip G., and Bonwoo Koo. "Biotechnology and Genetic Resource Policies." International Food Policy Research Institute (January 2003). Available online. URL:http://www.ifpri.org/pubs/rag/br1001.pdf. Accessed June 9, 2009. Takes a worldwide view of GM crops and the organizations that oversee programs to grow them, with eye toward eliminating hunger. Analyzes the progress made by various programs.

Perseley, G. J., and M. M. Lantin. "Agricultural Biotechnology and the Poor." (1999). CGIAR. Available online. URL: http://www.cgiar.org/biotech/rep0100/contents.htm. Accessed June 9, 2009. Detailed report from CGIAR's international conference, with papers submitted by international experts in agricultural biotechnology. Explores how GMOs may be used to alleviate poverty in sub-Saharan Africa, and looks candidly at the risks involved. Includes a chapter on South Africa.

Prakash, C. S. "Gene Revolution and Food Security." *AgBioWorld* (3/2/00). Available online. URL: http://www.agbioworld.org/biotech-info/topics/dev-world/revolution.html. Accessed June 9, 2009. Prakash is the director of the Center for Plant Biotechnology Research at Tuskegee University. He believes that agricultural biotechnology is essential in solving India's food crisis and reducing the country's reliance on pesticides.

Pschorn-Strauss, Elfrieda. "Bt Cotton in South Africa: The Case of the Makhathini Farmers." (April 2005). Available online. URL: http://www.grain.org/seedling/?id=330. Accessed June 9, 2009. Detailed report of a five-year study undertaken by Biowatch South Africa on the impact of Bt cotton on small farmers in a region of the country, which concludes that Monsanto's GM seed has failed to help farmers transcend their poverty and improve their lives.

Vandana Shiva Lecture at Michigan State University. (4/7/05). Available online. URL: http://www.archive.org/details/vandana-shiva-msu-04072005-mm. Accessed June 9, 2009. Video of a 75-minute speech in which Shiva comments on water rights, Monsanto, biotechnology, and other topics.

Other Media

Bad Seed: The Truth about Our Food. DVD. Directed by Timo Nadudvari, 2006. Documentary about GMOs that takes a grassroots approach. States that GMOs are considered dangerous and are not adequately tested. Features interviews with leading researchers and activists, including Jeffrey Smith, author of *Seeds of Deception*; Ignacio Chapela, who researched GM maize transgenic contamination in Mexico; and Percy Schmeiser, who was sued by Monsanto for patent violation when canola plants from Bt seed were found in his fields.

Annotated Bibliography

Beyond Closed Doors: Science, Ethics and Politics of Farm Animal Welfare. DVD. Directed by Hugo Dorigo. Sandgrain Films, 2006. Documentary about factory farming, with interviews from officials at the American Humane Society and other animal welfare organizations.

The Corporation. DVD. Directed by Mark Achbar and Jennifer Abbott. New York: Zeitgeist Films, 2005. Documentary about the rise of the corporation in the United States and its legal status as an individual with full rights. Among other examples, it explores Monsanto's marketing of rBGH as Posilac. Two investigative journalists, Jane Akre and Steve Wilson, were fired for their negative report on Posilac and a subsequent whistle-blowing trial upheld that decision. The case exemplifies, according to the directors, the legal protection given to large corporations by the government and their ability to suppress dissenting viewpoints.

The Future of Food. DVD. Directed by Deborah Koons Garcia. New York: Arts Alliance America, 2004. Documentary that explores the unlabeled GM substances in the U.S. food supply and the rise of the multinational corporations that influence government policy toward new seed technology. Interviews farmers in Canada and Mexico whose livelihoods have been destroyed by unintended patent infringement and discusses the effects of and alternatives to large-scale industrial, monoculture agriculture.

How to Save the World. 2007. Written, directed, and produced by Barbara Sumner-Burstyn and Tom Burstyn. Hawkes Bay, New Zealand: Cloud South Films, 2007. Documentary about 78-year-old New Zealand farmer Peter Proctor, who traveled to India to promote biodynamic farming among subsistence farmers. Using an array of natural processes that run counter to the "bio colonial" practices of corporate agricultural conglomerates, Proctor helps farmers reduce their debt, increase their yields, and raise their standard of living.

King Corn. DVD. Directed by Aaron Woolf. London, England: Mosaic Films, 2007. Documentary starring two young, urban Americans, Ian Cheney and Curt Ellis, who decide to grow an acre of corn on farmland in Iowa to demonstrate aspects of the farming life and agricultural policy in the United States. They explore how corn yields are greater than at any time in history and how this glut of corn has promoted the high-fructose corn syrup industry, which may be responsible for at least part of the obesity epidemic in the United States.

"Seeds of Suicide: India's Desperate Farmers." *Frontline* (7/26/05). Available online. URL: http://www.pbs.org/frontlineworld/rough/2005/07/seeds_of_suicid.html. Chad Heeter travels to Andhra Pradesh to investigate the phenomenon of "suicide by pesticide." Since 1997, more than 25,000 indigent farmers have committed suicide by drinking a pesticide they were told would benefit their crops and farms after transitioning to Bt cotton.

The World According to Monsanto. DVD. Directed by Marie-Monique Robin, 2007. Documentary by a French independent filmmaker that examines the revolving door between government officials at the FDA and Monsanto. The result is that government agencies have green-lighted GM products and procedures because of their close relationships with key individuals at Monsanto, without undergoing the process of due diligence.

BIOTECHNOLOGY, MEDICINE, STEM CELL RESEARCH, AND BIOWARFARE
Books

Alibek, Ken, and Stephen Handelman. *Biohazard: The Chilling True Story of the Largest Covert Biological Weapons Program in the World—Told from Inside by the Man Who Ran It.* Concord, Calif.: Delta, 2000. Alibek was a microbiologist and leader of the Soviet Union's biological weapons program in the 1980s and talks about the former superpower's plans to target U.S. cities with warheads loaded with anthrax, with an intended death toll in the millions. He eventually defected to the United States.

Broad, William, Judith Miller, and Stephen Engelberg. *Germs: Biological Weapons and America's Secret War.* New York: Simon & Schuster, 2001. Best-selling book by three *New York Times* journalists on how fringe groups could obtain bioweapons and unleash them on the public. Gives real-life examples of incidents, such as the intentional *Salmonella* poisoning of 1,000 people in 1984.

Garrett, Laurie. *The Coming Plague: Newly Emerging Diseases in a World Out of Balance.* New York: Penguin, 1995. Pulitzer Prize–winning journalist Garrett examines the failure of public health programs worldwide in preventing and limiting a number of diseases once believed to be under control, as well as the continuing evolution of microbes that thwart scientists' best efforts to eradicate them. AIDS, especially, threatens to destabilize governments, communities, health systems, and the world economy in areas of widespread poverty.

Gunn, Moira A. *Welcome to Biotech Nation: My Unexpected Odyssey into the Land of Small Molecules, Lean Genes, and Big Ideas.* New York: AMACOM, 2007. Gunn is a medical doctor and host of the public radio show "Tech Nation." She interviews leading figures in the biotech industry and explores current research into stem cells, GMOs, DNA, and pharmaceuticals from a business perspective.

Herold, Eve. *Stem Cell Wars: Inside Stories from the Frontlines.* New York: Palgrave Macmillan, 2007. Herold is a journalist who travels to the hot spots associated with stem cell research: South Korean laboratories, the halls of U.S. Congress, and the bedsides of patients who are waiting for cures that such research may provide.

Mangold, Tom, and Jeff Goldberg. *Plague Wars: The Terrifying Reality of Biological Warfare.* New York: St. Martin's Press, 1999. The authors are television correspondents who tackle the prospect of weaponized anthrax, plague, smallpox, and ebola in hot spots around the globe, including Iraq, North Korea, and sub-Saharan Africa.

Parson, Ann B. *The Proteus Effect: Stem Cells and Their Promise for Medicine.* Washington, D.C.: Joseph Henry Press, 2004. A pro–stem cell research book in which the author advocates stem cell research as the best hope for those suffering from a myriad of conditions.

Preston, Richard. *The Demon in the Freezer: A True Story.* New York: Random House, 2002. Story of the last remaining smallpox virus in the post–9/11 world, and possible attempts to harness the disease and others like it by rogue groups and governments to create weaponized viruses.

Ruse, Michael, and Christopher A. Pynes, eds. *The Stem Cell Controversy: Debating the Issues.* Amherst, N.Y.: Prometheus Books, 2006. A collection of essays written by experts on stem cell issues and divided into sections on science, morals, religion, and policy. Examines both pro–stem cell research opinions and anti–stem cell opinions, and leads off with the words of President George W. Bush, who limited federal funding for stem cell research early in his administration.

Stephens, Trent, and Rock Brynner. *Dark Remedy: The Impact of Thalidomide and Its Revival as a Vital Medicine.* New York: Perseus, 2001. The authors blame the thalidomide tragedy on the under-regulated pharmaceutical industry of the 1950s and trace trade in the drug on the black market in subsequent years, along with its re-emergence as an approved treatment for many conditions, including cancer, AIDS, and multiple sclerosis.

Articles, Reports, and Papers

Arnold, Wayne. "Singapore Acts as Haven for Stem Cell Research." *New York Times* (8/17/06). Article about the widespread availability on the Internet of stem cells originating from ES Cell International in Singapore and that city-state's ability to attract top researchers from around the world, whose work may be hindered by restrictions on stem cell research in their home countries.

Daemmrich, Arthur. "A Tale of Two Experts: Thalidomide and Political Engagement in the United States and West Germany." *Social History of Medicine* 15.1: 137–158. Traces the thalidomide tragedy in West Germany and the prevention of a similar tragedy in the United States by comparing the actions of Widukind Lenz and Frances Kelsey, the German and American officials, respectively, responsible for approving the drug in their countries.

Fackler, Martin. "Risk Taking Is in His Genes." *New York Times* (12/11/07). Profile of Shinya Yamanaka, who, independently from James Thomson, turned a human skin cell into an induced pluripotent stem cell at Kyoto University in Japan.

Gentleman, Amelia. "India Nurtures Business of Surrogate Motherhood." *New York Times* (3/10/08). Article about the rising popularity of couples traveling to India to engage the services of surrogate mothers.

Kolata, Gina. "Scientists Bypass Need for Embryo to Get Stem Cells." *New York Times* (11/21/07). News story about James Thomson and Shinya Yamanaka's independent announcements on having created induced pluripotent stem cells from human skin cells. The cells can be differentiated into any kind of cell, with the advantage of not having to destroy embryos to obtain them.

Lederberg, Joshua. "Biological Warfare." *Emerging Infectious Diseases* 7.6 (Nov./Dec. 2001): 1,071. A summary of Lederberg's testimony before the U.S. Congressional Committee on Foreign Relations following the anthrax attacks in the fall of 2001. He states that biological warfare is the gravest security threat the country faces, due to the fact that bioweapons are relatively cheap and easy to produce, can easily be obtained by rogue groups, and can inflict serious casualties.

———. "Getting in Tune with the Enemy—Microbes." *Scientist* 17.11 (6/2/03). Lederberg, a Nobel Prize–winner and leading authority on genetics, writes about how microbes and infectious disease will never be conquered, and how medical science must learn to make sure they remain in check through studying what he calls the "microbiome."

Maria, Augustin, Joel Ruet, and Marie-Helene Zerah. "Biotechnology in India." Report commissioned by the French Embassy in India (2002). Detailed, lengthy report that highlights India's efforts to create a biotechnology industry that will make the country competitive on the world stage. Provides figures for research and development efforts, and names the regulatory agencies that pursue the development of the new business sector.

Motari, Marion, et al. "South Africa—Blazing a Trail for African Biotechnology." *Nature Biotechnology* 22, supplement (December 2004): DC 37–41. Detailed article complete with graphs and charts on the prospects of South Africa's biotechnology sector. Highlights the country's leading role among African nations and its research and development efforts in HIV/AIDS vaccine trials.

Purkitt, Helen, and Virgen Wells. "Evolving Bioweapon Threats Require New Countermeasures." *Chronicle of Higher Education* 53.7 (10/6/06). Examines South Africa's defunct Project Coast, a bioweapons program of the Apartheid era, and more recent programs in Iraq prior to the U.S. invasion in 2003, in order to understand the threat of bioweapons including anthrax, ebola, smallpox, and other substances from nations and terrorist groups.

Ratner, Lizzy. "Brave New Boutique: Baby Sex Selection Sold on East Side." *New York Observer* (7/24/06). Article about fertility specialists in New York who offer clients sex selection for their baby-to-be. Ratner notes that the practice is still rare, but demand is growing, possibly leading to a future awash with designer babies.

Weiss, Rick. "The Power to Divide." *National Geographic* 208.1 (July 2005): 2. In-depth cover story on stem cell research that interviews key researchers, including the now-disgraced Hwang Woo-Suk, and explores why top researchers are moving to Singapore. Illustrates how a patient's life was saved with an umbilical cord blood transfusion, which eliminated leukemia from his bone marrow by replacing it with healthy stem cells. Explores both sides of the issue in detail.

Yuan, Robert. "Nurturing a Biotechnology Industry in China." *Genetic Engineering News* 21.3 (2/1/01): 10. Describes China's nascent biotechnology industry, which involves a small but growing group of academic researchers, special economic zones, and help from the stock market.

Web Documents

Burton, Stephanie G. "Development of Biotechnology in South Africa." *Electronic Journal of Biotechnology* (2002). Available online. URL: http://www.ejbiotech nology.info/content/vol5/issue1/issues/03/index.html. Accessed December 15, 2009. Summarizes the state of biotechnology in South Africa, acknowledging the country's vast social and economic problems and also its areas of potential, which include its natural resources and its active research community.

Committee on Research Standards and Practices to Prevent the Destructive Application of Biotechnology. *Biotechnology Research in an Era of Terrorism.* Washington, D.C.: National Academies Press, 2004. Available online. URL: http://www.nap.edu/catalog.php?record_id=10827#toc. Accessed June 9, 2009. Online book that charts the march of the life sciences into the realm of bioweapons and makes recommendations to safeguard U.S. citizens against the misuse of biotechnology from the standpoint of national defense.

Convention on the Prohibition of the Development, Production and Stockpiling of Bacteriological (Biological) and Toxin Weapons and on Their Destruction. United Nations Office at Geneva (1972). Available online. URL: http://www. opbw.org/convention/documents/btwctext.pdf. Accessed December 15, 2009. Text of the Biological Weapons Convention, which was entered into force in 1975 and which provides a framework for the United Nations General Assembly for dealing with states that manufacture, use, or store biological weapons for any purpose.

The Joshua Lederberg Papers. Available online. URL: http://profiles.nlm.nih. gov/BB/. Accessed June 9, 2009. A collection of the Nobel Prize–winning scientist's papers at the National Library of Medicine's Profiles in Science Web site. Includes information on his research in bacterial genetics and artificial intelligence and his pioneering role in biomedical research.

National Biotechnology Strategy for South Africa. Department of Science and Technology, Pretoria (June 2001). Available online. URL: http://www.pub. ac.za/resources/docs/biotechstrategy_2002.pdf. Accessed June 9, 2009. Detailed plan of the government's various initiatives to build upon current biotechnology efforts.

National Institutes of Health. "Stem Cell Basics." (8/11/08). Available online. URL: http://stemcells.nih.gov/staticresources/info/basics/StemCellBasics.pdf.

Accessed June 9, 2009. Document that outlines what stem cells are, the difference between embryonic stem cells and adult stem cells, their potential use, and sources for further information. Includes full-color graphics.

Russell, Alan. "Why Can't We Grow New Body Parts?" Video of lecture at TED Conference (February 2006). 19 minutes. Available online. URL: http://www. ted.com/index.php/talks/alan_russell_on_regenerating_our_bodies.html. Accessed June 9, 2009. Russell talks about regenerative medicine, in which the body can be engineered to grow new tissue and organs in a medically relevant time frame.

United Nations Conference on Trade and Development. "Key Issues in Biotechnology." (2002). Available online. URL: http://www.unctad.org/en/docs/ poitetebd10.en.pdf. Accessed June 9, 2009. Examines the impact of GM crops on the environment and human health, the beneficiaries of GMOs, terminator technology, food security, DNA vaccines, the Human Genome Project, and the policy challenges presented by complex matters of biosafety, bioethics, and intellectual property rights.

Zuckerman, M. J. "Biosecurity: A 21st Century Challenge." Carnegie Corporation of New York (2005). Available online. URL: http://www.carnegie.org/pdf/bio security_challenge_paper.pdf. Accessed June 9, 2009. A broad overview of the issues facing national security leaders as biotechnology increasingly infiltrates the realm of defense. Draws heavily from Gerald R. Zink's report "Biotechnology Research in an Era of Terrorism."

Other Media

Mapping Stem Cell Research: Terra Incognita. Directed by Maria Finitzo, 2008. The very personal story of Jack Kessler, a neurologist involved in using stem cells to regenerate damaged spinal cord tissue. His daughter was partially paralyzed in a skiing accident and could benefit from his research. The film follows the travails of Kessler's lab researchers as they conduct a crucial experiment on mice to see if they can regenerate damaged axons to facilitate the transformation of stem cells into spinal cord tissue. Their experiment is successful and will result in further studies that they hope will build on their findings. The film also gives voice to those opposed to stem cell research and explores the ethical questions involved, from both religious and humanistic viewpoints.

20th Century with Mike Wallace: Outbreak! The New Plague. DVD. New York: History Channel, 1997. Wallace recounts the details of history's worst plagues, including the 1918 influenza pandemic, and more recent scourges—AIDS, ebola, and hantaviruses, and interviews individuals who believe that public health services are guilty of malpractice in not doing a better job of containing future threats and keeping people safe from current ones.

GENETIC ENGINEERING/HUMAN GENOME PROJECT

Books

Acharya, Tara, and Neeraja Sankaran. *The Human Genome Sourcebook.* Westport, Conn.: Greenwood Press, 2005. An in-depth reference on genetics and the science of the human genome, with many visuals of cells, genes, chromosomes, and diseases.

Aldridge, Susan. *The Thread of Life: The Story of Genes and Genetic Engineering.* Cambridge: Cambridge University Press, 1996. A scientific guide to molecular biology, DNA, genetic engineering, and biotechnology, and their practical applications.

Cooke-Deegan, Robert. *Gene Wars: Science, Politics, and the Human Genome.* New York: W. W. Norton, 1995. Cooke-Deegan is a medical doctor and member of the National Academy of Science Institute of Medicine who writes of the Human Genome Project's early days under the leadership of James Watson, when the project was dominated by the United States and its leaders seemed united in their goals. Soon, as other countries joined the race, the line between public good and private gain became much more murky—and controversial.

Davies, Kevin. *Cracking the Genome: Inside the Race to Unlock Human DNA.* Baltimore, Md.: Johns Hopkins University Press, 2002. Davies is a journalist and science editor who interviews James Watson, Francis Collins, J. Craig Venter, and others involved in the Human Genome Project to explain the importance of the project.

Dennis, Carina, and Richard Gallagher, eds. *The Human Genome.* Houndsmill, Basingstoke, Hampshire, England: Palgrave, 2001. Full-color guide to the scientific history of DNA and the Human Genome Project, with information on the mechanics of mapping. Includes a foreword by James D. Watson.

DeSalle, Rob, and Michael Yudell. *Welcome to the Genome: A User's Guide to the Genetic Past, Present, and Future.* Hoboken, N.J.: Wiley-Liss, 2005. Lavishly illustrated textbook that charts the history of mapping the genome and scientists' understanding of it.

Duncan, David Ewing. *The Geneticist Who Played Hoops with My DNA . . . and Other Masterminds from the Frontiers of Biotech.* New York: William Morrow, 2005. Duncan is a journalist, columnist, and television news correspondent who focuses on current science issues. The book profiles seven of the world's leading biotech scientists, including Francis Collins, James Watson, and Craig Venter.

Enriquez, Juan. *As the Future Catches You: How Genomics and Other Forces Are Changing Your Life, Work, Health, and Wealth.* New York: Three Rivers Press, 2005. Enriquez is a professor at Harvard Business School who writes about the vast ramifications in all sectors of society of the revolution launched by the Human Genome Project.

Hall, Stephen S. *Invisible Frontiers: The Race to Synthesize a Human Gene.* New York: Oxford University Press, 2002. First published in 1987, this book is a blow-by-blow account of the 1976–1978 race between Eli Lilly, Genentech, and others to create insulin using recombinant DNA techniques, one of the first crucial biotech pharmaceutical breakthroughs.

Hill, Walter E. *Genetic Engineering: A Primer.* Amsterdam, the Netherlands: Harwood Academic, 2000. Begins with chapters on cell biology, proteins, nucleic acids, bacteria, and viruses. Other chapters cover recombinant DNA, IVF, cloning, gene therapy, and ethics.

Kohler, Robert E. *Lords of the Fly: Drosophila Genetics and the Experimental Life.* Chicago, Ill.: University of Chicago Press, 1994. Kohler demonstrates how the lowly fruit fly—*Drosophila melanogaster*—emerged as the laboratory standard species for conducting genetics research, beginning with Thomas Hunt Morgan in 1910. *Drosophila* proved uniquely suited to the laboratory environment and became the cornerstone of research experiments around the world.

Oakley, Barbara A. *Evil Genes: Why Rome Fell, Hitler Rose, Enron Failed, and My Sister Stole My Mother's Boyfriend.* Amherst, N.Y.: Prometheus, 2008. Part memoir and part exploration of history's most notorious evil geniuses, Oakley's book investigates the genetic basis for borderline personality disorder, using examples from her own life (her sister) and world history (Mao Zedong, Slobodan Milošević).

Ridley, Matt. *Genome: The Autobiography of a Species in 23 Chapters.* New York: HarperPerennial, 2006. Best-selling author Ridley examines what is known about each chromosome in the human body as a result of the Human Genome Project and other research, noting which traits are associated with which genes. Easy-to-read exposé of scientific findings, along with a primer on various controversies and bioethics debates, including the new eugenics, genetic determinism, and free will.

Silver, Lee M. *Remaking Eden: How Genetic Engineering and Cloning Will Transform the American Family,* New York: Ecco Press, 1998. Silver takes an optimistic look at the future of genetics and aims to demystify genetic engineering, cloning, stem cell research, and other techniques that cause consternation among the public.

Stock, Gregory. *Redesigning Humans: Choosing Our Genes, Changing Our Future.* New York: Mariner Books, 2003. Stock is an ethicist who writes on issues related to genetics. He believes that cloning is not the central issue of genetics. Rather, it is a host of other developments that will enable parents to select the traits of their offspring. Stock believes this process cannot be stopped—parents will always want to give their children an edge in physical appearance, talent, and intelligence. Rather than attempting to derail the process, he believes society should prepare for it.

Annotated Bibliography

Articles, Reports, and Papers

Baird, Stephen L. "Designer Babies: Eugenics Repackaged or Consumer Options?" *Technology Teacher* 66.7 (April 2007): 12. Cover story on the arguments for and against genetic engineering aimed at teachers who may want to address the issue in class. Includes additional bibliography and a link to a two-day lesson plan.

Dickinson, Boonsri. "The Jiffy Lube of Genome Decoding." *Discover* (9/20/08). Short article about Pacific Bioscience in Menlo Park, California, which aims to be able to sequence an individual's entire genome in a few minutes for just a few hundred dollars by 2013.

Kruvand, Marjorie, and Sungwood Hwang. "From Revered to Reviled: A Cross-Cultural Narrative Analysis of the South Korean Cloning Scandal." *Science Communication* 29.2 (December 2007): 177–197. Exposé of South Korean cloning researcher Hwang Woo-Suk, who claimed to have cloned human embryos in 2005. The authors relate how the subsequent scandal, in which Hwang was shown to have forged his data, unfolded differently in the Western media and the Korean media.

Lander, Eric S. "Genomics: Launching a Revolution in Medicine." *Journal of Law, Medicine & Ethics* 28.4 (Winter 2000): S3. Text of a speech by the author, one of the world's foremost experts on genomics, that traces the development of the field through advances at the start of the 20th century that led to the 21st-century biomedicine revolution, which he calls "one of the most remarkable revolutions in the history of mankind."

Lemonick, Michael D. "The Iceland Experiment: How a Tiny Island Nation Captured the Lead in the Genetic Revolution." *Time* 167.8 (2/20/06): 50. Article that interviews Kari Stefansson, founder of deCODE Genetics, who can trace his ancestry back 1,000 years through good national recordkeeping and genetic testing.

Nature. "Double Helix: 50 Years of DNA." Available online. URL: http://www.nature.com/nature/dna50/index.html. Accessed June 9, 2009. Special issue commemorating the 50th anniversary of Watson and Crick's paper outlining the structure of DNA.

Park, Alice. "The Perils of Cloning." *Time* 168.2 (7/10/06): 56. Lengthy article that tracks progress in animal cloning in the decade since the birth of Dolly the sheep. The process of somatic cell nuclear transfer has had both successes and failures, the latter including reprogramming errors that are transferred to the genome. Author interviews Ian Wilmut, the scientist behind Dolly, for his opinion on where the field of cloning is headed.

Pinker, Steven. "My Genome, My Self." *New York Times* magazine (1/11/09). Lengthy article in which Pinker, a cognitive psychologist, discusses the process of having his genome sequenced and the result uploaded to the Internet as part of the Personal Genome Project. The PGP aims to be a public database containing the genomes of 100,000 people that will help scientists make further refinements to mapping specific traits to specific genes. Pinker concentrates on the debate of nurture v. nature, which he believes is impacted by the element of chance.

"The Proper Study of Mankind." *Economist* (US) 356.8177 (7/1/00): 11. Article traces the work of Allan Wilson at the University of California at Berkeley into mitochondrial DNA, which has revolutionized the study of anthropology and our understanding of human origins, evolution, and similarity.

Web Documents

Enriquez, Juan. "Decoding the Future with Genomics." Video of lecture at TED conference (February 2003). 22 minutes. Available online. URL: http://blog. ted.com/2007/04/juan_enriquez_o.php. Accessed June 9, 2009. Enriquez, author of the bestselling *As the Future Catches You,* talks about the world of microbes that are just beginning to be understood, thanks to recent research in genomics and its monumental importance for the future of science and world economics.

Freudenrich, Craig. "How Cloning Works." *How Stuff Works.* Available online. URL: http://science.howstuffworks.com/genetic-science/cloning.htm. Accessed June 9, 2009. Article aimed at high school students outlining the steps scientists take to create clones of animals—or possibly humans. Includes photographs, diagrams, and video.

"Genetic Tests: Insurers Should Pay." *Business Week* (January 2008). Available online. URL: http://www.businessweek.com/debateroom/archives/2008/01/genetic_tests_i.html. Accessed June 9, 2009. Presents a pro and a con view of insurers paying for patients' genetic testing, both written by experts in the field.

The Genographic Project. Available online. URL: https://genographic.national-geographic.com/genographic/index.html. Accessed June 9, 2009. A Web site hosted by *National Geographic* where participants can purchase a kit that enables them to trace their genetic lineage back 60,000 years. The results are entered anonymously into a database that becomes a constantly updated profile of a person's genetic history. The project is headed by Spencer Wells, a geneticist and anthropologist.

HHMI Virtual Transgenic Fly Lab. Howard Hughes Medical Institute. Available online. URL: http://www.hhmi.org/biointeractive/vlabs/transgenic_fly/index. html. Accessed June 9, 2009. Gives a virtual tour of the transgenic fly lab at the Howard Hughes Medical Institute, where scientists study genomics using fireflies and *Drosophila melanogaster* as model organisms. Geared toward teachers and students, the program allows users to navigate around the lab and watch animations that simulate the making of transgenic flies and their use in experiments.

Human Genome Project Information Web site. Available online. URL: http://www. ornl.gov/sci/techresources/Human_Genome/home.shtml. Accessed June 9, 2009. Main U.S. government portal to all human genome-related Web sites. Contains links to sites that focus on medicine, education, social and ethical issues, the gene gateway, the research archive, and new developments in genomics.

Indian Department of Biotechnology, Ministry of Science and Technology. *Biotechnology: A Vision.* Available online. URL: http://dbtindia.nic.in/uniquepage.asp?id_pk=102. Accessed June 9, 2009. Detailed outline of the department's mission, which includes harnessing human resources to create a strong biotech infrastructure that will lead to advances in biotechnology and initiatives in agriculture, medicine, environmental conservation, and industry.

Judson, Olivia. "Resurrection Science." (11/25/08). Available online. URL: http://opinionator.blogs.nytimes.com/2008/11/25/resurrection-science. Accessed December 15, 2009. Judson is an evolutionary biologist who writes The Wild Side blog for the *New York Times.* In this post, she outlines the scientific hurdles that would need to be overcome before extinct animals, such as the mastodon, could be resurrected. She advocates saving currently endangered species over bringing back extinct ones through genome sequencing and cloning.

Kitcher, Philip. "Manipulating Genes: How Much Is Too Much?" Available online. URL: http://www.pbs.org/wgbh/nova/genome/manipulate.html. Accessed June 9, 2009. Interview with Kitcher by the producers of the documentary series *Nova,* in which Kitcher expounds on the future of molecular medicine and gene therapy, in particular, along with the ethical implications of altering a person's genes. He is also concerned about the ethics of the free market on technologies that could be used to improve and save the lives of millions of disadvantaged people.

National Library of Medicine. *Chromosome Map.* Available online. URL: http://www.ncbi.nlm.nih.gov/books/bv.fcgi?rid=gnd.chapter.272. Accessed June 9, 2009. In-depth information, including graphs and charts, of each chromosome in the human body, along with the known traits and conditions associated with each.

President's Council on Bioethics. *Human Cloning and Human Dignity: An Ethical Inquiry.* Washington, D.C. (July 2002). Available online. URL: http://www.bioethics.gov/reports/cloningreport/index.html. Accessed June 9, 2009. Extensive, detailed executive report outlining the history, ethics, morality, and public policy recommendations on therapeutic and reproductive cloning that guided President George W. Bush's policy for most of his administration.

"Revised Version of the German Genetic Engineering Act." (12/16/93). Available online. URL: http://web.uni-frankfurt.de/si/gentech/GenTGengl10-95c.pdf. Accessed June 9, 2009. In English. The goal of the act is to protect the health of all people, plants, and animals and "the symbiotic structure of the environment at large." Also provides a legal framework for regulating genetic engineering technology.

Rule, Stephen, and Zakes Ianga. "Public Understanding of Biotechnology in South Africa." February 16, 2005. Available online. URL: http://www.pub.ac.za/

resources/docs/survey_pub_feb2005.pdf. Accessed June 9, 2009. Findings of a detailed survey of 7,000 representative South Africans, which found that "nearly eight out of ten respondents interviewed did not know or had any knowledge about biotechnology. An equal number of respondents did not know what to think when they heard the words biotechnology, genetic engineering, genetic modification or cloning."

Venter, Craig. "A Voyage of DNA, Genes and the Sea." Video lecture. (February 2005). Available online. URL: http://www.ted.com/index.php/talks/craig_venter_on_dna_and_the_sea.html. Accessed June 9, 2009. Lecture given at the 2005 TED conference by Craig Venter. He talks about the imminent transition from reading the genomic code to writing it.

Watson, James. "The Double Helix and Today's DNA Mysteries." Video of lecture given at the TED conference. (February 2005). Available online. URL: http://www.ted.com/index.php/talks/james_watson_on_how_he_discovered_dna.html. Accessed June 9, 2009. Watson talks about how he and Francis Crick became interested in researching the structure of DNA, at the annual Technology, Entertainment, Design (TED) conference in California.

Winstead, Edward R. "In South Africa, the Quagga Project Breeds Success." *Genome News Network* (10/20/00). Available online. URL: http://www.genomenewsnetwork.org/articles/10_00/Quagga_project.shtml. Accessed June 9, 2009. Article about Reinhold Rau's 20-plus year experience with the Quagga Project, which breeds zebras with the physical qualities of the extinct Southern African quagga.

World Health Organization. "A Dozen Questions (and Answers) on Human Cloning." Available online. URL: http://www.who.int/ethics/topics/cloning/en/print.html. Accessed June 9, 2009. Detailed information for the layperson on how cloning occurs, the difference between reproductive and non-reproductive cloning, ethics, and regulations. Provides links to international regulations on cloning.

Other Media

Charlie Rose interview with Craig Venter. (12/25/07). 26 minutes. Available online. URL: http://www.charlierose.com/home. Accessed June 9, 2009. Rose interviews Venter about the scientist's own genome project.

Charlie Rose interview with Francis Collins. (7/29/08). 31 minutes. Available online. URL: http://www.charlierose.com/home. Accessed June 9, 2009. Interview with the outgoing director of the National Human Genome Research Institute.

Charlie Rose interview with Lee Silver. (11/26/01). 14 minutes. Available online. URL: http://www.charlierose.com/home. Accessed June 9, 2009. Rose talks with Silver, a noted molecular biologist, about human cloning.

Conversations in Genetics. 15-DVD set (1997). Produced by Rochelle Easton Esposito and distributed by the Genetics Society of America and Cold Spring

Harbor Laboratory. Each DVD contains a one-on-one interview with a leading figure in genetics research. The series is designed to be an oral history of scientific advancement.

Fed Up!: Genetic Engineering, Industrial Agriculture and Sustainable Alternatives. DVD. Directed by Angelo Sacerdote. San Francisco, Calif.: Microcinema, 2006. Documentary that explores the infiltration of GMOs into the U.S. food supply. Includes interviews with farmers, scientists, and activists.

NOVA: Cracking the Code of Life. DVD. Directed by Betsey Arledge. WGBH Boston (2001). Documentary that examines the race between government researchers and private biotech firms to map the human genome. Interviews people with inherited diseases that could be helped by the Human Genome Project and discusses legal and privacy issues involved.

NOVA: DNA—The Secret of Photo 51. DVD. WGBH Boston (2003). Rosalind Franklin's X-ray crystallography image, photo 51, which was shown to Watson and Crick without her knowledge and which indicated the double-helix structure of DNA, which would earn Watson and Crick the Nobel Prize. Includes interviews with Franklin's Ph.D. student, Raymond Gosling, who created photo 51 with Franklin, and Maurice Wilkins, the third member of the team that won the Nobel Prize.

NOVA: Ghost in Your Genes. DVD. WGBH Boston (2009). Documentary that examines the mystery of identical twins—who share identical DNA—whose health diverges in life. Explores the idea that environment and lifestyle play a more important role in disease than DNA.

BIOTECHNOLOGY HISTORY AND BIOGRAPHY
Books

Barry, John M. *The Great Influenza: The Epic Story of the Deadliest Plague in History.* Rev. ed. New York: Penguin, 2005. How many died during the flu epidemic of 1918? Between 50 and 100 million, according to Barry, who traces the worst epidemic in human history from a medical and social standpoint while warning that another influenza epidemic is unavoidable.

Berg, Paul, and Maxine Singer. *George Beadle, an Uncommon Farmer: The Emergence of Genetics in the 20th Century.* Woodbury, N.Y.: Cold Spring Harbor Laboratory Press, 2005. Beadle was a Nobel Prize–winning scientist who helped found modern molecular biology by understanding how genes interact with proteins. The biography was written by Berg, himself a towering figure in the history of genetics, and noted science writer Singer.

Binder, Gordon, and Philip Bashe. *Science Lessons: What the Business of Biotech Taught Me about Management.* Cambridge, Mass.: Harvard Business School Press, 2008. Binder was CEO of Amgen, a major biotechnology firm that began as Applied Molecular Genetics in 1980 and developed many of the life-saving pharmaceutical products on the market today. Binder, at the helm from

1988 to 2000, recounts the difficulties of shepherding the company through the travails involved in both science and business.

Black, Edwin. *War against the Weak: Eugenics and America's Campaign to Create a Master Race.* New York: Four Walls Eight Windows, 2003. Black is a journalist who conducted extensive research into American eugenics and brought to light documents that proved American eugenicists influenced Nazi eugenicists. He draws parallels between eugenics and human genetics and believes a solid understanding of history is necessary for those currently involved in genetic engineering.

Bruinius, Harry. *Better for All the World: The Secret History of Forced Sterilization and America's Quest for Racial Purity.* New York: Vintage, 2007. Account of eugenics in early 20th-century America, which includes the story of Carrie Buck and explores the links between American and Nazi eugenicists.

Bud, Robert. *The Uses of Life: A History of Biotechnology.* New York: Cambridge University Press, 1994. Survey of the relationship between biology and engineering through the years—in the Americas, Europe, and Japan—that led to the modern biotechnology industry and the promise it holds for the future.

Carson, Rachel. *Silent Spring.* New York: Fawcett, 1962. Still in print more than 40 years after its initial publication, *Silent Spring* was the first book to sound the alarm about pesticides in the environment. Though DDT was intended to slow the spread of malaria and prevent destruction of agricultural crops, Carson showed how it affected the environment, with all species accumulating the toxin in their bodies, passing it up the food chain through the process of bioaccumulation.

Crick, Francis. *What Mad Pursuit: A Personal View of Scientific Discovery.* New York: Basic Books, 1990. Crick's memoir is also a scientific primer that traces the Crick/Watson/Wilkins race to discover the structure of DNA in the 1950s. Far from a methodical pursuit in the hallowed halls of academia, the process was messy and full of trial and error. Along the way he introduces other key figures in science, from Linus Pauling to Sidney Brenner, and describes the research into brain chemistry he undertook after receiving the Nobel Prize.

Dawkins, Richard. *The Selfish Gene: 30th Anniversary Edition.* New York: Oxford University Press, 2006. The classic work of evolutionary biology, which introduced the concept of the "meme," the main engine of the evolutionary process. Dawkins is a leading Darwinian theorist and proponent of atheism, which makes him a controversial figure.

Gold, Hal. *Unit 731 Testimony.* North Clarendon, Vt.: Tuttle Publishing, 2004. Exposé of the infamous Japanese medical unit during World War II that experimented on live prisoners of war and developed weaponized bacteria and other substances as part of its biowarfare research program.

Hager, Thomas. *The Demon under the Microscope: From Battlefield Hospitals to Nazi Labs, One Doctor's Heroic Search for the World's First Miracle Drug.* New York: Harmony, 2006. Hager traces the discovery of sulfa drugs, the precursors of modern antibiotics, in the years between the world wars. German scientist Gerhard Domagk, working at Bayer, developed sulfa drugs in 1932 in an effort to eradicate malaria and tuberculosis.

Harris, Robert, and Jeremy Paxman. *A Higher Form of Killing: The Secret History of Chemical and Biological Warfare*. New York: Random House, 2002. The story begins with gas warfare during World War I, continues with the stockpiling of biological weapons during World War II, and ends with the black market trade in bioweapons after the collapse of the Soviet Union and their popularity among terrorist organizations.

Harris, Sheldon. *Factories of Death: Japanese Biological Warfare, 1932–1945 and the American Cover-Up*. 2nd edition. New York: Routledge, 2002. Harris provides evidence from the archives of China, the United States, and Russia of biological warfare research by Unit 731 during Japan's occupation of Manchuria. He states that the leaders of Unit 731 passed on information at the end of the war in exchange for immunity to criminal prosecution.

Henig, Robin Marantz. *The Monk in the Garden: The Lost and Found Genius of Gregor Mendel, the Father of Genetics*. New York: Mariner, 2001. Biography of Mendel, who Henig regards as an unsung genius for how his quiet diligence changed the course of science. Mendel's life was full of religion, science, and determination, but his work failed to attract the attention of scientists of the day.

Jones, James H. *Bad Blood: The Tuskegee Syphilis Experiment*. Rev. ed. New York: Free Press, 1993. Best-selling nonfiction account of the Tuskegee experiment from a professor of ethics, first published in 1981.

Maddox, Brenda. *Rosalind Franklin: The Dark Lady of DNA*. New York: HarperCollins, 2002. Maddox's biography restores Franklin's reputation as a stellar scientist who was just as much responsible for the discovery of DNA's double-helix structure as Watson and Crick. She conducted important research at laboratories where she was not particularly appreciated and died early, possibly as a result of her work with X-rays.

Olmstead, Alan L., and Paul W. Rhode. *Creating Abundance: Biological Innovation and American Agricultural Development*. New York: Cambridge University Press, 2008. The authors trace the history of agriculture in the United States, providing a revisionist time line demonstrating that farmers have been involved in biotechnology from early on in their quest to increase yields, improve soil conditions, deal with climate issues, improve livestock breeds, and protect human health.

Plotz, David. *The Genius Factory: The Curious History of the Nobel Prize Sperm Bank*. New York: Random House, 2006. Humorous exposé of the sperm bank founded by Robert Graham, designed to acquire the genetic material of Nobel Prize winners and other smart people. In the course of its 20-year history no woman ever became impregnated with sperm from a Nobel Prize winner and Plotz sets out to find out why.

Shreeve, James. *The Genome War: How Craig Venter Tried to Capture the Code of Life and Save the World*. New York: Ballantine, 2005. Shreeve is a science journalist who unravels the complicated relationship between public and private interests in the quest to unlock the "code of life" by sequencing the human genome. He examines the issue of whether or not the genome should be patented and explains the science behind the controversy.

Sturtevant, A. H., and Edward B. Lewis. *A History of Genetics.* Woodbury, N.Y.: Cold Spring Harbor Laboratory Press, 2001. Reprint of Sturtevant's 1965 book in which he relates his role in early genetics, working with Thomas Hunt Morgan in his infamous "Fly Room" at Columbia University, where they conducted experiments on *Drosophila melanogaster* to show how genetic mutations occur.

Vettel, Eric J. *Biotech: The Countercultural Origins of an Industry.* Philadelphia: University of Pennsylvania Press, 2008. The author traces the rise of the biotechnology industry in Northern California during the 1960s, driven by ambitious young scientists who demanded that their research be taken seriously, and a political, social, and economic climate that fostered new ideas.

Venter, J. Craig. *A Life Decoded: My Genome, My Life.* New York: Viking, 2007. Venter founded the private company Celera Genomics to compete with the government-funded Human Genome Project to map the human genome using a different method. Venter sees himself as a heroic scientist battling larger government forces in his effort to profit from patented sections of DNA, much to the consternation of government-funded scientists. Beyond Venter's story, the politics of big science come to the fore.

Watson, James D. *The Double Helix: A Personal Account of the Discovery of the Structure of DNA.* New York: Atheneum, 1968. Never out of print since its publication, Watson's tale of the race to uncover the structure of DNA reveals as much about him as it does about the scientific jet-set of the 1950s. Far from being a straight-laced environment of erudition, the laboratory was full of petty backbiting and personal mistrust.

West, Darrell M. *Biotechnology across National Boundaries: The Science-Industrial Complex.* New York: Palgrave Macmillan, 2007. The author argues that biotechnology has transcended national boundaries to become a deregulated, global science-industrial complex that operates independently of state control. He provides case studies regarding stem cell research, cloning, and GM food to show how the interactions between businesses, universities, and nongovernmental organizations work.

Wilkins, Maurice. *The Third Man of the Double Helix: The Autobiography of Maurice Wilkins.* New York: Oxford University Press, 2005. Very readable firsthand account of one of the major scientific breakthroughs of the 20th century. Though Wilkins received the Nobel Prize for discovering the structure of DNA along with Watson and Crick, his reputation has been overshadowed by his colleagues'. Written partly as a response to Watson's book, *The Double Helix,* which Wilkins thought was misleading, he tells the story from his own perspective and relates his thoughts on Rosalind Franklin, whose work in crystallography led to the discovery.

Annotated Bibliography

Articles, Reports, and Papers

Berg, Paul. "Asilomar 1975: DNA Modification Secured." *Nature* 455 (9/18/08): 290–291. Essay in which Berg looks back at the 1975 conference and affirms that it was the right move at the right time in order to create a standardized framework for scientists conducting genetic research. He believes that holding such a conference today to shore up agreement over current topics such as stem cell research would be difficult, due to the number of public and private scientists now working in the field.

Ditz, Susan. "Biotech Beginnings." *Business Journal* 17.36 (12/24/99): 26. States that the epicenter of modern biotechnology is Silicon Valley, which with its abundance of brain power, venture capital, and research infrastructure has remained the world leader in medical breakthroughs for decades. The foundation of "life science innovation," according to Ditz, was the discovery of the rDNA process.

Johnson, Carolyn. "The History of Biotechnology." *Boston Globe* (5/6/07). Short article that contains a biotech time line from 1953 to 2007 and focuses on events relating to the Boston/Cambridge area, one of the epicenters of biotech innovation.

Leaf, Clifton, and Doris Burke. "The Law of Unintended Consequences: Twenty-five Years Ago a Law Known as Bayh-Dole Spawned the Biotech Industry." *Fortune* (9/19/05): 250. Lengthy article about the legal battles of the modern biotech era. The authors trace these to the 1980 Bayh-Dole Act, which revolutionized the relationship between university research and intellectual property rights. The result is an expensive, inefficient system of sparse biotech breakthroughs that result in higher drug prices, higher college tuition rates, and higher taxes.

Louw, Johann. "Social Context and Psychological Testing in South Africa, 1918–1939." *Theory and Psychology* 7.2 (1997): 235–256. Historical article about eugenics in South Africa, whose leaders held views similar to those of Western nations regarding the "feeble-minded" and the genetic inferiority of black Africans. Traces the organizations and individuals that were influential.

Web Documents

Clinton, Bill. "Remarks by the President in Apology for Study Done in Tuskegee." (5/16/97). Transcript available online. URL: http://clinton4.nara.gov/textonly/New/Remarks/Fri/19970516-898.html. Accessed June 9, 2009. Clinton apologizes for the ethical violations perpetrated by the U.S. Public Health Service during its decadeslong study of syphilis on 400 African-American men who were never informed about their disease. He honors the eight subjects who are still alive and tells them "the United States government did something that was wrong—deeply, profoundly, morally wrong. It was an outrage to our commitment to integrity and equality for all our citizens."

Darwin, Charles. *On the Origin of Species.* London: Murray, 1859. Available online. URL: http://www.gutenberg.org/etext/1228. Accessed June 9, 2009. Darwin's groundbreaking text on natural selection, which explained how animals adapt to their environment.

Galton, Francis. *Hereditary Genius: An Inquiry into Its Laws and Consequences.* London: Macmillan, 1892. Available online. URL: http://galton.org. Accessed June 9, 2009. Facsimile of the second edition of Galton's book, in which he looks at the transmission of intellect through heredity, and his ideas about eugenics are strengthened.

Japanese Ministry of Health and Welfare. "Number of Sterilizations Reported from 1955 to 1967, by Sex, and Stated Reason." Available online. URL: http://www.bioethics.jp/licht_genetics.html. Accessed June 9, 2009. Table that provides at-a-glance information on eugenics sterilization in Japan.

Murphy, Ann, and Judi Perrella. "Overview and Brief History of Biotechnology." Woodrow Wilson Biology Institute (1993). Available online. URL: http://www.woodrow.org/teachers/bi/1993/intro.html. Accessed June 9, 2009. Geared toward teachers, the site includes a short article, a detailed time line of major biotech innovations, and a teacher's guide with lesson suggestions.

Pasteur, Louis. *The Physiological Theory of Fermentation.* Translated by F. Faulkner and D.C. Robb. Available online. URL: http://biotech.law.lsu.edu/cphl/history/articles/pasteur.htm#paperI. Accessed June 9, 2009. Originally published in 1879, the paper outlined Pasteur's landmark experiments on fermentation, controlling yeast's exposure to microbes and oxygen, and laying the foundation for germ theory.

Profiles in Science: The Joshua Lederberg Papers. National Library of Medicine. Available online. URL: http://profiles.nlm.nih.gov/BB. Accessed June 9, 2009. Robust archive of Lederberg's papers; a stellar collection of primary documents that spans the renowned scientist's entire career. Includes abstracts, autobiographies, bibliographies, correspondence, diaries, essays, interviews, and more on all his major research topics, including bacterial genetics, exobiology, and artificial intelligence.

Watson, James, and Francis Crick. "Molecular Structure of Nucleic Acids: A Structure for Deoxyribose Nucleic Acid." *Nature* (4/25/53). Available online. URL: http://www.nature.com/nature/dna50/watsoncrick.pdf. Accessed June 9, 2009. The landmark paper that first illustrated the double-helix structure of DNA.

Other Media

American Experience: Influenza 1918. Washington, D.C.: PBS Home Video (2005). 60 minutes. Produced by Katy Mostoller and narrated by David McCullough. Documentary that focuses on the Spanish Flu epidemic after World War I, particularly in the United States, where 675,000 people died in 18 months. The epidemic started at an army base in Kansas, quickly spread worldwide, and disappeared almost as quickly as it appeared.

Annotated Bibliography

Edward Jenner: The Man Who Cured Smallpox. London: ArtsMagic DVD (2007). 60 minutes. Documentary on Jenner's experiment that led to a successful vaccination against smallpox in 1798.

History: Modern Marvels Brewing. New York: A & E Television Networks (2008). 50 minutes. Documentary that traces the origins of beer in civilizations as diverse as ancient Sumeria, China, and Finland. Outlines brewing practices from the earliest times to now, from the most technologically rudimentary to the most sophisticated.

Homo Sapiens 1900. DVD. Directed by Peter Cohen. New York: First Run Features (2004). Documentary about the history of eugenics at the turn of the century, with archival footage from the American Eugenics Society and information about the first U.S. sterilization laws and their impact on "race hygiene" as practiced by the Nazi Party in Germany.

In Search of the Polio Vaccine: Modern Marvels. New York: A & E Home Video (2005). 50 minutes. Documentary on the scourge of polio in the early 20th century and efforts to eradicate it, which included backing from President Franklin Roosevelt, himself crippled by the disease. Traces the formation of the March of Dimes and the Salk Institute and the aggressive vaccination campaign waged as soon as an inoculation became available.

Josef Mengele: The Final Account. DVD. Directed by Dan Setton. Santa Monica, Calif.: Direct Cinema Limited (2007). Documentary on Joseph Mengele, Nazi Germany's Angel of Death, who conducted experiments on prisoners and ultimately led 400,000 of them to their deaths at concentration camps during World War II. After the war he escaped to Brazil and lived for decades as Jewish survivors attempted to track him down.

Miss Evers' Boys. DVD. Directed by Joseph Sargent. New York: HBO Productions (1997). Written by Walter Bernstein, based on the Pulitzer Prize–winning play by David Feldshuh, starring Alfre Woodard and Laurence Fishburne. This Emmy Award–winning production tells the story of "The Tuskegee Study of Untreated Syphilis in the Negro Male," begun in 1932, from the point of view of Nurse Eunice Evers, an African American who was torn between comforting the men whom she knew had syphilis and telling them the truth about their condition.

NOVA: Percy Julian: Forgotten Genius. DVD. Directed by Llewellyn M. Smith. WGBH Boston (2007). Tribute to African-American chemist Percy Julian (1899–1975), who discovered a way to turn soybeans into synthetic steroids, which made cortisone a widely available medical treatment for a variety of diseases. In the midst of violent racist attacks on him, he persevered in his research and isolated many substances from plants that were later developed into widely used drugs, including the birth control pill.

Chronology

10,000 B.C.E.

- Communities in the Mesopotamian region develop agriculture by selecting and collecting seeds to replant in the spring. They also develop rudimentary selective breeding techniques to domesticate animals and livestock.
- Smallpox, an extremely lethal virus, first surfaces and becomes one of humanity's greatest health threats, killing hundreds of millions of people worldwide before being eradicated in the 20th century.

9000 B.C.E.

- People use yeast for brewing beer, fermenting wine, and making bread.

8000 B.C.E.–3000 B.C.E.

- Several cultures use lactic-acid producing bacteria to make yogurt and cheese.

500 B.C.E.

- The Chinese use moldy bean curd paste as a rudimentary topical antibiotic.
- Greek mathematician Pythagoras believes that inherited traits are passed down only by the males of the human species.

304 C.E.

- Chinese physician Ko Hung writes *Handy Therapies for Emergencies*, which contains the first written account of smallpox and ways to treat it.

CA. 1300

- Muslim physician Ibn Khatima theorizes that bubonic plague is caused by "minute bodies" that infect a host. His hypothesis is proven correct about 600 years later.

Chronology

1516

- The German beer purity law (Reinheitsgebot) is established; it mandates that beer must be brewed from only three ingredients: water, barley, and hops. The role of yeast in the brewing process is not yet understood.

1546

- Italian physician Girolamo Fracastoro theorizes that epidemic diseases are caused by tiny "spores" that can be transmitted by direct or indirect contact.

1590

- Dutch eyeglass maker Sacharias Jansen makes improvements to glass lenses that lead to the invention of the telescope and the microscope.

1665

- British scientist Robert Hooke creates a powerful microscope, and on viewing a slice of cork he describes its structure as being composed of "cells." He publishes his observations in the book *Micrographia.*

1668

- German apothecary Friedrich Jacob Merck buys the Angel Pharmacy in the town of Darmstadt. It passes from father to son for generations, ultimately becoming Merck KgaA, one of the largest pharmaceutical companies in the world.

1677

- Antoni van Leeuwenhoek, a Dutch scientist dubbed the father of microbiology, discovers microorganisms through his microscope, the most powerful ever built. He witnesses fertilization of an egg, thereby debunking the widely held notion of spontaneous generation.

1796

- On May 14, English physician Edward Jenner successfully inoculates eight-year-old James Phipps with a smallpox vaccine created from weakened cowpox virus. In 1803, he founds the Jennerian Institution for the purpose of eradicating smallpox; it later becomes the Royal Society of Medicine.

1803

- British rulers in India outlaw *tikah,* a centuries-old smallpox vaccine in which pus from smallpox blisters is injected into the skin.

1839

- German scientists Matthias Schleiden and Theodor Schwann develop the cell theory, which states that cells are the basis of all living organisms.

1845

- The Irish Potato Famine kills a million people in Ireland and causes another million to emigrate. The famine is caused by a potato blight, which ruins the country's sole subsistence crop.

1850s

- Louis Pasteur's experiments with the fermentation process show that the production of beer, wine, cheese, and buttermilk rely on microorganisms.

1859

- Charles Darwin publishes *On the Origin of Species,* outlining his theory of natural selection.

1862

- The Land-Grant College Act of 1862 provides each state with 30,000 acres of federal land per representative in Congress. Each parcel of land is sold and the proceeds are used to endow a college that teaches agriculture and engineering.
- Louis Pasteur and Claude Bernard complete the first experiment in pasteurization—the process of heating fermentable liquids such as milk in order to kill bacteria and mold.

1863

- The German pharmaceutical company Bayer is founded and becomes the world's leading producer of aspirin.

1866

- Gregor Mendel publishes "Experiments on Plant Hybridization," a paper summarizing decades of research on inherited characteristics in pea plants; it is largely ignored.

1869

- Swiss biochemist Friedrich Miescher is the first to isolate human DNA.
- Francis Galton publishes *Hereditary Genius,* which outlines his ideas about eugenics and the importance of nature over nurture.

1875

- Walther Flemming discovers chromosomes.

1881

- Louis Pasteur develops the first vaccine for anthrax.

Chronology

1884

- Takahashi Yoshio publishes *A Treatise on the Improvement of the Japanese Race*, in which he proposes eugenic policies for Japan.

1885

- Chemist John Pemberton creates Coca-Cola, a patent medicine that contains cocaine, a legal substance.

1899

- Martinus Beijerinck discovers the tobacco mosaic virus.

1901

- John Queeny founds Monsanto in St. Louis, Missouri. The pharmaceutical company's first products are saccharine and caffeine, which they sell to Coca-Cola.
- Japanese scientist Ishiwata discovers *Bacillus thuringensis* (Bt).

1905

- The Indian Agricultural Research Institute is founded to guide the country's agricultural industry. During the 1970s, it becomes central to the country's Green Revolution.

1906

- Upton Sinclair's novel *The Jungle* highlights the horrendous conditions of the Union Stockyards in Chicago, prompting the U.S. government to create the Food and Drug Administration (FDA) to regulate slaughterhouse practices and ensure the safety of the food supply.

1910

- Charles Davenport and Harry H. Laughlin found the Eugenics Record Office at Cold Spring Harbor, New York. Its eugenic mission is abandoned by the 1940s, but Cold Spring Harbor Laboratory continues to be an important research center for genetics into the 21st century.

1912

- On July 24–29, the British Eugenics Education Society hosts the first International Eugenics Conference in London.

1913

- Father and son team William and Lawrence Bragg develop the technique of X-ray crystallography, which later leads to the first 3-D views of DNA.

BIOTECHNOLOGY AND GENETIC ENGINEERING

- Alfred Sturtevant, a student of Thomas Hunt Morgan at Columbia University, becomes the first researcher to create a genetic map of a chromosome.

1915

- Thomas Hunt Morgan publishes *The Mechanism of Mendelian Heredity*, a landmark in the history of genetics, which outlines "Mendelian-chromosome theory" and becomes the basis for future genetic research.

1917

- Hungarian inventor Karl Ereky coins the term *biotechnology* in his book *Biotechnologie*, which explores how technology can be used to transform living substances into products that are more useful than in their natural state.

1918

- The Spanish flu epidemic sweeps the globe, killing upward of 100 million people. It remains the most lethal epidemic in human history.

1919

- Phoebus Levene identifies adenine, guanine, thymine, and cytosine as the components of DNA, but he does not believe that DNA is the source of an organism's genetic code.

1921

- Margaret Sanger forms the American Birth Control League in New York as an organization to promote safe, effective birth control for women in hopes of limiting family size and improving the health of the nation's existing children.

1923

- *Frye v. United States* results in the Frye rule, which gives judges the right to determine if evidence based on new scientific techniques will be admitted in court. The rule favors admission of evidence gained by techniques that have "gained general acceptance" among scientists.

1925

- The Geneva Protocol outlaws bioweapons in warfare.

1926

- Harry H. Laughlin organizes the American Eugenics Society, a professional organization with more than 1,200 members. The society organizes "Fittest Family" contests and promotes forced sterilization of "undesirable" people.

Chronology

1927

- On May 2, in the U.S. Supreme Court decision *Buck v. Bell*, Justice Oliver Wendell Holmes upholds the practice of forced sterilization, stating that "three generations of imbeciles is enough," despite flimsy evidence that Buck, her mother, or her daughter are mentally impaired.

- Margaret Sanger organizes the first World Population Council in Geneva, Switzerland, as an international conference to promote birth control and eugenics.

1928

- Journalist Ikeda Shigenori becomes a central figure in Japan's eugenics movement after traveling to Germany and observing the eugenics movement there firsthand. He sponsors "Blood Purity Day," in which citizens can obtain a free blood test at the Tokyo Hygiene Laboratory.

- Scottish biologist Alexander Fleming discovers a type of mold that stops the growth of bacteria, which results in the world's most effective antibiotic, penicillin.

1930

- The Plant Patent Act takes effect in the United States; it allows plant breeders to prohibit cloning of hybrid plant varieties they have created.

- The Japanese Society of Health and Human Ecology is founded.

- The South African Association for the Advancement of Science establishes a eugenics committee, headed by H. B. Fantham, that recommends voluntary sterilization for "undesirables" but never advocates forced sterilization.

1932

- "The Tuskegee Study of Untreated Syphilis in the Negro Male" is inaugurated in Alabama with 400 test subjects who are not informed they have syphilis. The experiment continues until 1972, long after penicillin becomes a quick, effective, and standard treatment for the disease.

1933

- Thomas Hunt Morgan, who began his genetics research at Columbia University in New York and later moved to the California Institute of Technology, wins the Nobel Prize for his work with *Drosophila melanogaster* (fruit flies), which he breeds to express certain characteristics.

- The Law for the Prevention of Hereditarily Diseased Offspring is enacted in Germany. Harry H. Laughlin assists German officials in drawing up the guidelines that are ultimately used to sterilize 350,000 people.

BIOTECHNOLOGY AND GENETIC ENGINEERING

1935

- The Nuremburg Laws, also known as the Law for the Protection of German Blood and German Honor, are passed in Germany. They forbid Germans from marrying Jews and strip Jewish people of their German citizenship.

1938

- Congress passes the Federal Food, Drug, and Cosmetic Act (FFDCA), which gives the Food and Drug Administration (FDA) authority to set limits on the amounts of foreign substances in food, including herbicide residues.

1940

- Japan forms the Epidemic Prevention and Water Purification Department of the Kwantung Army, better known as Unit 731. It performs gruesome medical experiments on prisoners of war in World War II, which may have resulted in 200,000 deaths.

- Japan passes the National Eugenic Law under Prime Minister Fumimaro Konoe, which mandates sterilization of mentally defective citizens, limits access to birth control, and promotes genetic screening.

1941

- George Beadle and Edward L. Tatum discover that each gene in a DNA molecule codes for one enzyme.

1944

- At the Rockefeller Institute, Oswald Avery, Colin McLeod, and Maclyn Macarty discover that DNA, not protein, contains an organism's genes and chromosomes.

1945

- Alexander Fleming, Ernst Chain, and Howard Florey win the Nobel Prize in physiology or medicine for their development of penicillin.

1946

- The Communicable Disease Center is founded in Atlanta, Georgia, by the U.S. government for the purpose of eliminating malaria worldwide through the liberal application of the pesticide DDT.

1947

- The Nuremburg Code is established to outlaw human experimentation of the type perpetrated by Josef Mengele during World War II.

Chronology

- The Federal Insecticide, Fungicide, and Rodenticide Act (FIFRA) is passed to regulate pesticide testing and use. Since its inception, it has been expanded to include genetically engineered crops, such as Bt corn, as pesticides.

1948

- Japan passes the Eugenic Protection Law, which replaces its National Eugenic Law that was passed eight years earlier. The new law allows sterilization and abortion with the consent of a woman and her husband.

1949

- Germany's 1933 Law for the Prevention of Hereditarily Diseased Offspring is replaced by the Grundgesetz, or Basic Law, which refutes the idea of a "master race" and introduces a code of human rights.

1952

- Alfred Hershey and Martha Chase conduct the experiment that proves Oswald's theory that an organism's genetic material is contained within DNA, not protein.

- Rosalind Franklin, a crystallographer at King's College in London, takes Photograph 51, an X-ray diffraction image of DNA that reveals its double-helix structure. The image is shared with Watson and Crick, unbeknownst to Franklin. Watson and Crick recognize the photo's importance and use it to prove the structure of DNA the following year, for which they ultimately win a Nobel Prize.

- Jonas Salk begins wide-scale testing of his polio vaccine, which will almost eradicate the common and debilitating disease within a few years.

1953

- On April 25, James D. Watson and Francis Crick publish "A Structure for Deoxyribose Nucleic Acid" in the journal *Nature*, which reveals the double-helix structure of DNA for the first time. It is considered one of the most important moments in the history of science.

1957

- Thalidomide hits the market in more than 50 countries, without having undergone a thorough trial phase. Sold as a sedative, it causes birth defects in children whose mothers take it during pregnancy. It is pulled off the shelves in 1962 after roughly 10,000 children are born with flipper-like appendages instead of arms and legs.

- Alick Isaacs and Jean Lindenmann, British and Swiss virologists, respectively, are credited with discovering interferon, proteins that are crucial to the immune system's process of fighting viruses and other pathogens. It becomes an effective treatment for hepatitis C, many forms of cancer, and multiple sclerosis. It is eventually proved that two Japanese researchers, Nagano Yasuichi and Kojima Yasuhiko, discovered it three years earlier.

1958

- George Beadle and Edward L. Tatum win the Nobel Prize for their discovery that each gene in a DNA molecule codes for one enzyme.

1962

- Watson and Crick, along with their colleague Maurice Wilkins, receive the Nobel Prize in physiology or medicine for "their discoveries concerning the molecular structure of nucleic acids and its significance for information transfer in living material."

- Molecular biologist John B. Gurdon of Oxford University announces that he has cloned the South African clawed frog from a differentiated adult intestinal cell. It is the first verified instance of animal cloning.

- Marine biologist Rachel Carson publishes *Silent Spring*, which warns of the dangers of the pesticide DDT and the process of bioaccumulation, in which concentrated amounts of the chemical are passed up the food chain, with a variety of negative results, to human and animal life.

1963

- The Codex Alimentarius is established by the United Nation's Food and Agriculture Organization to ensure a healthy and safe food supply. It is amended in 2000 to take biotechnology advances into account.

1964

- The Declaration of Helsinki is established by the World Health Organization to prohibit unethical human experimentation.

1965

- Law professor Koichi Bai introduces the German notion of informed consent to the Japanese medical system.

1967

- On December 3, South African cardiologist Christiaan Barnard performs the first successful human-to-human heart transplant.

1968

- The 22nd World Medical Assembly meets in Sydney, Australia, and adopts the Declaration of Sydney on the Determination of Death and the Recovery of

Organs, which enumerates the conditions necessary for declaring brain death and harvesting organs for transplants.

1970

- The U.S. Environmental Protection Agency is formed to protect the health of American citizens and to safeguard the environment.
- The Plant Variety Protection Act is passed, which expands the 1930 Plant Patent Act in forbidding sexual reproduction of plant varieties without the permission of the breeders who developed them.

1971

- The first large-scale genetic screening takes place when Michael Kaback tests 1,800 people of Ashkenazi Jewish ancestry to see if they carry the gene for Tay-Sachs disease. The test proves highly successful and results in a drastic decline of the disease within a few years.
- Earl Butz becomes secretary of agriculture and changes the USDA's long-standing policy of paying farmers to limit their corn production. Under Butz's policy of "get big, or get out," farmers are given subsidies to produce as much corn as possible, which leads to a decline in food prices, a glut of corn, and vast changes in agriculture commerce and food production, including factory farms, monoculture, and the use of corn as an ingredient in an ever-growing number of food items.

1972

- Paul Berg conducts the first recombinant DNA experiments at Stanford University.
- The Sickle Cell Anemia Act is passed; it initiates a nationwide, voluntary screening program to identify carriers of sickle cell anemia that is not linked to eligibility for federal services.
- Scientists discover that the DNA of chimpanzees and gorillas is 99 percent similar to that of humans.
- The pesticide DDT is banned in the United States, largely as a result of Rachel Carson's book *Silent Spring,* and the environmental movement is spawned.

1973

- Stanley Cohen and Herbert Boyer invent a procedure to transfer genes from one organism to another through a process called recombinant DNA.

1974

- The Recombinant DNA Advisory Committee (RAC) is formed by the National Institutes of Health to ensure the safety of experiments using recombinant DNA.

- Following the disclosure of the ethics violations of the Tuskegee Syphilis Study, Congress forms the National Committee for the Protection of Human Subjects of Biomedical and Behavioral Research.

- The Japanese Brain Wave Society draws up guidelines on diagnosing brain death, which are necessary for establishing whether a patient's organs may be removed for transplant purposes. The guidelines are based on the World Medical Association's 1968 Declaration of Sydney.

1975

- The Asilomar Conference on Recombinant DNA is held in California. It brings together genetic researchers from across the country for the purpose of creating guidelines to ensure the safety of their research. The guidelines are an early example of the "precautionary principle" that later become standard in international agreements on biomedical research.

1976

- Genentech, the first company to profit from recombinant DNA technology, is founded by Herbert Boyer and Robert Swanson.

- The Genetic Diseases Act passes, which states that screening to identify carriers of inherited diseases will be done strictly on a voluntary basis and will not influence eligibility for federal services.

- The Toxic Substances Control Act is passed, which gives the Environmental Protection Agency the right to regulate new chemicals.

1977

- Frederick Sanger becomes the first scientist to sequence the DNA-based genome of a living organism, the bacteriophage phi X 174. He maps the genome of a bacteriophage by hand, which has a relatively uncomplicated 11 genes and 5,386 base pairs.

1978

- On July 25, Louise Brown is born in Manchester, England, becoming the first child born from the process of in vitro fertilization. She is dubbed a "test-tube baby," and the process sparks outrage among many, although subsequently IVF becomes a common procedure.

- David Rorvik publishes *In His Image: The Cloning of a Man*. A science writer, he claims to have witnessed a scientific cloning experiment in which a wealthy man had himself cloned. The book is later denounced as a hoax.

- Walter Fiers at the University of Ghent in Belgium reveals the complete nucleotide sequence of the RNA bacteriophage SV40 through the newly derived process of shotgun sequencing.

Chronology

- The South African Committee for Genetic Experimentation is formed to advise the National Department of Agriculture.

1979

- Scientists with the World Health Organization announce the global eradication of smallpox following a concerted effort to vaccinate millions of people worldwide. Two vials of the virus are kept in quarantine—one in the United States and one in Russia—as insurance against future events.
- An anthrax outbreak in Rhodesia kills 182 people and sickens 10,000. Many believe it is an act of biowarfare.

1980

- In *Diamond v. Chakrabarty,* the U.S. Supreme Court rules in favor of Ananda Chakrabarty, a scientist for General Electric, allowing him to patent a genetically engineered organism—a bacteria—that digests petroleum.
- Martin Cline's experiments to use gene therapy to cure thalassemia are denied by the Recombinant DNA Advisory Committee. He carries out the procedure in Italy and Israel, but it is unsuccessful.
- India's sixth Five-Year Plan creates a national strategy for steering commerce toward biotechnology by creating the National Biotechnology Board and implementing policies to encourage genetics research and development.

1981

- South Africa's covert Project Coast program develops biological weapons, violating the UN Biological and Toxic Weapons Convention, which entered into force in 1975.

1982

- Synthetic human insulin becomes the first bioengineered medicine to become widely available in the United States. Engineered by Genentech and marketed by Eli Lilly, Humulin, as it is named, becomes a common treatment for diabetes.

1984

- John Moore, a leukemia patient, sues his doctor David Golde for profiting from products derived from his discarded tissue without his knowledge or consent. In 1990, the California Supreme Court rules that Moore has no ownership rights to the tissue in question.
- A cow on a farm in West Sussex, England, dies of bovine spongiform encephalopathy (BSE), more commonly known as mad cow disease. The cause of death is not determined until two years later, a missed opportunity that throws the

British beef industry into crisis after dozens of people die of the disease's human form, variant Creutzfeldt-Jacob disease, a neurological condition caused by eating contaminated meat, for which there is no cure.

- The largest bioterror attack in U.S. history takes place in rural Oregon, where members of the Rajneeshee religious sect intentionally poison 10 local salad bars with *Salmonella*. Some 750 people are sickened and 45 are hospitalized, but there are no deaths.

1985

- Alec Jeffreys at the University of Leicester in England invents the process of DNA fingerprinting, or profiling, which forensic scientists use to link suspects to crimes.
- Belgian company Plant Genetic Systems develops the first genetically engineered seed, a Bt tobacco plant.

1986

- The Toxic Substances Control Act is amended to include genetically engineered organisms.

1987

- The "ice minus" bacteria developed by Steven Lindow at the University of California at Berkeley is the first genetically altered organism to be released into nature.
- Colin Pitchfork is the first person to be convicted of a crime based on DNA fingerprinting in the rape and murder of two teenagers in Leicestershire, England.
- The Plant Pest Act is passed; it gives the U.S. Department of Agriculture's Animal and Plant Health Inspection Service the power to regulate all organisms, including genetically engineered organisms, that can be considered pests.

1988

- Tommie Lee Andrews is the first American to be convicted through DNA identification testing in the burglary and rape of a woman in Florida.
- The Harvard Oncomouse is the first genetically altered animal to be patented. Designed to be susceptible to cancer, it is used in medical research.

1989

- Steven A. Rosenberg is the first scientist to insert altered genes into a human being to treat cancer.
- Harold Varmus and J. Michael Bishop win the Nobel Prize for their work on the genetics of cancer.

Chronology

- On September 14, four-year-old Ashanthi DeSilva, who suffers from severe combined immunodeficiency (SCID), becomes the first patient to undergo successful gene therapy, which results in a marked improvement in her condition, although she is not cured.

- The U.S. Department of Energy launches the Human Genome Project through the newly created National Center for Human Genome Research at the National Institutes of Health. The ambitious project is headed by James D. Watson and aims to unite scientists worldwide in sequencing all the genes in the human genome by 2005.

- Britain passes the Fertilisation and Embryology Act, which bans reproductive cloning and permits research cloning.

1991

- Germany passes the Embryo Protection Law, which protects all human embryos from destruction and prohibits embryonic stem cell research. It is one of the most restrictive such laws in the world.

1992

- In June, the Convention on Biological Diversity is opened for signature at the Earth Summit in Rio de Janeiro. The document aims to retain as many species of plants and animals on the planet as possible and discourages biopiracy—the practice of wealthy nations and corporations profiting from the commercialization of a less well-developed country's unique species of plants and animals.

- The Human Genome Diversity Project is launched by Stanford geneticist Luca Cavalli-Sforza. The project will collect DNA samples from 500 groups of endangered indigenous peoples and analyze them with regard to migration patterns and susceptibility to disease.

1993

- Jerry Hall and Robert Stillman at George Washington University clone early-stage human embryos from cells that are scheduled for destruction. Their work is protested by many who oppose the cloning of humans and research on embryos.

- UNESCO establishes the International Bioethics Committee to monitor advances in the life sciences.

1994

- Kary Mullis receives the Nobel Prize in chemistry for his discovery of the polymerase chain reaction.

- The DNA Identification Act is passed; it creates standards for DNA testing in criminal cases and authorizes the FBI to fund the Combined DNA Index System (CODIS) as a repository for DNA profiles obtained during criminal investigations.
- Calgene's Flavr Savr tomato is the first genetically modified food to be marketed to consumers. Released with great fanfare, a number of problems result in its discontinuation shortly thereafter.
- India passes the Transplantation of Human Organs Act, which forbids for-profit harvesting of organs. Despite this, a large black market in organs and "transplant tourism" evolve.
- Germany passes the Genetic Engineering Act, which legally protects farmers whose crops become contaminated with genetically modified seed.

1995

- Marya Norman-Bloodsaw sues the Lawrence Berkeley Laboratory for conducting genetic testing on her blood and urine without her knowledge or consent. Claiming violations of her rights under the Fourth Amendment and the 1964 Civil Rights Act, she wins.
- Christiane Nusslein-Volhard of Germany wins the Nobel Prize in physiology or medicine for her work on genetic control of embryo development. She is director of the Max Planck Institute for Developmental Biology in Tübingen and also leads its genetics department.

1996

- Joseph Vlakovsky and John C. Mayfield III, two U.S. Marines, are court-martialed for refusing to obey a direct order to provide samples of their DNA for archiving, believing it to be a violation of their right to privacy.
- U.S. Congress passes the Health Insurance Portability and Accountability Act (HIPAA), which forbids health insurers from denying coverage to those in employer group plans who have preexisting genetic conditions.
- Monsanto launches Roundup Ready soybeans. The seeds are genetically engineered to withstand the company's Roundup herbicide, meaning that spraying crops with Roundup will kill weeds without harming the soybeans.

1997

- The United Nations General Assembly adopts the Universal Declaration on the Human Genome and Human Rights and the World Medical Association passes its Resolution on Human Cloning. Both documents state that reproductive cloning of human beings is an affront to human dignity and should be banned.

- On February 27, Ian Wilmut of the Roslin Institute in Scotland announces in the journal *Nature* that he has cloned a sheep, named Dolly, from a mature body cell, the first successful somatic cell nuclear transfer experiment. The announcement sparks vociferous debate over the ethics of cloning.
- Monsanto Corporation, headquartered in St. Louis, Missouri, reorganizes as a life-sciences company specializing in agriculture and food. Its patented GM seed is used by thousands of farmers worldwide.
- Perry Adkisson and Ray F. Smith receive the 1997 World Food Prize for their work in developing integrated pest management, which eschews chemical pesticides and herbicides in favor of more natural processes for protecting crops.
- South Africa passes the Genetically Modified Organisms Act, which regulates the planting of GM crops.

1998

- On August 12, Árpád Pusztai announces on British television that he would not eat potatoes modified with snowdrop lectin, which his research shows is harmful to rats. Pusztai is fired from the Rowett Research Institute in Aberdeen, Scotland, where he has worked for decades, and his research files are seized. The high-profile controversy involves top members of the British government and launches a spirited public debate over the safety of GM food.
- James Thomson, a leading geneticist at the University of Wisconsin–Madison, isolates embryonic stem cells, which hold great promise for developing cures for many diseases. The use of stem cells in research proves to be controversial, as it necessitates the destruction of human embryos.

1999

- On September 17, 18-year-old Jesse Gelsinger dies from complications of gene therapy to treat his ornithine transcarbamylase deficiency. His death halts many further experimental gene therapy treatments.
- Ingo Potrykus, at the Institute of Plant Sciences at the Swiss Federal Institute of Technology, and Peter Beyer of the University of Freiburg create Golden Rice, which contains genes from daffodils and bacteria that expresses beta carotene. The hope is that the rice will help children in developing countries stave off malnutrition, blindness, and infectious diseases.
- Scientists at the National University of Singapore create the genetically engineered GloFish, a fluorescent zebrafish that contains a jellyfish gene for

phosphorescence. Intended to glow in the presence of environmental pollutants, the fish instead becomes the first genetically engineered animal to be marketed as a pet.

2000

- Starlink Bt corn, manufactured by Aventis Crop Science and supposedly produced only for livestock feed, is found in taco shells in the United States, to which some people claim to have had an allergic reaction. Starlink corn is subsequently discontinued, although no adverse health reactions are proven.
- The Durban Declaration is signed by more than 5,000 physicians and scientists at the International AIDS Conference in Durban, South Africa, affirming that HIV is the cause of AIDS. The declaration is made in response to the country's high-ranking AIDS denialists, including President Mbeki and Minister of Health Manto Tshabalala-Msimang.

2001

- In September–October, weaponized anthrax mailed to members of Congress and the media kill five and sicken 22, in the most significant event of bioterrorism in the United States to date. The case goes unsolved until 2008, when the FBI closes in on U.S. Army biologist Bruce Ivins, who commits suicide before he is charged with the crime.
- The President's Council on Bioethics is established by George W. Bush to address ethical issues raised by biomedical science and technology.
- Germany establishes the Bio-Seigel, a seal of approval granted to certified organic products. Within several years, more than 35,000 products brandish the seal, making Germany's organic food industry one of the largest in the world.
- Scientists at the University of Guelph in Canada trademark the Enviropig; it is genetically engineered to produce low-phosphorus manure, which will result in less soil and water pollution.
- The Human Proteome Organization is founded to build upon the findings of the Human Genome Project. The organization hopes to advance pharmocogenomics, in which a person's genetic make-up is used to create customized medical treatment.
- Ignacio Chapela and David Quist, professors at the University of California at Berkeley, publish a study in *Nature* that says U.S.–grown GM corn has contaminated native maize varieties in rural Oaxaca, Mexico, despite the country's ban on GM seed. The paper sparks controversy; Chapela is initially

denied tenure, and *Nature* ultimately retracts the paper—the first time it has done so in its history.

2002

- Japan's Biotechnology Strategy Council drafts the Biotechnology Strategy Guidelines, intended to guide the economic development of the life sciences in the country in the 21st century.
- Bayer CropScience is spun off from Bayer USA, a division of the German-based pharmaceutical conglomerate Bayer. The new company will focus on agricultural products such as GM seed.

2003

- In April, the Human Genome Project completes its goal of sequencing the human genome two years ahead of schedule.
- On May 13, the U.S. government files a challenge with the World Trade Organization stating that the European Union's anti-GM food policy violates international agreements.
- On September 11, the Cartagena Protocol on Biosafety enters into force, with the goal of protecting existing biodiversity from organisms modi-fied through modern biotechnology. The protocol allows countries to ban GMOs if they believe they pose a threat to their well-being, and requires all countries to label GMOs in the international marketplace so people can remain informed about what their food contains. It also stipulates that the precautionary principle should be used to guide all research involving "living modified organisms."
- The Center for Biomedical Ethics and Law is founded at the University of Tokyo.
- On November 27, the Icelandic Supreme Court bars the implementation of the Icelandic Health Sector Database—which contains tissue of and genetic information on Iceland's 300,000 citizens—by the biotech company deCODE Genetics due to widespread concerns about privacy and informed consent.
- Japan's Osaka Brewing Society, founded in 1923, becomes the Society for Biotechnology.
- In an effort to protect citizens from genetic discrimination by health insurers and employers, the German National Ethics Council states that individuals should not be forced to undergo genetic testing against their will.
- Monsanto successfully sues Oakhurst Dairy in Maine for labeling its milk as having come from cows not treated with bovine growth hormone. Monsanto

claims that the "reverse labeling" practice of stating what a product does not contain insinuates that the bovine growth hormone in its milk is not safe.

2004

- In *Monsanto Canada Inc. v. Schmeiser,* the Supreme Court of Canada rules in favor of Monsanto, stating that Saskatchewan farmer Percy Schmeiser deprived Monsanto of its monopoly on its proprietary GM canola seed by inadvertently and unknowingly storing and planting Roundup Ready canola seeds in his fields.

2005

- On June 17, South Korean geneticist Hwang Woo-Suk publishes an article in the journal *Science* in which he claims to have cloned 11 human embryonic stem cells. His claim is later disproven and Hwang is forced to resign his position at Seoul National University due to numerous ethics violations and allegations of fraud.

- The United Nations General Assembly approves the Declaration on Human Cloning, a nonbinding resolution that calls for a ban on all forms of reproductive and therapeutic cloning. The United States and Germany both vote in favor of the ban, but countries such as Great Britain, which have a significant industry in stem cell research, vote against it.

2007

- On September 4, Celera Genomics publishes the complete human genome, consisting of a sequence of 6 billion nucleotides of the company's founder, Craig Venter.

- In November, two researchers, James Thomson of University of Wisconsin–Madison and Yamanaka Shinya at Kyoto University in Japan, announce independently that they have created induced pluripotent stem cells (iPSC), which are stem cells created from human skin cells that can be used to create any kind of cell without destroying an embryo.

2008

- Stemagen scientists Andrew French and Samuel Wood announce that they have cloned several human embryos from adult skin cells using the somatic cell nuclear transfer technique. The embryos are later destroyed.

- The Svalbard Global Seed Vault opens in the Arctic Circle in Norway. It is the largest seed repository in the world, housing 4 million packets of seeds and is managed by the Norwegian government, the Nordic Genetic Resource Center, and the Global Crop Diversity Trust.

Chronology

- On May 21, the Genetic Information Nondiscrimination Act of 2007 is signed into law by President George W. Bush, which prohibits the improper use of genetic information by health insurers and employers.
- India's Prime Minister Manmohan Singh calls for a second Green Revolution in the country, which will use agricultural biotechnology to drastically expand the country's crop yields and eliminate hunger.

2009

- In March, President Obama reverses George W. Bush's policy on stem cell research, allowing hundreds of new embryonic stem cell lines to be used in federally funded research.

Glossary

adenine a purine nucleobase that binds with thyamine in DNA and with uracil in RNA.

adult stem cells undifferentiated cells found in the human body after the embryonic stage that replenish dying cells. Unlike embryonic stem cells, research with adult stem cells does not require the destruction of an embryo. Same as somatic stem cells.

allele portions of a DNA sequence that may be expressed in an organism's genotype as a specific trait.

amino acid a molecule that is a building block of protein and is present in every living organism.

antibiotic a microorganism that is derived from another living microorganism that is intended to kill bacteria.

antibody a protein produced by the blood that the immune system uses to fight bacteria and viruses. Also called immunoglobulins.

antiretroviral drugs used to treat retroviruses, such as HIV.

autosome a non-sex chromosome. Humans have 22 pairs of autosomes. The sex chromosomes, a combination of either XX or XY, comprises the 23rd pair.

***Bacillus thuringensis* (Bt)** a bacteria that kills most pest insects but does not harm most non-pest insects. The gene that produces Bt has been used in organic insecticides and has been inserted into genetically engineered corn, cotton, soy, and other types of seed.

bacteriophage a virus that infects bacteria. Ubiquitous in all life forms on earth. Scientists are exploring the use of bacteriophages in light of the increasing number of antibiotic-resistant pathogens.

bacterium a type of single-celled organism that lacks a nucleus. Bacteria are ubiquitous on the planet and in all living objects. Many types of bacteria are crucial to vital biological processes, while others cause serious diseases.

base pairs   the two nucleotides on a strand of DNA or RNA that are linked via hydrogen bonds. In DNA, adenine and thyamine are a base pair, as are guanine and cytosine. In RNA, adenine and uracil are a base pair.

bioethics   the study of ethics and controversies brought about by scientific advances in biology and medicine. Bioethics involves various fields, including medicine, politics, law, philosophy, and theology.

biochemistry   the branch of science that studies chemical processes in living organisms, especially those concerned with proteins, nucleic acids, and enzymes.

biodiversity   the variation of plant and animal species within a given region. A high level of biodiversity corresponds to a healthy environment.

bioinformatics   the practice of using information technology to collect and interpret data obtained from molecular biology research; bioinformatics is expected to become a leading growth industry in the coming years. DNA mapping is a type of bioinformatics.

biopharming   using genetic engineering to grow plants with pharmaceutical properties.

bioprospecting   the appropriation of indigenous biological processes, knowledge, or resources by public or private corporations without adequate compensation or cooperation with its original practitioners; often obtained by patenting intellectual property and acquiring legal rights to it. Also called biopiracy.

bioremediation   the process of using naturally occurring microorganisms to restore a contaminated environment to its original condition.

biosafety   the practice of safe transfer, handling, and use of organisms modified by biotechnology in order to limit risks to human health and the environment.

biosphere   the global sum of all ecosystems.

biosynthesis   the process by which chemical compounds necessary to metabolism are created in living organisms through the work of enzymes.

biotechnology   defined by the United Nations Convention on Biological Diversity as "any technological application that uses biological systems, living organisms, or derivatives thereof, to make or modify products or processes for specific use." Biotechnology is technology used in agriculture, food science, and medicine.

bioterrorism attack/biowarfare   the intentional release of biological agents (viruses, germs, bacteria, or toxins) intended to harm people, animals, or plants. The agents can be naturally occurring or modified to increase their potency.

blastocyst an early stage of fetal development shortly after fertilization in which an inner cell mass is surrounded by a trophoblast. The inner cell mass later becomes an embryo and the trophoblast the placenta.

bovine growth hormone (BGH), bovine somatotropin (bST) a protein hormone produced naturally in the pituitary gland of cattle. When it is produced artificially through recombinant DNA technology, it is called recombinant bovine growth hormone (rBGH) or recombinant bovine somatotropin (rBST), and is injected into cows to increase milk production.

bovine spongiform encephalopathy (BSE) commonly known as mad cow disease, it is a fatal, untreatable degenerative neurological disease that strikes cattle. It can be spread by cannibalistic feeding practices and spread to humans who consume infected meat. In humans, the disease is known as new variant Creutzfeldt-Jakob disease (vCJD or nvCJD). An outbreak of BSE in Great Britain in the 1990s resulted in the deaths of 163 people and the slaughter of 4.4 million cattle that may have been infected with BSE.

Cartagena Protocol on Biosafety an international agreement on biosafety entered into force on September 11, 2003, that applies the precautionary principle in developing standards for the safe handling of living modified organisms (LMOs). Also called the Biosafety Protocol.

cell the basic unit of all living things. A cell is self-containing and maintaining. It takes in nutrients, converts them into energy, carries out specialized functions, and reproduces.

chimera a living organism containing DNA from two different zygotes. This can happen naturally, as in the case of a mule, which is the offspring of a male donkey and a female horse, or through genetic engineering, such as in attempts to create human-chimpanzee hybrids.

chromosome an organized structure of DNA and proteins found in cells. Humans have 46 chromosomes, each of which contains many specific genes and nucleotide sequences.

clone a genetically identical copy of an organism.

Combined DNA Index System (CODIS) the DNA database authorized by the DNA Identification Act of 1994 that created the standards for DNA testing in law enforcement. It is funded by the Federal Bureau of Investigations (FBI) and stores DNA profiles obtained from evidence collected by local, state, and federal crime laboratories.

Creutzfeldt-Jakob disease a fatal brain disease that affects humans. It is a transmissible spongiform encephalopathy caused by prions, abnormal cellular proteins.

cytosine one of the four bases in DNA and RNA. Cytosine always pairs with guanine.

DNA (deoxyribonucleic acid) a double-helix-shaped string of two antiparallel nucleotides held together by a hydrogen bond that contain an organism's genetic material and which resides in the nucleus of a cell.

DNA fingerprinting a forensic technique of identifying an individual based on extracting and identifying the unique base pairs of his or her DNA.

DNA sequencing the process by which the order of nucleotide bases in a molecule of DNA is determined.

Drosophila melanogaster a genus including more than 1,500 species of flies commonly known as fruit flies. The flies that are used by biologists and geneticists as model organisms because they are easily cultured, have short generations, and readily express genetic mutations that are valuable for experimental purposes.

ecosystem the state of interconnectedness between all living organisms in a given region.

enzyme a complex protein produced by living cells that is important in producing biochemical reactions that affect body temperature.

epidemiology the study of infectious disease.

eugenics the science of improving the overall health or qualities of a race of people by controlling their ability to reproduce. Positive eugenics is the practice of encouraging people possessing esteemed genetic traits to reproduce; negative eugenics is the practice of discouraging or forbidding (sometimes by force) those possessing undesirable traits from reproducing.

eukaryote a organism comprised of complex cells with defined membranes and nuclei.

fermentation the chemical conversion of carbohydrates into alcohol or acids. Commonly, yeast is used to convert sugar to alcohol in turning juice into wine or grain into beer.

gene a specific sequence of nucleotides in DNA or RNA that is located on a specific chromosome and codes for a specific trait. A regulatory gene produces proteins that control (regulate) the expression of a structural gene. A structural gene contains the code for amino acids or ribosomal or transfer RNA.

gene mapping the process of creating a diagram of DNA sequences that belong to an organism's chromosomes.

gene splicing the process of manipulating an organism's genes. Also called genetic modification.

gene therapy a medical procedure in which an altered gene is inserted into a patient's DNA to treat or cure a disease caused by a malfunctioning gene. The two types of gene therapy are germ-line gene therapy and somatic cell gene therapy. Germ-line gene therapy involves altering reproductive cells—sperm or eggs—and will effect changes that can be passed on to future generations. Somatic cell gene therapy makes changes to a person's non-reproductive cells, which will have no effect on future generations.

genetic code   the uniform sequence of nucleotides in DNA and RNA that is the biochemical basis of heredity for a given organism.

genetic determinism   the belief that genes determine an organism's physical and behavioral characteristics.

genetic engineering   the manipulation of an organism's genes. Same as gene splicing.

genetic marker   a known DNA sequence that can be copied and transferred to another organism in order to replace a faulty DNA sequence.

genetic modification   intentional manipulation of an organism's genes.

genetics   the science of heredity and variation in living organisms.

genetic use restriction technology (GURT)   commonly called "terminator technology," it refers to restricting genetically modified plants (which are typically patented) by engineering the seed's second generation to be sterile. As of 2009, no such seed is on the market anywhere in the world.

genome   the complete set of genes of a particular species.

genomics   the study of an organism's entire genome, including DNA sequencing and gene mapping, as opposed to the study of specific genes, which falls under the aegis of molecular biology.

genotype   the inherited instructions contained in an organism's genetic code.

germ-line genes   genes that are included in the sex cells and can be passed on to offspring.

guanine   one of the four bases in DNA and RNA. Guanine is always paired with cytosine.

HIV (Human Immunodeficiency Virus)   a transmissible retrovirus that causes AIDS, acquired immunodeficiency syndrome.

horizontal gene transfer (HGT)   the process of an organism incorporating genetic information from an organism other than its parent. Significantly less common than vertical gene transfer, in which genetic information is passed on through an organism's offspring, horizontal gene transfer may be the mechanism by which GM seed contaminates nearby non–GM crops. Also called lateral gene transfer.

human genome   the genome of *Homo sapiens,* which contains more than 3 billion DNA base pairs and between 20,000 and 25,000 genes.

Human Genome Project   international research effort implemented by the U.S. Department of Energy and the National Institutes of Health in 1990 to identify the base pair sequences and genes of the human genome. The project was completed in 2003 and has made its data available to the public.

hybrid   the offspring resulting from the cross-breeding of plants or animals. Plant hybrids are relatively common and have been cultivated for centuries. Animals hybrids between different breeds of dogs, for example, are also common; but less common are hybrids between felines, such as the liger, a cross between a lion and a tiger.

induced pluripotent stem cell (iPS)   a stem cell created from an adult somatic cell (such as a skin cell) that is able to be transformed into any kind of cell (i.e., a pluripotent cell). Unlike embryonic stem cell research, iPS research does not require the destruction of an embryo.

infectious disease   a disease spread by pathogens of any kind, including bacteria, viruses, fungi, prions, and parasites, that can be transmitted from one organism to another, as opposed to an inherited disease.

inoculate   introducing a substance into the body for the purpose of boosting the body's immune system to a certain disease.

in vitro fertilization (IVF)   a form of assisted reproductive technology in which a female's egg and a male's sperm are extracted and fertilized outside the womb (in vitro). The resulting zygote is implanted in the female's uterus and the pregnancy continues normally.

junk DNA   the common term for the vast portions of an organism's DNA or genome for which no purpose has yet been discovered. Up to 95 percent of the human genome is comprised of junk DNA, which scientists think may be the detritus of millions of years of evolution.

karyotype   the chromosomal characteristics of a cell, often depicted in a karyogram showing the 23 numbered pairs of chromosomes.

lateral gene transfer   same as horizontal gene transfer. The process by which an organism obtains genetic material from an organism other than the one from which it descended. Common among bacteria.

Mendelian genetics   the laws of inheritance outlined by Gregor Mendel that explain the transmission of characteristics from parents to offspring. First published in 1865, they form the foundation of modern genetics. Also known as Mendelian inheritance and Mendelism.

messenger RNA (mRNA)   RNA produced by transcription and carrying the code for a particular protein.

microbiology   the study of microorganisms.

mitochondria   the part of a cell responsible for energy production.

mitochondrial DNA (mtDNA)   a circular molecule of DNA found in the mitochondria of a cell—not the cell nucleus—that is maternally inherited. It has only about 16,500 base pairs and is inherited without change (except for mutations) from mother to child. It can be used to trace matrilineage back thousands of years.

molecular biology   the study of biology at the molecular level. As a discipline, it overlaps greatly with biochemistry and genetics. All study proteins, DNA, RNA, and cells.

monoculture   the practice of growing one crop exclusively over a large area. Long-term monoculture frequently depletes soil of nutrients and can have a negative effect on a region's biodiversity.

mutation   a change in a DNA sequence. Mutations can be due to DNA copying errors, exposure to radiation, chemical mutagens, or viruses. Germ line mutations can be passed on to an organism's offspring, but somatic mutations cannot—at least in organisms that reproduce sexually. Mutations create variety in the gene pool and can influence the processes of natural selection and evolutionary adaptation.

nanobiotechnology (also bionanotechnology)   the science of devising small technical devices incorporating biological and biochemical agents that are used in the realm of atoms and molecules.

natural selection   the perpetuation of organisms that are genetically best suited for their environment through continued reproduction and the gradual elimination of organisms that fail to adapt to their environment.

nucleic acid   a large molecule comprised of a chain of nucleotides that carries an organism's genetic information. All living things contain nucleic acid; the most common nucleic acids are DNA and RNA.

nucleotide   the basic structural unit of the nucleic acids RNA and DNA.

Pasteur effect   the phenomenon, discovered in controlled experiments by Louis Pasteur in 1857, explaining how oxygen increases the growth of yeast in yeast-containing substances while decreasing fermentation. Conversely, depriving yeast-containing liquids of oxygen (that is, putting them in an anaerobic environment) increases fermentation.

pasteurization   the process of heating liquids to destroy harmful bacteria and pathogenic (illness-causing) microbes.

phagocyte   cells found in blood, bone marrow, and other bodily tissue that consume foreign microorganisms; they are central to the immune system and help fight infections.

pharmacogenetics   the science of tailoring medicine to genetic traits. Also called pharmacogenomics.

pharming   inserting genes into plants or animals so they express new characteristics that make them useful pharmaceuticals. Biopharming refers to creating pharmaceuticals by genetically altering plants.

phenotype observable characteristics of an organism that are the result of both the organism's genotype and its environment. Phenotype variations are necessary for evolution by natural selection.

plasmid a double-stranded, circular molecule of DNA that does not contain chromosomes and can replicate autonomously. Most commonly found in bacteria.

pluripotent cell an undifferentiated cell that can become any kind of cell in the body. Embryonic stem cells are pluripotent.

polymerase chain reaction (PCR) a useful laboratory technique in which a segment of DNA can be replicated quickly and abundantly in-vitro, making genetic research more efficient.

precautionary principle the idea that harm must be anticipated and mitigated before it occurs. In science, it means researchers must assume the burden of proof that their experiments will not harm people or the environment. As a legal concept, the precautionary principle is designed to foster transparency, communication, and consensus in the scientific community to protect the rights of the public at large.

prion an abnormal cellular protein that causes brain disease.

prokaryote a one-celled organism that lacks a nucleus and membrane-bound organelles. Includes bacteria and archaea.

protein a substance comprised of amino acids arranged in an order determined by a gene. They perform many essential functions within a cell.

proteomics the study of proteins.

recombinant DNA (rDNA) synthetic DNA derived from inserting DNA sequences from one organism into the genome of another organism, creating a new sequence that expresses new characteristics.

retrovirus a type of virus comprised of an RNA (single-stranded) genome that uses the process of reverse transcription to transform its RNA into DNA, which is then integrated into a host cell's DNA. HIV is a retrovirus.

reverse transcriptase a type of DNA enzyme that converts single-stranded RNA into double-stranded DNA. Normal transcription creates RNA from DNA; reverse transcription creates DNA from RNA.

ribonucleic acid (RNA) a nucleic acid that controls the chemical activities of a cell. Includes ribosomal RNA (rRNA), the main component of ribosomes that synthesize proteins; transfer RNA (tRNA), which transmits amino acids to the ribosomes; and messenger RNA (mRNA), which carries the code for a particular protein.

ribosomal RNA (rRNA) a main structural component of ribosomes, which synthesize proteins in a cell.

ribosome   a particle in a cell's cytoplasm that binds messenger RNA and transfer RNA to synthesize proteins and polypeptides.

somatic cell   any cell other than a reproductive cell.

somatic-cell nuclear transfer (SCNT)   a cloning process, sometimes used in stem cell research, in which the nucleus of a non-sex cell (a somatic cell) is removed and inserted into an egg cell that has had its nucleus removed. The egg cell is then stimulated to divide until it becomes a blastocyst.

somatic stem cell   an undifferentiated cell found in any post-embryonic organism that multiplies by cell division. Also adult stem cell.

spontaneous generation   the belief that life can arise from nonliving matter. Also called abiogenesis.

sustainable agriculture   the process of farming land so that it retains its fertility and ability to yield the same amount of food year after year without harming the surrounding ecosystem or resulting in a loss of biodiversity.

stem cell   a cell than can renew itself through cell division and can differentiate itself into one of a number of specialized cells. There are two types of stem cells: embryonic stem cells, which are found in blastocysts; and adult stem cells, which are found in adult tissues.

terminator technology   formally known as genetic use restriction technology (GURT), it refers to restricting genetically modified plants (which are typically patented) by engineering the seed's second generation to be sterile. As of 2009, no such seed is on the market anywhere in the world.

thymine   one of the four bases in DNA. Thymine is always paired with adenine.

transgenic   an organism that has been altered by having genes from another organism inserted into its DNA.

transcription   the process by which DNA copies one of its nucleotide sequences into a molecule of messenger RNA, which will then be transmitted to the site of protein synthesis in the cell.

transfer RNA (tRNA)   RNA that transfers a certain amino acid to the site of protein synthesis during the process of translation.

vaccine   a biological preparation derived from microorganisms given to a patient in order to increase his or her immunity to a given pathogenic disease.

vector   in genetics, a substance used to transmit genetically modified DNA into another organism. Plasmids and viruses are often used as vectors.

vertical gene transfer   the process by which an organism receives genetic material from its ancestor or parent organism.

virus   a very small microorganism that cannot grow or replicate outside a host organism. Ubiquitous in history and in all known forms of life; frequently pathogenic.

Glossary

X-ray crystallography a valuable technique for obtaining information about biological substances by subjecting a crystallized version of the substance to an X-ray and recording data on how the beam diffracts.

yeast a single-celled microorganism that reproduces asexually and is crucial in baking and fermentation.

zygote a cell fertilized via sexual reproduction.

Index

Page numbers in **boldface** indicate major treatment of a subject. Page numbers followed by *c* indicate chronology entries. Page numbers followed by *f* indicate figures. Page numbers followed by *g* indicate glossary entries. Page numbers followed by *m* indicate maps.

333

Index

Index